APPLIED MATHEMATICS FOR
SCIENCE AND ENGINEERING

APPLIED MATHEMATICS FOR SCIENCE AND ENGINEERING

LARRY A. GLASGOW
Department of Chemical Engineering
Kansas State University

Published by John Wiley & Sons, Inc., Hoboken, New Jersey
Published simultaneously in Canada

For general information on our other products and services or for technical support, please contact our Customer Care Department within the United States at (800) 762-2974, outside the United States at (317) 572-3993 or fax (317) 572-4002.

Wiley also publishes its books in a variety of electronic formats. Some content that appears in print may not be available in electronic formats. For more information about Wiley products, visit our web site at www.wiley.com.

Library of Congress Cataloging-in-Publication Data:

Glasgow, Larry A., 1950– author.
    Applied mathematics for science and engineering / Larry A. Glasgow, Department of Chemical Engineering, Kansas State University.
      pages cm
    Includes index.
    ISBN 978-1-118-74992-0 (hardback)
    1. Engineering mathematics.   2. Technology–Mathematical models.   I. Title.
    T57.G53   2014
    510–dc23
                                        2014008366

Printed in the United States of America

10   9   8   7   6   5   4   3   2   1

CONTENTS

PREFACE

This book, *Applied Mathematics for Science and Engineering*, is the culmination of many years of experience preparing upper-division students in engineering and the physical sciences for graduate-level work (and particularly for subjects such as advanced transport phenomena). We strive to acquire key competencies that can be utilized to solve important practical problems of the type found in advanced coursework as well as those that may arise in a student's research program. The course is intended for engineers and scientists in the science/technology/engineering/mathematics (STEM) fields, and this book is my effort to fit the presentation of the material to the dominant learning styles of such students, many of whom are visual and tactile learners.

In the fall of 2013, the Programme for International Student Assessment (PISA) released its key findings from the 2012 examinations administered to 15-year-old students around the globe. The performance of students from the United States ranked about 26th in mathematics, far below the levels seen in Shanghai, Singapore, Hong Kong, Korea, Japan, and other countries. In the *Country Notes*, PISA observed that "students in the United States have particular weaknesses in performing mathematic tasks with higher cognitive demands, such as taking real-world situations, translating them into mathematical terms, and interpreting mathematical aspects in real-world problems." This result has occurred *despite* the fact that the United States spends the most money per student on education according to a survey conducted by the Organization for Economic Cooperation and Development (OECD). There are a number of possible explanations.

The life experiences of contemporary adolescents are very different from those of previous generations. Educators in the STEM fields have found—and employers of young technical professionals confirm—that the disconnect between contemporary students and the physical world is staggering. Even when our domestic students have basic mathematical tools, rarely do they possess a frame of reference or necessary judgment that might allow them to critically evaluate a result obtained from solution of a model or from a computational simulation. For the applied sciences, this is a perilous situation.

It has also been suggested that a factor contributing to this crisis is the lack of integration between science and mathematics in secondary education. Focus in public education in the United States is usually based on content coverage and not on contextual understanding. This is not a new problem; for generations, the lack of a demonstrable connection between, say, algebra and the world (as perceived by a 15-year-old) has been an obstacle to learning. The author distinctly recalls his impression that trigonometry was the *only* important math subject in high school because we used it in physics to solve problems that *looked like they might have bearing on something that actually mattered*. The failure of typical secondary-school course structure to relate mathematical subjects to problems in context makes it extremely difficult for students to fully appreciate the significance of the material.

With these observations in mind, I have tried to present these topics as I do in class, with frequent attention paid to applications of obvious importance. An overarching goal is to demonstrate why a particular mathematical method is worthy of study, and we do this by relating it to things that the student of applied science can appreciate. Consequently, this book contains many examples of important applications in biology, chemistry, physics, and engineering, and most include graphical portrayals of model results and computations. This book also covers some topics rarely treated in similar texts; these include integro-differential equations, interpretation of time-series data, and an introduction to the calculus of variations. My hope is that students will find their interest piqued by the approach we have taken and the topics we have covered, and that they will turn to the literature of mathematics to learn more than we can possibly provide here.

L. A. G.

1

PROBLEM FORMULATION AND MODEL DEVELOPMENT

INTRODUCTION

Our purpose in this course is to review some mathematical techniques that can be used to solve important problems in engineering and the applied sciences. We will focus on problem types that are crucial to the analysis and simulation of real, physical phenomena. Sometimes, our objective will be to predict the future behavior of a system and sometimes it will be to interpret behavior that has already occurred. We want to stress that *the author and the readers are collaborators in this effort*, and whether this text is being used in a formal setting or for self-study, the ultimate goal is the same: We want to be able deal with problems that arise in the applied sciences and do so efficiently. And—this is important—we do not want to rely on calculation software unless we know something about the method(s) being employed. Too often, real problems can have multiple solutions, so it is essential that the analyst be able to exercise some judgment based on understanding of the problem and of the algorithm that has been selected.

Many of the problems we will be solving will come from both transient and equilibrium balances, and they will involve forces, fluxes, and the couplings between driving force–flux pairs. Examples of the latter are Newton's, Fourier's, and Fick's laws:

$$\tau_{yx} = -\mu \frac{\partial v_x}{\partial y}, \quad q_y = -k \frac{\partial T}{\partial y}, \quad \text{and} \quad N_{Az} = -D_{AB} \frac{\partial C_A}{\partial z},$$
(1.1)

where τ_{yx} is the shear stress (acting on a y-plane due to fluid motion in the x-direction), q_y is the flux of thermal energy in the y-direction, and N_{Az} is the molar flux of species "A" in the z-direction. Note that these three fluxes are *linearly related* to the velocity gradient, the temperature gradient, and the concentration gradient, respectively. Each driving force–flux pair has, under ideal conditions, a constant of proportionality (the viscosity, μ; the thermal conductivity, k; and the diffusivity, D_{AB}); these constants are *molecular properties of the medium* that can be determined from first principles if the right conditions are met. Unfortunately, it is also possible for viscosity to depend on velocity, for thermal conductivity to depend on temperature, and for diffusivity to depend on concentration. In such cases, the driving force–flux relationships are no longer linear as indicated by eq. (1.1).

The *balances* we speak of usually come from some statement of conservation; and this could be conservation of mass, energy, momentum, and so on. For an example, consider heat transfer occurring in an electrical conductor, perhaps a copper wire. The conductor is carrying an electric current so thermal energy will be produced in the interior by dissipation (I^2R heating) and thermal energy will be lost to the surroundings at the wire's surface. We will construct a thermal energy balance on a volume element, an annular region extracted from the wire of length L that extends from r to $r + \Delta r$; this is shown in Figure 1.1.

Applied Mathematics for Science and Engineering, First Edition. Larry A. Glasgow.
© 2014 John Wiley & Sons, Inc. Published 2014 by John Wiley & Sons, Inc.

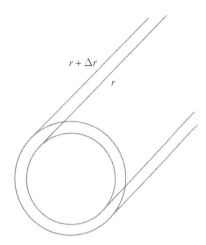

FIGURE 1.1. Annular volume element extracted from conductor for the thermal energy balance. The thickness of the annular shell is Δr and the length is L.

We express the balance verbally in the form

{Rate of thermal energy in at r} − {Rate of thermal

energy out at $r + \Delta r$} + {Production of thermal energy

by dissipation} = {Accumulation}.   (1.2)

Since the temperature in the conductor may vary continuously with *both* position and time, the result of this balance will be a partial differential equation. We can rewrite the balance (eq. 1.2) introducing the appropriate symbols:

$$+(2\pi r L q_r)_r - (2\pi r L q_r)_{r+\Delta r} + 2\pi r L \Delta r P_e = 2\pi r L \Delta r \rho C_p \frac{\partial T}{\partial t}.$$
(1.3)

Now we divide by $2\pi L \Delta r$, take the limit as $\Delta r \rightarrow 0$, apply the definition of the first derivative, and substitute Fourier's law for q_r (we also divide by r):

$$k\left[\frac{1}{r}\frac{\partial}{\partial r}\left(r\frac{\partial T}{\partial r}\right)\right] + P_e = \rho C_p \frac{\partial T}{\partial t}.$$
(1.4)

Note that we have assumed that the volumetric rate of thermal energy production, P_e, is a constant; this is not strictly correct since the resistance of copper wire (e.g., AWG 12) is 1.65 ohms/1000 ft at 25°C, but increases to 3.08 ohms/1000 ft at 250°C. In our model, we neglected the temperature dependence of the conductor's resistance; this would probably be acceptable if the temperature change in the wire is modest. For steady-state conditions, the solution for eq. (1.4) is simply

$$T = -\frac{P_e}{4k}r^2 + C_1 \ln r + C_2.$$
(1.5)

If T is finite at the center, then clearly, $C_1 = 0$. One question that arises in such problems concerns the speed of approach to steady state: When might we use eq. (1.5) and when must we proceed with the solution for eq. (1.4)? We can illustrate this concern using 2 AWG bare copper wire ($d = 0.6544$ cm) with a constant surface temperature of 30°C (this is an example of a *Dirichlet* boundary condition). We set $P_e/\rho C_p = 5950$°C/s and let the wire have a uniform initial temperature of 30°C. Because copper has a very large thermal diffusivity, $\alpha = k/\rho C_p = 1.14$ cm²/s, the approach to steady state should be quick:

t (s)	0.005	0.01	0.02	0.05	0.075	0.100	0.175
T center (°C)	59.67	86.87	124.51	162.43	168.01	169.20	169.52

As we anticipated, steady-state conditions are attained rapidly, suggesting that for many similar applications eq. (1.5) could be used to find $T(r)$.

The previous example is a microscopic balance, that is, we are modeling a *distributed parameter* system. We will also have occasion to use macroscopic balances for lumped-parameter systems in which the field (or dependent) variable does not vary with position. For the electrical wire with dissipation discussed earlier, this would mean that the temperature would *not vary* in the r-direction. This is clearly not valid for the case we just examined where $T(r = R)$ was forcibly maintained at 30°C by removing heat at the surface. We will discuss the circumstances under which the temperature might be (nearly) independent of position a little later.

In modern industrial production, applied scientists and engineers constantly struggle to meet product specifications, satisfy regulatory constraints, increase output, and maximize the return for investors and stakeholders. A reality of modern industrial operations is that economic survival is often predicated on continuous process improvement. And because of the scale of industrial processes, even incremental improvements can be very significant to the bottom line. In the early twentieth century, process tweaking was carried out empirically by trial and error; this usually worked since margins were wide, there was less global competition, and product specifications were often loose. Since there was little automatic process control, skilled operators quickly learned through experience how to make adjustments to improve production. That era has passed, and now operational decisions and control strategies are often based on models or process simulations. As Hanna and Sandall (1995) point out, contemporary economic reality dictates that modeling and simulation be favored over labor-intensive experimental investigations. In this introduction, we will examine a few of the possible ways models can be formulated, and we will look at some examples illustrating the underlying principles that are key to modeling and simulation.

Before we do that, however, we need to recognize that a model—however complex—is merely a representation of reality. Though we may understand the governing physical principles thoroughly, our mathematical formulation will never be in perfect fidelity with the "real" world. This is exactly what Himmelblau and Bischoff (1968) referred to when they noted, "the conceptual representation of a real process cannot completely encompass all of the details of the process." Nearly always in real processes, there are random events, stochastic elements, or nonlinear couplings that simply cannot be anticipated. Nowhere does this become more apparent than in the examination of engineering or industrial catastrophes; the actual cause is almost always due to a chain or cascade of events many of which are quite improbable taken individually. In cases of this kind, the number of state variables may be *very* large such that no mathematical model—at least none that can be realistically solved—will account for every contingency. And even in relatively simple systems, quite unexpected behavior can occur, such as a sudden jump to a new state or the appearance of an aperiodic oscillation. Examples of real systems where such behaviors are observed include the driven pendulum, the Belousov–Zhabotinsky chemical reaction, and the Rayleigh–Bénard buoyancy-driven instability. Real systems are always dissipative; that is, they include "frictional" processes that lead to decay. Where we get into trouble is in situations that include both dissipation and at least one mechanism that acts to sustain the dynamic behavior. In such cases, the dynamic behavior of the system may evolve into something much more complicated, unexpected, and possibly even dangerous.

There is an area of mathematics that emerged in the twentieth century (the foundation was established by Henri Poincaré) that can provide some *qualitative* indications of system behavior in some of these cases; though what has become popularly known as *catastrophe theory* is beyond the scope of our discussions, it may be worthwhile to describe a few of its features. In catastrophe theory, we concern ourselves with systems whose normal behavior is smooth, that is, that possess a stable equilibrium, but that may exhibit abrupt discontinuities (become unstable) at instants in time. Saunders (1980) points out that catastrophe theory applies to systems governed by sets (even *very large* sets) of differential equations, to systems for which a variational principle exists, and to many situations described by partial differential equations. In typical applications, the number of state variables may indeed be very large, but the number of *control* variables may be quite small. Let us explain what we mean by *control variable* with an example: Suppose we wished to study the flow of water through a cylindrical tube. We impose a particular pressure gradient (or head, Δp) and then measure the resulting flow rate. The head is the control variable and the flow rate through the tube is established in response to Δp. If the number of

control variables is less than or equal to four, then there are only *seven elementary types of catastrophes*. The beauty of catastrophe theory is that it makes it possible to predict the *qualitative* behavior of a system, even for cases in which underlying differential equations are unknown or hopelessly complicated. An excellent review of this field with numerous familiar examples (including biochemical reactions, population dynamics, orbital stability, neural activity in the brain, the buckling of structures, and hydrodynamic instability) has been provided by Thompson (1982). His book is a wonderful starting point for students interested in system instabilities.

The principal fact we wish to emphasize as we conclude this introduction is that *every model is an idealization* and when we rely on a mathematical analysis (or a process simulation), it is prudent to keep its limitations in mind. We would do well to remember the statistician George E. P. Box's admonition, "essentially, all models are wrong but some are useful." In the modern practice of applied science, we must add a corollary: Not only can models be useful, but sometimes they are also absolutely necessary even when they are wrong in some limited sense.

Let us now look at just a few examples of how problems of the types we wish to solve are actually developed. We will begin with a situation involving equilibrium between gas and liquid phases; this problem requires solution of a set of algebraic equations.

ALGEBRAIC EQUATIONS FROM VAPOR–LIQUID EQUILIBRIA (VLE)

Problems in VLE require solution of mass balances, but in cases where the temperature (T) is unknown (as in this instance), a trial-and-error process can be employed. We will assume that we have a vapor consisting of an equimolar mixture of light hydrocarbons, ethane (1), propylene (2), propane (3), and isobutane (4). The vapor phase mole fractions are all ¼, that is, $y_1 = y_2 = y_3 = y_4 = 0.25$. The constant total pressure is 14.7 psia (1.013 bars), and the vapor phase is cooled slightly until the first drop of liquid is formed (this temperature is the *dew point*). Our objective is to find the temperature, T, at which this occurs, and the composition of the liquid that forms (in equilibrium with the vapor). We will solve this problem in two different ways and then compare the results.

First, we will use the Antoine equation to get the vapor pressures of all four species as functions of temperature:

$$\log_{10} p^* = A - \frac{B}{C+T}. \qquad (1.6)$$

The necessary constants will be obtained from Lange's *Handbook of Chemistry* (1961).

	A	B	C
Ethane	6.80266	656.4	256
Propylene	6.8196	785	247
Propane	6.82973	813.2	248
Isobutane	6.74808	882.8	240

Keep in mind that T must be in Celsius and $p*$ is in millimeters of mercury. We will assume that Raoult's law is applicable such that

$$y_1 P = x_1 p_1 * (T). \qquad (1.7)$$

P is the total pressure and it is 760 mmHg. Therefore, the liquid-phase mole fractions are determined from $x_i = y_i P / p_i*(T)$, and a solution will be found when $\Sigma x_i = x_1 + x_2 + x_3 + x_4 = 1$. Such problems are amenable to machine computation and a simple strategy suggests itself: Estimate T, compute the vapor pressures with the Antoine equation, then calculate the liquid-phase mole fractions and check their sum. If the $\Sigma x_i \neq 1$, adjust T and repeat. A very short program was written for this purpose, and it shows $T = -34.2°C$ ($-29.6°F$), with

$$x_1 = 0.0273 \quad x_2 = 0.1406 \quad x_3 = 0.1789 \quad x_4 = 0.6611$$
$$(\textstyle\sum x_i = 1.0079).$$

Of course, VLE for light hydrocarbons are enormously important and have been intensively studied. For an alternative procedure, we can use data published by DePriester (1953) which are in the form of nomograms relating the distribution coefficients, Ks ($K = y/x$), for light hydrocarbons to temperature and pressure. Again, we estimate T, then find the distribution coefficients and use them to calculate the liquid-phase mole fractions. By trial and error, we find for $T = -30°F$ (with $P = 14.7$ psia):

	K (y/x)	x
Ethane	7.95	0.031
Propylene	1.92	0.130
Propane	1.43	0.175
Isobutane	0.37	0.672

In this case, the summation of the liquid-phase mole fractions is 1.008, and the agreement with our first solution is reasonable (the worst case is ethane, with a difference between values of x_i of about 13%).

This VLE example illustrates a possible outcome when two different solution procedures are available; the results, particularly for the volatile constituents, are slightly different. It is important to note, however, that the results obtained from the two solution procedures for the *major components of the liquid* (propane and isobutane) are very close, differing only by about 2%.

MACROSCOPIC BALANCES: LUMPED-PARAMETER MODELS

We observed at the beginning of this chapter that the temperature of a copper conductor carrying electrical current *might not* vary with radial position. Such a situation could arise if there was large resistance to heat transfer at the surface; that is, if the heat generated by dissipation could not escape to the surrounding fluid phase. In the following example, we look at a case that might meet this stipulation, a hot metal casting that is being quenched in an oil bath. It is always good practice for the analyst to begin with a verbal statement of the balance:

{Rate of thermal energy in} − {Rate of thermal energy out}
  = {Accumulation}.

This is a *macroscopic* balance in which we *assume* that the temperature *throughout the casting is the same* (as we pointed out, this cannot be strictly correct, but under the right circumstances, it may be adequate). There is no {rate in} since the casting simply loses thermal energy to its surroundings, and we will assume that this loss is approximately described by Newton's law of cooling. Therefore,

$$-hA(T - T_\infty) = MC_p \frac{dT}{dt}. \qquad (1.8)$$

This first-order, *lumped-parameter* model is readily separated to yield:

$$-\frac{hA}{MC_p} dt = \frac{dT}{(T - T_\infty)}. \qquad (1.9)$$

We integrate and find

$$T - T_\infty = C_1 \exp\left(-\frac{hA}{MC_p} t\right). \qquad (1.10)$$

The casting has an initial temperature, T_i, at $t = 0$, and thus:

$$\frac{T - T_\infty}{T_i - T_\infty} = \exp\left(-\frac{hA}{MC_p} t\right). \qquad (1.11)$$

According to this first-order model, the temperature of the casting will follow an exponential decay, ultimately approaching the temperature of the oil bath. It is worth noting that the quotient, $hA/(MC_p)$, is an inverse time constant for this system, $1/\tau$. When $t = 1\tau$, about 63% of the ultimate change will have been accomplished; when $t = 2\tau$, about 86%; and at $t = 3\tau$, about 95%. Consequently, a common rule of thumb for simple first-order systems is that the dynamic behavior is nearly complete in 3τ.

We should wonder what to look for in more general applications that would indicate that a lumped-parameter model is acceptable. What are the conditions that might allow us to neglect the variation of the field (dependent) variable in the interior of the medium? We have pointed out that in a case like the metal casting considered earlier, we can answer this question by identifying where the main resistance to heat transfer is located. If the main resistance is in the fluid phase, then the temperature in the interior of the solid may be nearly uniform. But if the main resistance is in the solid material, then T may vary significantly with position. We can assess the location of the resistance to heat transfer through the use of the Biot modulus, $Bi = hR/k$, where h is the heat transfer coefficient on the *fluid side* of the interface and k is the thermal conductivity of the *solid* medium. If Bi is very small, then the fluid side of the interface offers the main resistance to heat transfer.

Let us look at a second macroscopic balance example that is slightly more complicated; we will model a perfectly mixed continuous stirred-tank reactor (CSTR) in which the reactant species, "A," is consumed by a first-order chemical reaction. We will have flow into the tank, flow out of the tank, and depletion of "A" by chemical reaction. In this case, a verbal statement of the mass balance on "A" will appear: {Rate in} − {Rate out} − {Depletion by reaction} = {Accumulation}. Symbolically, we write

$$\dot{v}C_{Ain} - \dot{v}C_A - k_1 V C_A = V \frac{dC_A}{dt}. \qquad (1.12)$$

We will divide by the volumetric flow rate, $\dot{v}$, obtaining

$$C_A(1 + k_1\tau) - C_{Ain} = -\tau \frac{dC_A}{dt}. \qquad (1.13)$$

In this instance, the time constant for this system (τ) is merely the volume of the reactor, V, divided by volumetric flow rate ($\tau = V / \dot{v}$); this is called the *mean residence time* of the reactor. For the initial condition, we take the concentration of "A" in the tank to be zero ($C_A = 0$ for $t = 0$), and once again the solution for this first-order model is elementary:

$$C_A = \frac{C_{Ain}}{1 + k_1\tau} \left[1 - \exp\left(-(1 + k_1\tau)\frac{t}{\tau} \right) \right]. \qquad (1.14)$$

Note that under steady-state conditions, the concentration, C_A, is attenuated from the inlet (feed) value by the factor, $1/(1 + k_1\tau)$. If the residence time, τ, in the reactor is large, and if the reaction rate constant, k_1, is large, then the exiting concentration will be *much smaller* than the feed, C_{Ain}.

Both of the previous examples immediately led to exponential solutions. This is not the only possible outcome for this model type; consider a *batch* kinetic study of a second-order chemical reaction in which the reactant species is consumed according to

$$r_A = -k_2 C_A{}^2. \qquad (1.15)$$

Since this is a batch process, there is no flow either into or out of the reactor volume, so the macroscopic mass balance yields

$$\frac{dC_A}{dt} = -k_2 C_A{}^2. \qquad (1.16)$$

Differential eq. (1.16) is nonlinear, but it is easily integrated:

$$-\frac{1}{C_A} + C_1 = -k_2 t. \qquad (1.17)$$

At $t = 0$, $C_A = C_{A0}$, so the dynamic behavior in the batch reactor is

$$\frac{1}{C_{A0}} - \frac{1}{C_A} = -k_2 t. \qquad (1.18)$$

If the initial concentration in the reactor is 1, then

$$C_A = \frac{1}{1 + k_2 t}; \qquad (1.19)$$

we have an inverse relationship between concentration and time. If this batch process is carried out for a *long* period of time, the reactant species, "A," will nearly disappear. Of course, the decay in an exponential process is more rapid; for example, taking $kt = 3$, we have $\exp(-3) = 0.0498$, whereas $1/(1 + 3) = 0.25$.

We conclude this section by noting the study of chemical kinetics in constant volume systems is fertile ground for the exploration of macroscopic balances with less familiar outcomes (i.e., other solutions that are *not* exponential). For example, we could have *parallel* first- and second-order reactions yielding

$$\frac{dC_A}{dt} = -k_1 C_A - k_2 C_A{}^2. \qquad (1.20)$$

This equation can also be integrated and the reader may want to show that

$$\frac{1}{k_1} \ln\left[\frac{k_2 C_A}{k_2 C_A + k_1} \right] = -t + C_1. \qquad (1.21)$$

Other possibilities include rate expressions with fractional orders and Michaelis–Menten kinetics for certain catalyzed systems, where

$$r_A = \frac{dC_A}{dt} = \frac{-K_1 C_A}{K_2 + K_3 C_A}. \tag{1.22}$$

If eq. (1.22) is integrated, we find that the sum of both $K_2 \ln(C_A)$ and $K_3 C_A$ terms is proportional to time, $-K_1 t$ (plus the constant of integration).

FORCE BALANCES: NEWTON'S SECOND LAW OF MOTION

Many problems in mechanics require us to set

$$Ma = \sum F. \tag{1.23}$$

For example, consider a mass, M, suspended vertically from the ceiling, through a spring (that exhibits Hookean behavior). We will take the *positive z*-direction to be down. The apparatus also has some kind of viscous damping provided by a dashpot or shock absorber:

$$M \frac{d^2 z}{dt^2} = Mg - Kz - A\frac{dz}{dt}. \tag{1.24}$$

In this case, we obtain a second-order, ordinary differential equation. We will divide by M and rewrite the equation:

$$\frac{d^2 z}{dt^2} + \frac{A}{M}\frac{dz}{dt} + \frac{K}{M}z - g = 0. \tag{1.25}$$

The response of this system to an initial perturbation can assume different forms depending on the values selected for the constants. Suppose, for example, that we let $A/M = 3$ and $K/M = 5$ ($g = 9.8\,\text{m/s}^2$, of course). We will initiate the dynamic behavior of the system by pulling the mass downward to a new, extended, position. The equilibrium position is $z = 9.8/5 = 1.96\,\text{m}$, so we will start by extending the assembly to $z = 5$. We will then repeat the solution but decrease the damping coefficient, A/M, to 0.5. For this second case, the response will be much more oscillatory as shown in Figure 1.2.

We will explore options for the solution of differential equations similar to eq. (1.25) in Chapters 5 and 6.

It is worth pointing out that the form of differential eq. (1.25) can arise in other phenomena, quite unrelated to $F = Ma$. For example, if we place a resistance (R), an inductance (L), and a capacitance (C) in series to form an *RLC* circuit, and if we supply sufficient energy to offset the dissipative losses, we can see sustained oscillation. The governing equation in this case is

$$\frac{d^2 I}{dt^2} + \frac{R}{L}\frac{dI}{dt} + \frac{1}{LC}I = a,$$

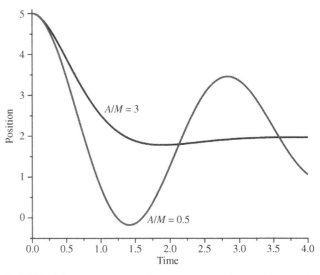

FIGURE 1.2. Displacement of a suspended weight attached to the ceiling with a spring. Viscous damping is reflected by the magnitude of A/M. The mass is moved to an initial position of $z = 5$ and then released.

which looks exactly like eq. (1.25). In fact, if we were to choose a large inductance and a small resistance, the circuit would be severely underdamped and oscillation in response to a forcing function would be guaranteed.

DISTRIBUTED PARAMETER MODELS: MICROSCOPIC BALANCES

We looked at some cases previously where lumped models were used to make difficult problems tractable; these included heat lost by a metal casting and the operation of stirred chemical reactors. We pointed out that in heat transfer we might be able to use the Biot modulus to assess the suitability of the lumped approach. Let us provide a little elaboration by way of additional examples. Consider a *tray* (*plate* or *stage*) in a distillation column; we often treat this type of separation as a cascade of equilibrium stages where the *conditions over the entire tray are the same*. This means that the temperature and the mole fractions in the liquid and vapor *do not vary* over the plate. This is incorrect as local fluid mechanics as well as spatially variable heat and mass transfer affect the approach to equilibrium for each stage. Sometimes, we can account for such variations with a *stage efficiency*, which is the fractional approach to equilibrium for that tray and often this approach works very well.

Similarly, in the case of a CSTR, we normally *assume* that the temperature and concentration are *completely uniform* throughout the reactor volume. This too cannot be correct as we know with certainty that there will be dead zones (in corners and perhaps near baffles) where little mixing takes place. In some instances, the idealized treatment of a chemical reactor will result in a suitable model. But what happens if fine-scale variations seriously impact process performance? In such cases, we must turn to distributed parameter models, and let us illustrate the process with an example from heat transfer.

Consider a *long* cylindrical rod, initially at some uniform temperature, T. At $t = 0$, the surface temperature of the rod is instantaneously changed to some new, elevated value. Our interest is the flow of heat into the interior of the rod. We will make a shell energy balance on an annular element of length, L, and thickness, Δr:

{Rate in at r} − {Rate out at $r + \Delta r$} = {Accumulation of thermal energy}.

Thus,

$$+(2\pi rLq_r)_r - (2\pi rLq_r)_{r+\Delta r} = 2\pi rL\Delta r\rho C_p \frac{\partial T}{\partial t}. \quad (1.26)$$

We divide by $2\pi L\Delta r$, take the limit as $\Delta r \rightarrow 0$, and apply the definition of the first derivative:

$$-\frac{1}{r}\frac{\partial}{\partial r}(rq_r) = \rho C_p \frac{\partial T}{\partial t}. \quad (1.27)$$

Since $q_r = -k(\partial T/\partial r)$, we can rewrite this partial differential equation in a more useful form:

$$\frac{\partial T}{\partial t} = \alpha\left[\frac{\partial^2 T}{\partial r^2} + \frac{1}{r}\frac{\partial T}{\partial r}\right], \quad (1.28)$$

where α is the thermal diffusivity of the medium, $\alpha = k/\rho C_p$. Equation (1.28) is a parabolic partial differential equation and it is a candidate for solution by a technique known as *separation of variables*. We will now illustrate some of the initial steps in this process, deferring the intricacies for Chapter 7. We take $T = f(r)g(t)$ and introduce this product into eq. (1.28):

$$fg' = \alpha\left[f''g + \frac{1}{r}f'g\right], \quad (1.29)$$

Dividing by the product, fg, we note that the left-hand side is a function of time and the right-hand side is a function only of r. The only way the two can be equal is if they are both equal to a constant:

$$\frac{g'}{\alpha g} = \frac{f'' + \frac{1}{r}f'}{f} = -\lambda^2. \quad (1.30)$$

Notice that we have purposefully chosen a negative constant—the necessity for this will become apparent in a moment. We have identified two ordinary differential equations (the second of the pair is a form of Bessel's differential equation):

$$\frac{dg}{dt} = -\alpha\lambda^2 g \quad \text{and} \quad \frac{d^2 f}{dr^2} + \frac{1}{r}\frac{df}{dr} + \lambda^2 f = 0, \quad (1.31)$$

and the solutions are

$$g = C_1\exp(-\alpha\lambda^2 t) \quad \text{and} \quad f = AJ_0(\lambda r) + BY_0(\lambda r). \quad (1.32)$$

Therefore, according to our initial assumption regarding the form of the solution

$$T = C_1\exp(-\alpha\lambda^2 t)[AJ_0(\lambda r) + BY_0(\lambda r)]. \quad (1.33)$$

However, $Y_0(0) = -\infty$, and since our temperature must be finite at the center of the cylindrical rod,

$$T = A\exp(-\alpha\lambda^2 t)J_0(\lambda r). \quad (1.34)$$

Let us emphasize that, at this point, we merely have *a solution* for partial differential eq. (1.28)—to get *the solution* of interest, we must apply a boundary condition at $r = R$ and an initial condition for $t = 0$. As we shall see later in the course, these final two steps often cause the analyst the most difficulty. It is entirely appropriate for us to wonder how well this model for transient heat transfer in a cylindrical rod represents physical reality. A comparison will be provided here in which an acrylic plastic rod was heated by immersion in a constant temperature bath; part of the discrepancy between the model and the experiment is caused by the fact that the experimental rod was of finite length, whereas the model is for an *infinitely long* cylinder. The rod was chilled to 3°C then immersed in a heated water bath maintained at 69°C. A thermocouple, embedded at the center of the rod, was used to measure the cylinder's (centerline) temperature as a function of time (Figure 1.3).

Of course, if the cylinder length-to-diameter ratio is not large, then axial (z-direction) transport would have to be included in the model; the term, $\partial^2 T/\partial z^2$, would be added to the right-hand side of eq. (1.28). This illustrates a case where the experience of the analyst comes into play. How might we determine if the axial transport term is required? Of course, we could solve the complete problem for a sequence of decreasing ratios (of L/d) until the solutions begin to show significant differences. We will actually try this later, but we also recognize that such an approach is time-consuming and computationally expensive, so we will seek another course of action as well. One possibility is to construct dimensionally correct representations for the second derivatives using

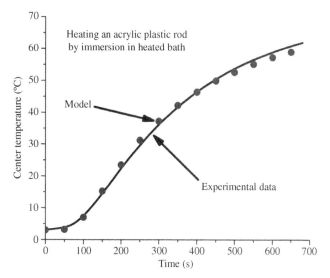

FIGURE 1.3. Comparison of a distributed parameter model (filled circles) with experimental data (solid curve) for transient heating of a cylindrical rod. The discrepancy between the two is mainly due to the fact that the experimental cylinder was not infinitely long (in fact, $L/d = 6$).

appropriate temperature differences and suitable characteristic lengths:

$$\frac{\partial^2 T}{\partial r^2} \approx \frac{\Delta T}{\delta^2} \quad \text{and} \quad \frac{\partial^2 T}{\partial z^2} \approx \frac{\Delta T}{L^2}.$$

If $\delta^2 << L^2$, then axial transport is almost certainly unimportant. Let us explore the effect of L/d in a concrete way by revisiting the previous example (transient heating of an acrylic plastic rod); we will solve eq. (1.28) numerically, but with the addition of the axial conduction term, $\partial^2 T/\partial z^2$.

Temperature at Center of Acrylic Plastic Rod of Finite Length

t (s)	$L/d = 10$	$L/d = 6$	$L/d = 4$	$L/d = 2$
50	3.164	3.164	3.164	3.164
100	7.365	7.365	7.365	7.365
150	15.952	15.952	15.952	15.955
200	25.107	25.108	25.108	25.131
250	33.242	33.244	33.245	33.322
300	40.040	40.041	40.043	40.208
400	50.112	50.114	50.117	50.482
500	56.702	56.704	56.708	57.214

It is clear from these data that for cases in which molecular transport is dominant, the "tipping" point for the assumption of infinite-length behavior occurs at about $L/d = 4$. In cylindrical geometries, if $L/d \leq 4$, axial transport must generally be accounted for.

Of course we understand that we can have distributed parameter problems where more than one dependent vari-

able changes with position. Consider the case of an energy balance drawn on a fluid contained within parallel, planar walls; the space between the walls extends from $y = 0$ to $y = B$. The lower wall is heated, the upper wall is insulated, and the fluid moves in the z-direction (fully developed, laminar flow). The appropriate energy balance is

$$\rho C_p \left(\frac{\partial T}{\partial t} + V_z \frac{\partial T}{\partial z} \right) = k \left[\frac{\partial^2 T}{\partial y^2} + \frac{\partial^2 T}{\partial z^2} \right]. \qquad (1.35)$$

In this case, the momentum balance can be solved separately to obtain V_z, and the velocity distribution for the fully developed laminar flow is

$$V_z = \frac{1}{2\mu} \frac{dp}{dz} \left(y^2 - By \right). \qquad (1.36)$$

Therefore, the temperature of the fluid between the parallel walls is governed by

$$\frac{\partial T}{\partial t} = -\frac{1}{2\mu} \frac{dp}{dz} \left(y^2 - By \right) \frac{\partial T}{\partial z} + \alpha \left[\frac{\partial^2 T}{\partial y^2} + \frac{\partial^2 T}{\partial z^2} \right], \qquad (1.37)$$

assuming that the fluid properties are constant. Finding a solution for eq. (1.37) will be a more difficult proposition than in our previous examples. However, if we restrict our attention to steady state and neglect axial conduction (i.e., omit the $\partial^2 T/\partial z^2$ term), we can solve the problem easily by forward-marching in the z-direction using only values from the previous z-row; some computed results are provided in the following table. We will take water as the fluid (it enters at a uniform temperature of $4.44°C$), assume the fluid properties are constant, and let the lower (heated) wall be maintained at $37.78°C$. The maximum (centerline) fluid velocity will be $15\,cm/s$ and we let $B = 2\,cm$. There is an important observation to make with respect to these computed results: Note how slowly thermal energy is transported in the transverse (y-) direction. This is characteristic of processes in which we rely on molecular transport. We will see additional examples of this explicit computational process in Chapter 8.

y/B	$z/B = 10$	$z/B = 100$	$z/B = 1,000$	$z/B = 10,000$
0.1	8.31	21.91	30.75	36.34
0.2	4.45	10.03	23.96	34.96
0.4	4.44	4.49	12.94	32.53
0.8	4.44	4.44	5.34	29.95

Using the Equations of Change Directly

In many of the previous examples, we formulated balances on, say, mass or thermal energy and then developed a governing equation from the balance statement. However, a model can be formulated directly from the *equations of change*, that is, from the complete balances for momentum, thermal energy, or concentration (of species "A"). To illustrate this process, we will consider mass transfer occurring in laminar flow in a cylindrical tube (of radius R). We assume species "A" is transported from the tube wall to the fluid, which is in motion in the positive z-direction. We have a highly ordered (laminar) flow and we can obtain the needed equations from Bird et al. (2007) or from Glasgow (2010). We will need the continuity (mass balance) equation for species "A" in cylindrical coordinates:

$$\frac{\partial C_A}{\partial t} + V_r \frac{\partial C_A}{\partial r} + \frac{V_\theta}{r}\frac{\partial C_A}{\partial \theta} + V_z \frac{\partial C_A}{\partial z}$$
$$= D_{AB}\left[\frac{1}{r}\frac{\partial}{\partial r}\left(r\frac{\partial C_A}{\partial r}\right) + \frac{1}{r^2}\frac{\partial^2 C_A}{\partial \theta^2} + \frac{\partial^2 C_A}{\partial z^2}\right] + R_A, \quad (1.38)$$

and we also need the z-component of the Navier–Stokes equation (momentum balance):

$$\rho\left(\frac{\partial V_z}{\partial t} + V_r \frac{\partial V_z}{\partial r} + \frac{V_\theta}{r}\frac{\partial V_z}{\partial \theta} + V_z \frac{\partial V_z}{\partial z}\right)$$
$$= -\frac{\partial p}{\partial z} + \mu\left[\frac{1}{r}\frac{\partial}{\partial r}\left(r\frac{\partial V_z}{\partial r}\right) + \frac{1}{r^2}\frac{\partial^2 V_z}{\partial \theta^2} + \frac{\partial^2 V_z}{\partial z^2}\right]. \quad (1.39)$$

The tube has a constant cross section, the flow is steady, and the fluid is incompressible. Therefore, $V_r = V_\theta = 0$, and $\nabla \cdot V = 0$ so that $\partial V_z/\partial z = 0$. Furthermore, $\partial p/\partial z$ is constant, and for eq. (1.39), we are left with

$$0 = -\frac{\partial p}{\partial z} + \mu\left[\frac{1}{r}\frac{\partial}{\partial r}\left(r\frac{\partial V_z}{\partial r}\right)\right]. \quad (1.40)$$

Integrating the first time,

$$r\frac{dV_z}{dr} = \frac{1}{2\mu}\frac{dp}{dz}r^2 + C_1. \quad (1.41)$$

Of course, $dV_z/dr = 0$ at $r = 0$, so $C_1 = 0$. Integrating a second time, we find

$$V_z = \frac{1}{4\mu}\frac{dp}{dz}r^2 + C_2. \quad (1.42)$$

We apply the no-slip condition at the tube wall: $V_z = 0$ at $r = R$, so

$$C_2 = -\frac{1}{4\mu}\frac{dp}{dz}R^2,$$

and

$$V_z = \frac{1}{4\mu}\frac{dp}{dz}(r^2 - R^2). \quad (1.43)$$

In addition to the simplifications already made, we assume that the molecular transport (diffusion) of species "A" in the flow direction is negligibly small (this is often quite reasonable), and we make a rather severe assumption that the flux of "A" from the wall into the fluid is *constant* (we could provide experimental conditions to make this approximately true over a finite tube length). Therefore, $\partial C_A/\partial z = \beta$, and for eq. (1.38), we are left with

$$V_z\frac{\partial C_A}{\partial z} = V_z\beta = D_{AB}\left[\frac{1}{r}\frac{d}{dr}\left(r\frac{dC_A}{dr}\right)\right]. \quad (1.44)$$

Since we have the parabolic velocity distribution given by eq. (1.43) for V_z, we can multiply by $r\,dr$ and integrate the first time:

$$r\frac{dC_A}{dr} = \frac{\beta}{4\mu D_{AB}}\frac{dp}{dz}\left(\frac{r^4}{4} - \frac{R^2 r^2}{2}\right) + C_1. \quad (1.45)$$

C_A must be finite at the center of the tube, so $C_1 = 0$. We integrate a second time, noting that $C_A = C_{A0}$ at $r = R$. It is also helpful to introduce a new concentration variable by letting $C = C_A - C_{A0}$; the result is

$$C = \frac{\beta}{4\mu D_{AB}}\frac{dp}{dz}\left(\frac{r^4}{16} - \frac{R^2 r^2}{4} + \frac{3}{16}R^4\right). \quad (1.46)$$

We can explore the general behavior of this result by making a few calculations as shown here:

r/R	$\left(\dfrac{r^4}{16} - \dfrac{R^2 r^2}{4} + \dfrac{3}{16}R^4\right)$
0	$0.1875R^4$
0.1	$0.1850R^4$
0.3	$0.1655R^4$
0.5	$0.1289R^4$
0.7	$0.0800R^4$
0.9	$0.0260R^4$
1	0

The reader may want to plot these data to get a better sense of the shape of the profile. Remember that we still have to multiply by

$$\frac{\beta}{4\mu D_{AB}}\left(\frac{dp}{dz}\right)$$

(which is negative) to get $C = C_A - C_{A0}$.

A CONTRAST: DETERMINISTIC MODELS AND STOCHASTIC PROCESSES

In much of the preceding discussion, we formulated (or simply selected) differential equations that expressed the continuous relationship between dependent and independent variables. The appeal of this approach is that given boundary conditions and the initial condition for the system, and the differential equation, the *future behavior of the system is set for all time, t*. This is what we mean when we say that a system is *deterministic*. Often in engineering and the applied sciences, systems of interest do behave exactly this way. But suppose, for example, that we have to concern ourselves with a population of animals. We know that the animals will reproduce and that some will die, whether it be from old age, disease, or predation. It is also possible that some animals will immigrate, and some may emigrate. Let N be the number (density) of animals, perhaps the number of animals per acre; a simplistic approach to the dynamic problem might be formulated:

$$\frac{dN}{dt} = cN^2 + bN + a, \qquad (1.47)$$

with the initial condition, $N = N_0$ at $t = 0$. Therefore, we have

$$\int \frac{dN}{cN^2 + bN + a} = t + C_1. \qquad (1.48)$$

The solution for the integral will depend on our choices for a, b, and c. We will arbitrarily select $c = 1/16$, $b = -1/4$, and $a = 1/2$, and assume we have nine animals per acre at $t = 0$. We take

$$q = 4ac - b^2 = (4)(1/2)(1/16) - (1/16) = 0.0625.$$

The left-hand side of eq. (1.48) can be found in any standard table of integrals, for example, Weast (1975), *CRC Handbook of Tables for Mathematics*; therefore,

$$\frac{2}{\sqrt{q}} \tan^{-1}\left(\frac{2cN + b}{\sqrt{q}}\right) = t + C_1, \qquad (1.49)$$

such that

$$N = \frac{\sqrt{q}\,\tan\left[\frac{\sqrt{q}}{2}(t + C_1)\right] - b}{2c}. \qquad (1.50)$$

For the initial condition we have selected, $C_1 = 10.34$, and the number of animals per acre, N, behaves as shown in Figure 1.4.

It is clear from these data that $N(t)$ will grow without bound if we employ the deterministic model, eq. (1.50).

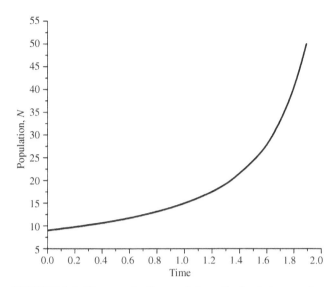

FIGURE 1.4. The growth of a population of animals (per acre) as described by eq. (1.50). The initial population was 9.

Obviously this is not correct. In biological systems, an explosive growth of numbers leads to collapse—the available food supply simply cannot sustain too many animals. We might even see extinction. This is an important lesson: Certain types of systems exhibit variabilities that should not be represented in familiar deterministic form. Biological systems particularly are affected by behavioral drives—greed, hunger, sex, and so on—that *might* be modeled deterministically if the total numbers are very large, but more generally must be treated stochastically. It will be far more appropriate in such cases to base our modeling on the *probability that the population may total N animals at some point in time*. We will discuss this type of formulation in greater detail in Chapter 9.

EMPIRICISMS AND DATA INTERPRETATION

A system of interest to us can be so complex that even though we might be able to write down the underlying physical laws, there will be absolutely no thought of actually obtaining a solution. Alternatively, we might be given data with the requirement that a framework for interpretation be developed—what do the data mean, and can they be used to formulate a coherent picture of what has taken place? In either case, the conventional approach of writing out some model equation and then seeking a solution will not be applicable. We are forced to adopt different viewpoints and we will spend just a little time here considering some of the alternatives.

Let us begin by thinking about heat transfer in a tube with turbulent flow. Since turbulence is always three-dimensional and time-dependent, we would need all three components of the Navier–Stokes equation, the continuity equation (conservation of mass), and the energy equation to completely describe the system. A solution—*without any approximation*—would therefore require that we handle all five of the listed partial differential equations simultaneously, including accurate boundary and initial conditions. This is an impossibility and it is likely to remain so for many decades to come, perhaps forever. Nevertheless, processes must be designed and their anticipated performance estimated. An approach that has often been taken historically in such situations is the development of a correlation based on experimental data. In the case of heat transfer in tubes, the Dittus and Boelter (1930) correlation provides the Nusselt number in terms of the Reynolds and Prandtl numbers:

$$\text{Nu} = \left(\frac{hd}{k}\right) = 0.023\,\text{Re}^{0.8}\,\text{Pr}^{0.4}, \qquad (1.51)$$

where

$$\text{Re} = \frac{d\langle V\rangle\rho}{\mu} \quad \text{and} \quad \text{Pr} = \frac{\mu C_p}{k} = \frac{\nu}{\alpha}.$$

Equation (1.51) allows us to estimate the heat transfer coefficient, h, and then use that value to find the rate at which thermal energy is transferred to the fluid. For example, if $\text{Re} = 50,000$ and if $\text{Pr} = 10$, then the Nusselt number is about 332. We find a very similar situation for mass transfer occurring in wetted-wall towers; the Gilliland and Sherwood (1934) correlation is

$$\text{Sh} = \left(\frac{Kd}{D_{AB}}\right) = 0.023\,\text{Re}^{0.83}\,\text{Sc}^{0.44}. \qquad (1.52)$$

In this case, the Sherwood number (Sh) is given as a function of the Reynolds and Schmidt numbers, and for the latter, $\text{Sc} = \nu/D_{AB}$. Equation (1.52) will allow us to obtain the mass transfer coefficient, K, and thereby to determine the rate of mass transfer. These two correlations illustrate a time-tested approach to the solution of extremely difficult problems in the applied sciences: Identify the pertinent dimensionless groups (either by dimensional analysis of the governing equations of change or by the use of a technique like the *Buckingham pi method*), use those dimensionless groups to form a correlation, and then use a statistical approach to select the "best" values for the coefficients and exponents using available experimental data. Please note that it is inappropriate for us to refer to such an exercise as modeling. What we are doing in such cases is fitting an empirical relationship (often a power law) to measured or observed data. There is nothing inherently wrong with this strategy and it has been used successfully in engineering practice for more than 100 years. But it is critical that we

remember that a correlation thus developed is *valid only for the conditions covered by the experimental work*. We cannot extrapolate these results to systems unlike the original experiments.

Unfortunately, there are cases where even the correlation approach may not be possible. Consider, for example, the frictional resistance experienced by a fluid flowing through a cylindrical tube. And furthermore, assume we are primarily interested in the transition region, that is, that range of Reynolds numbers in which the flow ceases to be laminar and becomes highly disordered. Although hydrodynamicists have some understanding of the mechanism by which a flow becomes turbulent, no model has been developed that can fully describe what happens in this process. What we can do, however, is to measure the pressure drop for flow through a specified tube at a given flow rate. The friction factor, f, is related to the pressure drop over a length of tube, L:

$$f = \frac{(P_0 - P_L)}{\rho V^2}\frac{R}{L}. \qquad (1.53)$$

In Figure 1.5 is a typical set of data obtained for flow through a polycarbonate plastic tube (which is hydraulically smooth) with a diameter of 0.375 in. (9.525 mm). The experimental effort was concentrated on Reynolds numbers ranging from about 2000 to 5000; since the tested fluid was water and since $d = 9.525$ mm, the approximate range of velocities was about 21–52 cm/s.

For a nominal Re of 3000, the friction factor appears to vary from possibly 0.006 to nearly 0.0107, a value which is 70% larger than the low end. Figure 1.5 illustrates

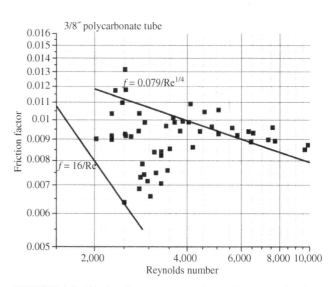

FIGURE 1.5. Friction factors measured experimentally for flow through a polycarbonate tube with a diameter of 0.375 in. (9.525 mm). Note particularly the scatter of the data in the range $2000 < \text{Re} < 4000$. There is no discernible functional relationship for these data.

a physically elementary situation for which there is no model nor is there a rational empirical correlation. Naturally, if we had a *sufficiently large data set for a particular apparatus*, we might be able to say something about the *probability* that f at Re = 3000 would be between, say, 0.008 and 0.009. This example (merely water flowing through a cylindrical tube) underscores the difference between a deterministic process and one that should be characterized as stochastic. In the transition region, the extreme sensitivity to initial conditions (SIC) yields such large variability in outcome that two trials carried out at the same velocity will rarely produce identical results. Furthermore, when such experiments are carried out with *viscoelastic* fluids, the results may exhibit *hysteresis*—the results obtained for a sequence of experiments with increasing flow rate may not be the same as those obtained from a sequence with decreasing flow rate. This is a consequence of molecular creep that may occur for certain non-Newtonian fluids.

We conclude that may be able to say something about the *likely range* for the friction factor, f, but we cannot say with certainty that $f = 0.0085$ at Re = 3000. Some physical phenomena are not well enough understood (or the governing mathematics are so intractable) that the conventional processes of model development and problem solution are simply inappropriate. Numerous examples—which we pointed out previously—originate from the large variability that is an inherent part of many biological processes.

Finally, we will think about a scenario in which some data are received by the analyst along with the requirement that an interpretation of those data be produced *very rapidly*. Speech recognition is a classic example of this problem type and we will examine the data shown in Figure 1.6,

which is a recording of a human voice speaking the word *integration*.

There are a few segments in these data where the signal is attractively simple. For example, the first *phone* (a single speech sound) appearing on the left-hand side of the figure is from the "in" part of the word (in'tə-grā'shən) and it exhibits a sharply defined frequency of about 175 Hz. Note, however, that much of the balance of the recorded signal is more difficult to characterize in such a direct way. One of the features that does stand out is the high-frequency burst that accompanies the start of the shən sound of the last syllable. The frequency here is certainly much higher, with components that appear to be around 1250 Hz. At the very end of the signal, the main oscillation returns to about 133 Hz with a small, higher-frequency component superimposed. The question we want to address is: Do these isolated features constitute a "model" or a "pattern" that would permit identification of the word *integration*? Probably not. We will need a better tool if we are to deal with this case successfully, and toward the end of this book (in Chapter 10), we will consider the treatment and interpretation of time-series data and we will learn how useful harmonic analysis can be for this type of situation.

CONCLUSION

We have looked at a few examples in which some type of balance (typically from a conservation principle) was used to develop a model for a physical situation of interest. We saw cases in which this technique led to algebraic equations and also to differential equations. This is entirely appropriate since our focus in this course is on mathematical methods that engineers and scientists routinely use to solve the problems thus formulated. We have also introduced the notion that certain types of problems will not lend themselves to this type of analysis—in some cases, whether it be due to system complexity or the nature of the data that must be interpreted, we will have to employ a different process. Some options for these problem types will be discussed later in this text.

Regardless of the exact nature of the problem, however, there is an important aspect to the overall process that we have not mentioned at all. Model development, data interpretation, and problem solution are merely elements of the larger decision-making process; we will conclude this introduction with an illustration of this broader context using a situation that arises frequently in industrial practice.

Suppose we have a requirement for gas compression. We can assume that the intake and final pressures will be specified, and that the mass flow rate of the compressed gas is dictated by process throughput. What types of questions would we need to address in this situation? One

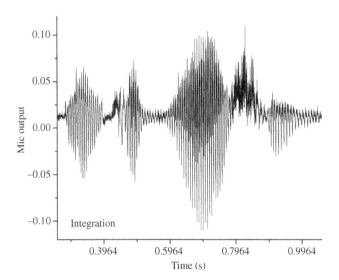

FIGURE 1.6. A recording of a human voice speaking the word "integration."

might presume that the critical issues concern the type of compressor and the required power input. However, there are many choices for the compressor (single-stage, multistage, with intercooling or without, reciprocating, centrifugal, etc.). Furthermore, the estimated power requirement will depend on the assumed thermodynamic pathway (isothermal or isentropic). We will now give some definitive shape to this discussion; suppose we need to compress air from 14.7 psia to 10 atm or 147 psia at a rate of 1000 ft³/min (based on inlet conditions). For a single-stage *isothermal* compression, the power requirement can be estimated:

$$P = (3.03 \times 10^{-5})(14.7)(144)(10^3) \ln\left(\frac{147}{14.7}\right) = 147.7 \text{ hp}$$
(1.54)

or 110 kW. For a single-stage *isentropic* compression, we find

$$P = \frac{(3.03 \times 10^{-5})(1.4)}{(1.4-1)}(14.7)(144)(10^3)\left[\left(\frac{147}{14.7}\right)^{0.4/1.4} - 1\right]$$
$$= 208.9 \text{ hp}$$
(1.55)

or 156 kW. And for a *three-stage isentropic* compression with *intercooling*,

$$P = \frac{(3.03 \times 10^{-5})(1.4)(3)}{(1.4-1)}(14.7)(144)(1000)$$
$$\left[\left(\frac{147}{14.7}\right)^{0.4/(1.4)(3)} - 1\right] = 165.1 \text{ hp}$$
(1.56)

or 123 kW. What compressor power should we specify? We have three solutions for (one aspect of) the problem, but we may be only a little closer to producing an actual answer. Consider that

- the thermodynamic paths chosen are merely idealizations; the real compression process will be neither isothermal nor isentropic
- since the estimates for required power are idealized, they do not account for any dissipative effects; the compressor efficiency may be much less than 100%
- other factors may weigh on the decision-making process, including capital cost, operating cost, reliability and maintenance, labor requirements, and safety issues.

This example underscores the fact that while we may be able to select and employ models for a real process, and then find solutions using those models, rarely will that process

result in a global answer for the questions being investigated. Most textbook problems and examples are far too "clean" in the sense that there is usually a narrow focus that will yield *one correct solution*. Real situations simply do not work out that way, and since models always entail idealizations, it is prudent to maintain a little skepticism until corroboration is at hand. It is essential that we try to understand the limitations of *both* the model and the solution procedure and, at the same time, not lose sight of the complications that distinguish a real-world problem from an artificially constructed textbook example.

PROBLEMS

Investigate the following scenarios and either formulate or describe an appropriate modeling approach or data analysis strategy for each. *Do not* try to solve the problem unless directed to do so.

1.1. A colony of prairie dogs numbers 47 individuals at time t. How many prairie dogs might there be at time $t = t + 30$ days?

1.2. A contaminant that is slightly soluble in water lies at the edge of a flowing stream. How might we predict the concentration distribution at a downstream distance of 1000 m?

1.3. How could we characterize the number of automobiles expected to arrive at a busy intersection over a 2-hour period during midday?

1.4. Students drop a tennis ball from the top of a building 375 ft high. What would the velocity of the tennis ball be at impact with the ground?

1.5. A equimolar stream of liquid hydrocarbons containing C_4s, C_5s, C_6s, and C_7s is flashed (partially vaporized) to split off the butanes. If the temperature and pressure are 130°F and 19 psi, how much vapor will be produced and what will the composition be?

1.6. An intravenous injection of a pharmacologically active agent gets distributed between plasma and tissue (and some is lost by elimination processes). How might we predict the concentrations in both plasma and tissue (as functions of time)?

1.7. Water flowing in a network of pipes reaches a "T" fitting and is split into two streams. One of the lines leaving the "T" has a diameter of 5 cm and the other line's diameter is 7.5 cm. What system of equations might be used to predict the distribution of flow at the "T" fitting?

1.8. A few data for ¼-mi acceleration times (drag racing) for stock cars and trucks are provided here. If a car weighs 3000 lb, what horsepower might be required to provide an ET (elapsed time) of 11.75 seconds?

Weight	Horsepower	ET/Speed
5680	300	17.3/81.2
5440	367	16.2/84.6
3883	426	13.0/109.8
3300	460	12.0/119
3220	330	13.5/107
2880	173	15.7/91
2680	170	15.8/90

1.9. Blood flows through a major artery ($d = 0.5$ cm) at a mean velocity of about 40 cm/s. How might we estimate the shear stress at the artery wall? Is blood a Newtonian fluid?

1.10. Air flows past a heated flat plate at 20 m/s. What data and equations should we employ to estimate the rate at which heat is transferred to the moving air stream? How might one determine whether or not the temperature dependence of the fluid properties, that is, $\mu = \mu(T)$ and $\rho = \rho(T)$, should be taken into account?

1.11. You have been directed to study neuron excitability and the generation of electric pulses (spikes) in nerve tissue. What characteristics must a model of this phenomenon have (you may want to start by looking at the FitzHugh–Nagumo model)?

1.12. Atherosclerosis is a disease of the arteries in which lesions form on the artery walls (actually the *arterial*, or *tunica*, *intima*). The process is not well understood, but as these lesions develop, they begin to obstruct blood flow, lead to plaque formation, and ultimately result in death by heart attack or stroke. Because atherosclerosis is a leading cause of death in much of the developed world, there is great interest in modeling lesion development. It is known that the lesions tend to form in curved arteries or near arterial branches, and it is also known that there is a strong correlation between cholesterol levels and the likelihood of atherosclerosis. What might a model of lesion formation and plaque growth look like?

1.13. It has been suggested oscillations might occur in the operation of nuclear reactor, perhaps when localized heating of a reactor component resulted in thermal expansion, changing the geometry of the reactor and thereby affecting reactivity. Critics of nuclear power have even suggested that a catastrophic event might be triggered by such a phenomenon. At least one model has been developed purporting to show this mathematically (see Thompson and Thompson, 1988). What are the essential elements of such models, and

are they capable of assessing the probability of a catastrophic event? What appears to be the major deficiency of the Thompson–Thompson model?

1.14. Uneven spots in the rails, or eccentricities in the wheels, can lead to side-to-side rocking of cars in normal railroad operations. Suppose you have been assigned the task of figuring out if such motions could ever be amplified, resulting in load-shifting, overstressing the car's structure, derailment, and so on. What kind of model might be formulated for this purpose, and what would its essential components be?

1.15. Recently, researchers from Haverford College (Pennsylvania) have successfully attached "head-cams" to falcons to study their hunting behavior. They have obtained video that reveals falcons carrying out attacks on flying crows. The video footage shows that falcons in pursuit of crows fly in such a way that the crow seems nearly stationary (and almost centered in the field of view). Can a mathematical model be formulated for a falcon's flight during pursuit of a flying crow, and what would such a model look like? *Hint: curve of pursuit.*

REFERENCES

Bird, R. B., Stewart W. E., and E. N. Lightfoot. *Transport Phenomena*, revised 2nd edition, John Wiley & Sons, New York (2007).

DePriester, C. L. Light-Hydrocarbon Vapor-Liquid Distribution Coefficients. *Chemical Engineering Progress, Symposium Series*, 49:1 (1953).

Dittus, F. W. and L. M. K. Boelter. Heat Transfer in Automobile Radiators of the Tubular Type. *University of California Publications, Engineering (Berkeley)*, 2:443 (1930).

Gilliland, E. R. and T. K. Sherwood. Diffusion of Vapors into Air Streams. *Industrial and Engineering Chemistry*, 26:516 (1934).

Glasgow, L. A. *Transport Phenomena: An Introduction to Advanced Topics*, John Wiley & Sons, New York (2010).

Hanna, O. T. and O. C. Sandall. *Computational Methods in Chemical Engineering*, Prentice Hall, Upper Saddle River, NJ (1995).

Himmelblau, D. M. and K. B. Bischoff. *Process Analysis and Simulation: Deterministic Systems*, John Wiley & Sons, New York (1968).

Lange, N. A., editor. *Handbook of Chemistry*, revised 10th edition, McGraw-Hill, New York (1961).

Saunders, P. T. *An Introduction to Catastrophe Theory*, Cambridge University Press, Cambridge (1980).

Thompson, A. S. and B. R. Thompson. A Model of Reactor Kinetics. *Nuclear Science and Engineering*, 100:83 (1988).

Thompson, J. M. T. *Instabilities and Catastrophes in Science and Engineering*, John Wiley & Sons, Chichester (1982).

Weast, R. C., editor. *CRC Handbook of Tables for Mathematics*, revised 4th edition, CRC Press, Cleveland, OH (1975).

2

ALGEBRAIC EQUATIONS

INTRODUCTION

Algebraic equations are commonly encountered in science and engineering, and a few examples of where they originate include material balances for separation processes, force resolution in structures, flow in pipe networks, application of Kirchoff's rules to electric circuits and networks, radiative exchange in enclosures, solution of discretized differential equations, and balances in chemical equilibria. Though such problems are often thought of as elementary, cases can arise that offer greater challenge than the analyst might expect.

It is impossible to know exactly when an algebraic equation was solved for the first time, but there is evidence indicating that quadratic equations were solved by the Babylonians perhaps 3700 years ago. Heath (1964, reprinted from the 1910 Edition) notes that the "father" of algebra, Diophantus of Alexandria, authored *Arithmetica* in 13 books in the third century AD. Six of the original 13 books still exist, and Heath produced English translations of them in 1885 (with the second edition published in 1910). In the *Arithmetica*, Diophantus solves determinate equations of the first and second degree; for quadratic equations he sought only rational, positive solutions in either integral or fractional form. For example, he gives

$$325x^2 = 3x + 18 \qquad (2.1)$$

and concludes that $x = 78/325$ or $6/25$. Given an equation of the form $ax^2 - bx + c = 0$, Diophantus would multiply

by a to obtain $a^2x^2 - abx + ac = 0$ and then write the solution as

$$ax = \tfrac{1}{2}b \pm \sqrt{\tfrac{1}{4}b^2 - ac}. \qquad (2.2)$$

In the twenty-first century, nearly any middle school student confronted by a quadratic equation such as

$$x^2 + 2x - 15 = 0 \qquad (2.3)$$

will immediately write the quadratic formula:

$$x = \frac{-b \pm \sqrt{b^2 - 4ac}}{2a}. \qquad (2.4)$$

Clearly,

$$x = \frac{-2 \pm \sqrt{4 + 60}}{2} = -1 \pm 4;$$

that is, $x = -5$ and $x = +3$. As we noted earlier, some determinate equations of first and second degree have been successfully dealt with for several thousand years. However, according to Heath, only one *cubic* equation is solved in *Arithmetica*:

$$x^2 + 2x + 3 = x^3 + 3x - 3x^2 - 1, \qquad (2.5)$$

Applied Mathematics for Science and Engineering, First Edition. Larry A. Glasgow.
© 2014 John Wiley & Sons, Inc. Published 2014 by John Wiley & Sons, Inc.

which is readily rewritten as

$$x^3 - 4x^2 + x - 4 = 0 \quad \text{or} \quad x(x^2 + 1) = 4(x^2 + 1). \quad (2.6)$$

Diophantus notes that $x = 4$, but it appears that he knew of no general method of solution for cubic equations. Such third-degree equations are much less mysterious today; consider the equation

$$x^3 - \tfrac{1}{2}x^2 - 6x - \frac{9}{2} = 0, \quad (2.7)$$

which we will write as $x^3 + px^2 + qx + r = 0$. If we take $x = y - p/3$, where $p = -1/2$, the second-degree term is eliminated, resulting in the form

$$y^3 + ay + b = 0. \quad (2.8)$$

The solution for the cubic equation can now be written in terms of a and b as described in any standard algebra reference book. For example, noting in eq. (2.7) that p, q, and r are $-1/2$, -6, and $-9/2$, respectively, then $a = (1/3)(3q - p^2)$ and $b = (1/27)(2p^3 - 9pq + 27r)$. A and B are now calculated from

$$A = \left[-\frac{b}{2} + \sqrt{\frac{b^2}{4} + \frac{a^3}{27}} \right]^{1/3} \quad \text{and} \quad B = \left[-\frac{b}{2} - \sqrt{\frac{b^2}{4} + \frac{a^3}{27}} \right]^{1/3},$$

and the solutions for eq. (2.8) are then $y = A + B$,

$$y = -\frac{A+B}{2} + \frac{A-B}{2}\sqrt{-3},$$

and

$$y = -\frac{A+B}{2} - \frac{A-B}{2}\sqrt{-3}.$$

The reader may want to show that for our example, eq. (2.7), $a = -6.08333$ and $b = -5.50926$, resulting in $x = -1$, $+3$, and $-3/2$.

In the types of applications that are of interest to us, it is not much of a stretch to arrive at more "interesting" cases where we have polynomials of higher degree, or products of functions, such as

$$x^4 - 3x^3 + 2x^2 - 5x + 9.76048 = 0 \quad (2.9)$$

or

$$x^4 \exp(-x) = 4. \quad (2.10)$$

Such problems arise in science and engineering all of the time; the solutions for eq. (2.10), by the way, are $x = -1.07967$, $x = 2.975924$, and $x = 5.23573$. Moreover,

in technological fields, *sets of simultaneous equations*—both linear and nonlinear—are encountered regularly. Our purpose in this chapter is to review some useful techniques for solving problems of these types.

ELEMENTARY METHODS

Newton–Raphson (Newton's Method of Tangents)

This technique was employed in the seventeenth century by Isaac Newton and Joseph Raphson, both of whom viewed it as an algebraic method to be applied to polynomials. The iterative form that is familiar to us, using derivatives of the function $f(x)$, appears to have been used in the eighteenth century by Thomas Simpson. Newton–Raphson is a powerful tool that, when it works, can be used to solve a variety of nonlinear algebraic equations.

Let us return to eq. (2.9). We wish to find a solution for this equation and we proceed in the following manner: We select an estimate for x, then construct a line tangent to the curve at that point, and extrapolate it to the x-axis where $f(x) = 0$. We use that point as our new estimate and repeat the process. The algorithm for Newton's method is easily obtained by fitting $f(x) = mx + b$ to two points: x_n, with $f(x_n)$, and x_{n+1}, where $f(x_{n+1}) = 0$. Of course, the slope, m, is simply $f'(x_n)$ and the intercept b is eliminated by subtraction resulting in

$$x_{n+1} = x_n - \frac{f(x_n)}{f'(x_n)}. \quad (2.11)$$

We illustrate with eq. (2.9):

$$f(x) = x^4 - 3x^3 + 2x^2 - 5x + 9.76048,$$

such that

$$f'(x) = 4x^3 - 9x^2 + 4x - 5.$$

We choose 3 for an initial estimate for x, and the following sequence of values emerges:

3
2.62469
2.39536
2.27549
2.31760
2.22515
2.225001
2.225001

This seems reasonable, but the polynomial we are trying to solve is fourth degree. Might there be other real solutions as well? An obvious way to explore this is to start the Newton–Raphson method at different values of x. If we try $x = 4$, we end up at the same place, but if we start the sequence at 2, we arrive at 1.875346 in five trials. For confirmation, we can use the program *cSolve* on a Texas Instruments™ TI-89 (or a comparable device) and obtain $-0.550174 \pm 1.42705i$, 1.87535, and 2.225.

Now let us suppose that the function of interest is $f(x) = 2 + \sin(2x)$. Of course, we see immediately that $f(x)$ can never be smaller than 1.0; Newton–Raphson does not know this, and if we start with 2 as our initial estimate, we get the following sequence:

2.000000
2.950975
2.074051
3.153893
2.141288
3.450326
1.868810
2.738003
1.814223
2.681012
1.686051
2.595879
1.389607
2.648622
1.592425
2.571719
1.265582
2.835763
1.965004
2.880478
etc.

Obviously, the technique does not always converge! With more complicated algebraic equations, such behavior will not be easily anticipated (as it was here). We must exercise care that we do not ask Newton–Raphson to do something it cannot do. The method is best employed in situations where the behavior of the function is well understood (of course, this is a trite observation since the statement is valid for just about *every* numerical method).

Let us conclude this section by returning to eq. (2.9) for a moment. Recall that we identified two real roots; we also pointed out the existence of a second (complex conjugate) pair of roots. It is occasionally necessary to locate complex zeros for polynomials, and in the case of eq. (2.9), it is possible to accomplish this by polynomial deflation. If we divide eq. (2.9) by $x^2 - 4.10035x + 4.17265$, we obtain

$$\frac{(x^4 - 3x^3 + 2x^2 - 5x + 9.76048)}{(x^2 - 4.10035x + 4.17265)} = x^2 + 1.10035x + 2.33917;$$

we can then use the quadratic equation on the second-degree polynomial that remains (you may want to verify that $x = 0.550175 \pm 1.42705i$). It is also possible to use a search procedure to identify complex zeros of functions and Hamming (1971) provides a useful discussion of this topic. We will illustrate a direct, brute force approach here using a real function of the complex variable, z (where $z = x + iy$). Let

$$w(z) = e^z - z^2 = e^x[\cos y + i \sin y] - (x^2 + 2ixy - y^2);$$
$$(2.12)$$

now we identify the real and imaginary parts, which are $e^x \cos y - x^2 + y^2$ and $e^x \sin y - 2xy$, respectively. We will use a *preplanned* search (a preplanned search is one in which the experiments or trials are set in advance of the calculations), looking over a range of x and y values: $-\pi \le x \le 2\pi$ and $0 \le y \le 2\pi$, and employing an interval of 0.2 in both directions. We let each (x, y) position be represented by discrete values for the indices j and k. Based on the signs of the computed real and imaginary parts, we assign the appropriate quadrant number for each (j, k) pair:

IF realp>0 AND imagp>0 THEN quadr(j,k)=1
IF realp<0 AND imagp>0 THEN quadr(j,k)=2
IF realp<0 AND imagp<0 THEN quadr(j,k)=3
IF realp>0 AND imagp<0 THEN quadr(j,k)=4

The result is a two-dimensional array of quadrant values that we can present as a contour plot, and this example is illustrated by the construction of Figure 2.1.

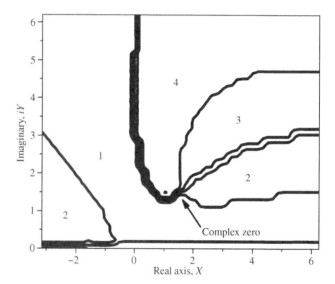

FIGURE 2.1. Results of the preplanned search in which the appropriate quadrant designations have been determined. The complex zero is located at the locus where quadrants 1, 2, 3, and 4 meet.

The direct approach that we have employed here is extremely simple, although the actual execution is a bit cumbersome. Naturally, if we needed a more accurate estimate for the complex zero, we would simply refine the search array. For example, in Figure 2.1, it is clear that $1 \leq x \leq 2$ and $1 \leq y \leq 2$. The reader is encouraged to try this to pin down the location of the complex zero.

Regula Falsi (False Position Method)

Suppose we wish to find a solution for an equation of the type

$$f(x) = \tfrac{1}{4}x^3 - x^2 + 3x - 6.50547. \qquad (2.13)$$

Furthermore, suppose we know that $f(x = 0)$ is negative (actually, -6.50547) and that $f(x = 5)$ is positive (actually, 14.7445). Clearly, there must be a sign change in the interval $(0, 5)$ that corresponds to a solution for this problem (we know that solution to be $x = 2.93333$). We could take these known endpoints, x_1 and x_2, and fit a straight line to them, since

$$f(x_1) = y_1 = mx_1 + b \qquad (2.14a)$$

and

$$f(x_2) = y_2 = mx_2 + b. \qquad (2.14b)$$

By subtraction we find the slope of this line: $m = (y_1 - y_2)/(x_1 - x_2)$, and the intercept can then be written as

$$b = y_1 - \left(\frac{y_1 - y_2}{x_1 - x_2}\right)x_1.$$

We are seeking the point where this straight-line approximation crosses the axis, that is, where $f(x_0) = y = 0$. From our equation for the linear approximation,

$$x_0 = x_1 - \left(\frac{x_1 - x_2}{y_1 - y_2}\right)y_1. \qquad (2.15)$$

If the product of the functions is $f(x_0)f(x_2) < 0$, then the sign change we are trying to identify still lies to the right, which means that we move the left-hand endpoint from the original value, x_1 to x_0. If the product is positive, then we move the right-hand endpoint inward. Let us apply this method to our sample equation using the following logic:

```
#COMPILE EXE
#DIM ALL
    REM *** Application of regula falsi method to an
        algebraic equation
```

```
    GLOBAL x1,x2,y1,y2,x0,f0,fx,x,trial,ZZ AS
        SINGLE
FUNCTION PBMAIN
    x1=0:x2=5:trial=0
    OPEN "c:regulaF.dat" FOR OUTPUT AS #1
100 REM *** continue
    x=x1
    GOSUB 300
    y1=fx
        x=x2
        GOSUB 300
        y2=fx
            x0=x1-(x1-x2)/(y1-y2)*y1
            x=x0
            GOSUB 300
            f0=fx
    IF f0*y2<0 THEN x1=x0 ELSE x2=x0
        trial=trial+1
            PRINT trial,x1,x2
                WRITE#1,trial,x1,x2
                IF trial>20 THEN 200 ELSE 100
200 REM *** continue
        INPUT "Shall we continue?";ZZ
        IF ZZ>0 THEN CLOSE
        END
300 REM *** continue
        fx=1/4*x^3-x^2+3*x-6.50547
        RETURN
END FUNCTION
```

This simple code produces the following results:

1,	1.53069877624512,	5
2,	2.17453002929688,	5
3,	2.53261733055115,	5
4,	2.72695684432983,	5
5,	2.82875323295593,	5
6,	2.88081741333008,	5
7,	2.90708756446838,	5
8,	2.92024779319763,	5
9,	2.92681646347046,	5
10,	2.93008899688721,	5
11,	2.93171787261963,	5
12,	2.93252825737,	5
13,	2.93293142318726,	5
14,	2.93313193321228,	5
15,	2.93323159217834,	5
16,	2.9332811832428,	5
17,	2.93330574035644,	5
18,	2.93331789970398,	5
19,	2.93332409858704,	5
20,	2.93332719802856,	5
21,	2.93332862854004,	

Notice that we have moved the left-hand endpoint very close to the actual solution ($x = 2.93333$) in 21 trials. *Regula falsi* is closely related to another linear approximation technique known as the secant method. With the secant method, the initial endpoints do not have to bracket the zero (the solution). The principal advantage of regula falsi over Newton–Raphson is that the former requires no evaluation of derivatives. A disadvantage of straight-line approximations is their inability to follow curvature; therefore, we might improve this approach by adding one more function evaluation (use an additional point so that the total is three) and fitting a parabola to them. This variation of the technique is known as Müller's method, and it can often provide a better first estimate for the root location. Müller's method will be the focus of one of the exercises at the end of the chapter.

Dichotomous Search

Search procedures can often be used to advantage in the solution of nonlinear algebraic equations and this is particularly evident for unimodal, one-dimensional problems. One of the simplest techniques available is the *dichotomous search*, and we will illustrate it by applying it to eq. (2.9). We are seeking an optimum, so we take eq. (2.9), square it, and invert it; that is, the merit function we wish to evaluate will be written as

$$f(x) = \frac{1}{[x^4 - 3x^3 + 2x^2 - 5x + 9.76048]^2}. \quad (2.16)$$

If problems are created by division by zero, we simply add a small constant to the denominator. The basic idea is that we select an interval to search, find the midpoint (MP) of the interval, and perform two function evaluations located at MP $+ \delta$ and MP $- \delta$. We reject the "worst" side and move the interval endpoint to that location. The logic will look something like this:

```
#COMPILE EXE
#DIM ALL
   REM *** dichotomous search for solution of an
      algebraic equation
            GLOBAL delta,MP,x1,x2,xl,xh,x,FL,FH,test,
               trial,FX,ZZ AS SINGLE
FUNCTION PBMAIN
         INPUT "Specify delta:";delta
         INPUT "Select x1:";x1
         INPUT "Select x2:";x2
            OPEN "c:dichoto2.dat" FOR OUTPUT AS #1
100 REM *** continue
         MP=(x1+x2)/2
            xl=mp-delta
```

```
            xh=mp+delta
            x=xl
               GOSUB 300
               FL=FX
            x=xh
               GOSUB 300
               FH=FX
            test=FH-FL
         IF test>0 THEN x1=xl ELSE x2=xh
         PRINT x1,x2
         WRITE#1,x1,x2
            trial=trial+1
               IF trial>25 THEN 200 ELSE 100
200 REM *** continue
         INPUT "Shall we continue?";ZZ
         IF ZZ>0 THEN CLOSE
         END
300 REM *** continue for function evaluation
         FX=1/(x^4-3*x^3+2*x^2-5*x+9.76048)^2
         RETURN
```

We will set $\delta = 0.001$ and choose the initial search interval to be $x = 0$ to $x = 22$. The results are

0,	11.0010004043579
0,	5.50150012969971
0,	2.7517499923706
1.37487494945526,	2.7517499923706
2.06231260299683,	2.7517499923706
2.06231260299683,	2.40803122520447
2.06231260299683,	2.23617172241211
2.14824223518372,	2.23617172241211
2.19120693206787,	2.23617172241211
2.21268939971924,	2.23617172241211
2.22343063354492,	2.23617172241211
2.22343063354492,	2.23080110549927
2.22343063354492,	2.22811579704285
2.22343063354492,	2.22677326202392
2.22343063354492,	2.22610187530518
2.2237663269043,	2.22610187530518
2.22393417358398,	2.22610187530518
2.22393417358398,	2.22601795196533
2.22397613525391,	2.22601795196533
2.22399711608887,	2.22601795196533
2.22399711608887,	2.22600746154785
2.22399711608887,	2.22600221633911
2.22399711608887,	2.22599959373474
2.22399711608887,	2.22599840164184
2.2239978313446,	2.22599840164184
2.2239978313446,	2.22599792480469

We repeat, again with $\delta = 0.001$, but select the initial interval to be $x = 1$ to $x = 21$.

1,	11.0010004043579
1,	6.00150012969971
1,	3.5017499923706
1,	2.25187492370605
1.62493741512298,	2.25187492370605
1.62493741512298,	1.93940627574921
1.78117179870605,	1.93940627574921
1.85928905010223,	1.93940627574921
1.85928905010223,	1.90034770965576
1.85928905010223,	1.88081848621368
1.86905372142792,	1.88081848621368
1.87393605709076,	1.88081848621368
1.87393605709076,	1.87837731838226
1.87393605709076,	1.87715673446655
1.87393605709076,	1.87654650211334
1.87424123287201,	1.87654650211334
1.87424123287201,	1.87639391422272
1.87431752681732,	1.87639391422272
1.87431752681732,	1.87635576725006
1.87433660030365,	1.87635576725006
1.87434613704681,	1.87635576725006
1.87434613704681,	1.87635099887848
1.87434613704681,	1.87634861469269
1.87434732913971,	1.87634861469269
1.87434792518616,	1.87634861469269
1.87434792518616,	1.87634837627411

Clearly, more numerical effort was required than needed with Newton–Raphson, but a search technique like this one can often be more revealing with regard to function behavior. In addition, no differentiation is required and for very complicated functions that can be a significant advantage. One might wonder if this approach could be improved with respect to numerical efficiency. For example, what if some of the function evaluations from previous stages could be reused? You will note that in the dichotomous procedure described earlier, we establish a new MP requiring two new accompanying evaluations for every stage. If we use the *golden section* search, we will achieve greater efficiency.

Golden Section Search

The *golden section* technique is based on the *golden ratio*, and its advantages can be made very clear by a simple example. Suppose we wish to locate the maximum value for a function, $f(x)$, over the interval $(0 < x < 10)$. Consider the impact of placing our initial function evaluations at $x = 6.18033989$ and $x = 3.81966011$; assume that the rightmost portion of the interval is rejected—that the extremum we seek is located between 0 and 6.18033989. If we multiply 6.18033989 by 0.618033989, we get 3.81966.... This, of course, is the location of one of our initial trials, so we can *reuse that existing function evaluation* in the next stage. Thus, the golden section technique can be used to solve an

algebraic equation with far fewer numerical operations than the dichotomous procedure.

For eq. (2.9), which we used previously, with the initial search interval $(0 < x < 20)$ we find

7.639319896698,	12.3606796264648
1.80339872837067,	2.22912335395813
2.22912335395813,	2.49223566055298
2.22912335395813,	2.32962322235107
2.22912335395813,	2.26751089096069
2.22912335395813,	2.24378609657288
2.22912335395813,	2.2347240447998
2.22566175460815,	2.22698402404785
2.22484469413757,	2.22515678405762
2.22484469413757,	2.22496390342712
2.22496390342712,	2.22503757476807

Once again, we discover that we must change the initial search interval if we wish to locate the other real-valued solution; setting $(1 < x < 21)$ we end up with the final pair, 1.875486 and 1.875679. Golden section searches are appealing because they are both efficient and easy to implement. But for *very complex equations* where execution time is critical, one can go one step farther and use the Fibonacci technique, which gives the most rapid reduction in the interval of uncertainty. Fibonacci searches are described in both Beveridge and Schechter (1970) and Shoup and Mistree (1987); the technique differs from the golden section in that the number of merit function evaluations is selected in advance and the interval of uncertainty can be computed from the Fibonacci numbers (Fs), 1, 1, 2, 3, 5, 8, 13, 21, 34, 55, 89, 144, 233, and so on. The very last evaluation uses the dichotomous technique since it cuts the final interval to 50% rather than the 61.8+% possible with the golden section. The initial function evaluations are located a distance inside the starting interval endpoints corresponding to

$$\frac{F_{n-1}}{F_n} + \frac{(-1)^n}{F_n}\delta.$$

Thus, for $n = 20$ with $\delta = 0.001$, this distance is $(6765/10{,}946) + (1/10{,}946)(0.001) = 0.618034$. Using the same values for n and δ, we can estimate the final interval of uncertainty:

$$I_u = \frac{1}{F_n} + \frac{F_{n-2}}{F_n}\delta = \frac{1}{10{,}946} + \frac{4181}{10{,}946}(0.001) = 0.000473.$$

SIMULTANEOUS LINEAR ALGEBRAIC EQUATIONS

Just about every student in science and engineering has seen a problem of this type:

$$2x_1 + 3x_2 - x_3 = 9 \tag{2.17}$$

$$-5x_1 + 2x_2 + 2x_3 = 4 \tag{2.18}$$

$$x_1 + x_2 + 3x_3 = 17. \tag{2.19}$$

Our goal, of course, is to find the correct values for the three variables, x_1, x_2, and x_3. We will do this using a technique familiar to many students, *Gaussian elimination*. The objective of this procedure is to obtain a triangular form (with 1s starting in the upper left-hand corner and proceeding diagonally down through the matrix, with zeros in the triangle underneath) where the unknowns are determined by back substitution. We begin by noting that this set of equations can be written equivalently as the *augmented coefficient matrix*:

2	3	−1	9
−5	2	2	4
1	1	3	17

We divide the first row (equation) by 2:

1	3/2	−1/2	9/2
−5	2	2	4
1	1	3	17

Divide the second row by −5 and subtract the first row from the second:

1	3/2	−1/2	9/2
0	−19/10	1/10	−53/10
1	1	3	17

Subtract the first row from the third and divide the second row by −19/10:

1	3/2	−1/2	9/2
0	1	−1/19	53/19
0	−1/2	7/2	25/2

Divide the third row by −1/2 and then subtract the second row from it:

1	3/2	−1/2	9/2
0	1	−1/19	53/19
0	0	−132/19	−528/19

Divide the third row by −132/19:

1	3/2	−1/2	9/2
0	1	−1/19	53/19
0	0	1	4

This is the triangular form we were seeking: The coefficient matrix has "1s" on the diagonal and zeros beneath. Therefore, $x_3 = 4$, and by back substitution, $x_2 = 3$ and $x_1 = 2$.

This direct elimination scheme is simple to understand and to execute, but there are several potentially serious problems. First, a machine computation will not produce the whole numbers and the fractions we see in the previous example. Each number will be represented only with the precision of the machine, which means that roundoff error will occur, and for large sets of equations, possibly accumulate. Second, if the system of equations is ill-conditioned, the number of simultaneous equations that can be solved will be quite limited. An ill-conditioned coefficient matrix is one in which small changes to a coefficient have enormous impact on the solution. Third, if a pivot element is zero and we do not interchange rows (employ partial pivoting), then division by zero occurs. Fourth, if a pivot element is very small relative to other coefficients in the matrix, the accuracy of the solution will be impaired. For these and other reasons, Gaussian elimination is not used to solve large sets of equations unless the coefficient matrix is sparse (contains many zeros). There are far better elimination methods available, and we will discuss one of those in the next section.

Crout's (or Cholesky's) Method

Crout's method is perhaps the most powerful of the direct elimination schemes—it can be used to solve fairly large sets of simultaneous linear algebraic equations and it can do so efficiently. Let us use the following problem as an example:

$$2x_1 + x_2 - x_3 + 1.26x_4 = 1.3467 \tag{2.20}$$

$$-x_1 + 3x_2 + x_3 + 0.53x_4 = 4.04 \tag{2.21}$$

$$1.082x_1 + 2x_2 + 3x_3 - x_4 = 5.3607 \tag{2.22}$$

$$3x_1 + 2.97x_2 - 1.48x_3 - x_4 = -0.82. \tag{2.23}$$

This set of equations can be written equivalently as

a_{11}	a_{12}	a_{13}	a_{14}	x_1		c_1
a_{21}	a_{22}	a_{23}	a_{24}	x_2	$=$	c_2
a_{31}	a_{32}	a_{33}	a_{34}	x_3		c_3
a_{41}	a_{42}	a_{43}	a_{44}	x_4		c_4

or more compactly as $\mathbf{A}\{\mathbf{X}\} - \{\mathbf{C}\} = 0$. If we could get the coefficient matrix in the triangular form we discussed previously with "1s" on the diagonal and zeros below, we could find the xs by back substitution. With Crout's method, a nonsingular n by n coefficient matrix is decomposed into lower ($\mathbf{L}$) and upper ($\mathbf{U}$) triangular matrices such that $\mathbf{A} = \mathbf{LU}$. $\mathbf{U}$, of course, is *exactly the form we seek*, with 1s (ones) on the diagonal (a_{11}, a_{22}, a_{33}, etc.) and zeroes below.

A typical structure for a Crout's (Cholesky's) procedure is shown as follows for the four-equation example. This routine has been adapted from one given by James et al. (1977):

```
#COMPILE EXE
#DIM ALL
    REM *** Crout's method for solution of simultaneous algebraic equations
        GLOBAL M,NEQ,J,I,II,SUM,JM1,K,IP1,IM1,JJ,NN,L AS SINGLE

    FUNCTION PBMAIN
  DIM A(10,11) AS SINGLE
  DIM X(10) AS SINGLE
    PRINT "****************************************************************"
    PRINT "      Crout's method example with four equations:            "
    PRINT "      2X1 +  X2 -  X3 + 1.26X4 = 1.3467"
    PRINT "      -X1 + 3X2  + X3 + 0.53X4 = 4.04"
    PRINT "      1.082X1 + 2X2 + 3X3 - X4 = 5.3607"
    PRINT "      3X1 + 2.97X2 - 1.48X3 - X4 = -0.82"
    PRINT "****************************************************************"

    REM *** Here's where you must input the coefficient matrix.
    A(1,1)=2:A(1,2)=1:A(1,3)=-1:A(1,4)=1.26:A(1,5)=1.3467
    A(2,1)=-1:A(2,2)=3:A(2,3)=1:A(2,4)=0.53:A(2,5)=4.04
    A(3,1)=1.082:A(3,2)=2:A(3,3)=3:A(3,4)=-1:A(3,5)=5.3607
    A(4,1)=3:A(4,2)=2.97:A(4,3)=-1.48:A(4,4)=-1:A(4,5)=-0.82

    REM *** Crout's method of matrix decomposition
    INPUT "SPECIFY THE NUMBER OF SIMULTANEOUS EQUATIONS:";NEQ
      M=NEQ+1
      FOR J=2 TO M
      A(1,J)=A(1,J)/A(1,1):NEXT J
    FOR I=2 TO NEQ
        J=I
        FOR II=J TO NEQ
        SUM=0!
        JM1=J-1
          FOR K=1 TO JM1
          SUM=SUM+A(II,K)*A(K,J):NEXT K
        A(II,J)=A(II,J)-SUM:NEXT II
        IP1=I+1
          FOR JJ=IP1 TO M
          SUM=0!
          IM1=I-1
            FOR K=1 TO IM1
            SUM=SUM+A(I,K)*A(K,JJ):NEXT K
        A(I,JJ)=(A(I,JJ)-SUM)/A(I,I):NEXT JJ
        NEXT I
      X(NEQ)=A(NEQ,NEQ+1)
      L=NEQ-1
        FOR NN=1 TO L
        SUM=0!
        I=NEQ-NN
        IP1=I+1
          FOR J=IP1 TO NEQ
          SUM=SUM+A(I,J)*X(J)
          NEXT J
        X(I)=A(I,M)-SUM:NEXT NN
          FOR J=1 TO NEQ
          PRINT X(J)
          NEXT J
      PRINT "Shall we continue with program?"
      INPUT "Respond with any positive number";NN
          IF NN>0 THEN 300
  300 REM *** continue
END FUNCTION
```

For the given example, the program produces the following results for x_1 through x_4: 0.3333475, 0.6666642, 1.66668, and 1.333349.

Matrix Inversion

Consider the following set of equations:

$$X_1 + 2X_2 + 3X_3 + X_4 = 8.98958 \qquad (2.24)$$

$$X_1 + X_2 - 9X_3 + X_4 = 0.65625 \qquad (2.25)$$

$$X_1 + 6X_3 - X_4 = 4.01042 \qquad (2.26)$$

$$5X_1 - 2X_2 - X_3 + 7X_4 = 2.14931, \qquad (2.27)$$

with the solution: 1.5, 3.0, 0.444444, and 0.15625 for X_1 through X_4, respectively. As we noted previously, we can write this set of equations alternatively as $\mathbf{AX} = \mathbf{C}$. If the coefficient matrix, $\mathbf{A}$, is nonsingular, then an inverse matrix exists such that $\mathbf{A^{-1}A} = \mathbf{I}$. The right-hand side is the identity matrix which consists of 1s on the diagonal and 0s elsewhere. This suggests the following multiplication: $\mathbf{A^{-1}AX} = \mathbf{A^{-1}C}$. Consequently, $\mathbf{X} = \mathbf{A^{-1}C}$, and the solution is at hand. So, how does one determine the inverse of $\mathbf{A}$?

Since $\mathbf{A^{-1}A} = \mathbf{I}$, we can obtain the result we seek by applying the Gauss–Jordan method (the forward and backward eliminations are combined into a single procedure) to the *augmented* matrix (note the form of the right half):

1	2	3	1	1	0	0	0
1	1	-9	1	0	1	0	0
1	0	6	-1	0	0	1	0
5	-2	-1	7	0	0	0	1

The required steps are illustrated by James et al. (1977), but one can also use software tools like Mathcad™ for this purpose (matrix toolbar, M^{-1}). The reader should verify that the inverse we are looking for is

-0.113	0.323	0.548	0.048
0.381	0.084	-0.077	-0.077
0.052	-0.090	0.00645	0.00645
0.197	-0.219	-0.413	0.087

Since $\mathbf{X} = \mathbf{A^{-1}C}$, we find that $X_1 = -0.113(8.98958) + 0.323(0.65625) + 0.548(4.01042) + 0.048(2.14931) =$ 1.497+, $X_2 = 0.381(8.98958) + 0.084(0.65625) - 0.077(4.01042) - 0.077(2.14931) = 3.0059$, and X_3 and X_4 are, respectively, 0.4481 and 0.1579. Note that the effect of roundoff error is apparent here. Matrix inversion is generally *not used* if one is only interested in a single set of simultaneous equations. It may become practical for cases with *numerous* sets of simultaneous equations.

Iterative Methods of Solution

Iterative methods are often used to solve the types of simultaneous algebraic equations that arise in the numerical solution of elliptic partial differential equations (PDEs). We will introduce one technique for dealing with such equations here but will postpone a more extensive discussion until we treat PDEs later in the book in Chapter 8. The *Gauss–Seidel* (GS) iterative method is distinguished from the *Jacobi* method by the fact that the latest iterative values are used immediately in subsequent calculations. GS is easy to understand and easy to apply, and particularly so in cases where the coefficient matrix is sparse. This is exactly what happens when we solve elliptic PDEs numerically. Let us look at an example and then reveal the required computational scheme in detail.

Suppose we have a square domain over which the equation

$$T_{i,j} = \tfrac{1}{4}[T_{i+1,j} + T_{i-1,j} + T_{i,j+1} + T_{i,j-1}] \qquad (2.28)$$

is to be applied at all interior points. This particular form has come about through *discretization* of the governing elliptic PDE and the *isolation* of the term with the largest numerical coefficient (which was 4). Let us assume that we have constant values for T on all four boundaries; in particular, let these edge values be (100, 200, 200, 300) for (L, R, T, B). In other words, along the left-hand side of the domain, the variable T has the value 100, and across the bottom, it is 300; the other two edges are maintained at 200.

This is a Dirichlet problem; we apply the computational algorithm (eq. 2.28) at each point across the bottom *interior* row successively, then we move up to the second row and repeat, and so on. We will assume that our discretization has resulted in a total of 400 mesh points of which 324 are interior points where eq. (2.28) is to be applied— we are solving 324 simultaneous linear algebraic equations. We let these interior points be initialized with the default value of 0. Obviously we could hasten the convergence of the GS scheme by making a better initial guess at the distribution, but that is not of much consequence here.

```
#COMPILE EXE
#DIM ALL
    REM *** solution of simultaneous algebraic equations
        by Gauss-Seidel iterative method
            GLOBAL i,j,iter,Told,eps AS SINGLE
FUNCTION PBMAIN
            DIM T(20,20) AS SINGLE
                iter=0
10 REM *** continue
    Told=T(10,10)
            FOR j=1 TO 20
                T(1,j)=100:T(20,j)=200
                NEXT j
            FOR i=1 TO 20
                T(i,1)=300:T(i,20)=200
                NEXT i
            FOR j=2 TO 19
            FOR i=2 TO 19
                T(i,j)=1/4*(T(i+1,j)+
                T(i-1,j)+T(i,j+1)+T(i,j-1))
                NEXT i:NEXT j
            iter=iter+1:eps=ABS(Told-T(10,10))
                PRINT iter,T(10,10),eps
                IF eps<1e-09 THEN 40 ELSE 10
40 REM *** continue
        OPEN "C:gausseid.dat" FOR OUTPUT AS #1
            FOR j=1 TO 20
            FOR i=1 TO 20
                WRITE#1,i,j,T(i,j)
                NEXT i:NEXT j
                    WRITE#1,iter
        CLOSE:END
END FUNCTION
```

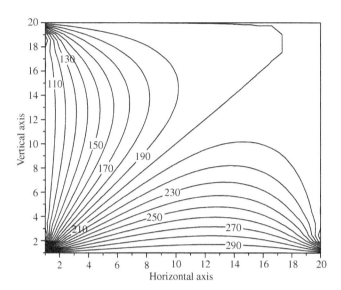

FIGURE 2.2. Contour plot of the results obtained through application of the Gauss–Seidel iterative method of solution for simultaneous linear algebraic equations. In this example, 324 simultaneous equations (resulting from the discretization of an elliptic PDE) are being solved and a satisfactory solution is obtained with less than 500 iterations. The bottom is maintained at 300 and the left-hand side at 100; the top and the right-hand side are set to 200.

Notice that progress toward convergence is being assessed by determining the change in value for a single, central point by comparing old and new iterates. For this example, 249 iterations are required to reach ε (*eps*) $= 0.01$ and 492 iterations are required to attain $\varepsilon = 1 \times 10^{-8}$. We can better examine the results of the computation by constructing an appropriate contour plot, which is provided in Figure 2.2.

The Gauss–Seidel iterative method provides a powerful tool for solving large sets of simultaneous linear algebraic equations, and we will have many opportunities to use the technique in our efforts to solve elliptic PDEs.

SIMULTANEOUS NONLINEAR ALGEBRAIC EQUATIONS

Unfortunately, we occasionally encounter problems such as the following set of equations:

$$3x_1{}^2 x_2 + x_3 = 25.032 \tag{2.29}$$

$$x_1 x_3 / x_2{}^{1.5} = 2.1896 \tag{2.30}$$

$$\sqrt{x_3}\,(x_1{}^3 + x_2) = 15.7171. \tag{2.31}$$

In this particular case, we do know that all three unknowns are positive (all greater than 0). The "easiest" way to attempt solution is through *successive substitution*. We first estimate (guess, really) two of the unknowns. We then rearrange the set of equations to solve for the third value. We use that estimate plus one of our guesses to get the second unknown, then use the two revised values to obtain the third; the entire process is repeated successively. For example, we might try to write the equations above as

$$x_1 = 2.1896 x_2{}^{3/2} / x_3 \tag{2.32}$$

$$x_2 = 15.7171 / \sqrt{x_3} - x_1{}^3 \tag{2.33}$$

$$x_3 = 25.032 - 3x_1{}^2 x_2. \tag{2.34}$$

We start by setting x_2 and x_3 equal to 2 and 2, say. The resulting sequence of computed values is

1	3.096561909	−18.578321	559.4556274
2	1.350865364	−1.8006176	34.88750458
3	0.203487575	2.652528763	24.70049858
4	0.382956296	3.106257439	23.66335106
5	0.506576121	3.100982904	22.64268303
6	0.528063238	3.155748367	22.3900528
7	0.548230529	3.156806707	22.18360329
8	0.553610981	3.167327881	22.1177845
9	0.558036625	3.16818738	22.0702343
10	0.559466541	3.170446396	22.05292511
11	0.560504615	3.170782566	22.04154968
12	0.560883105	3.171288967	22.03703499
13	0.561132371	3.171396494	22.03427315
14	0.561231256	3.171513081	22.03310585
15	0.561291933	3.171544313	22.03242874
16	0.561317503	3.171571732	22.03212929
17	0.561332405	3.171580315	22.03196144
18	0.561338961	3.171586752	22.03188705
19	0.561342597	3.171589136	22.03184509
20	0.561344266	3.171590805	22.03182602
21	0.56134522	3.171591282	22.03181458
22	0.561345637	3.171591759	22.03181076
23	0.561345816	3.171591759	22.03180885
24	0.561345875	3.171591997	22.03180695
25	0.561345994	3.171591997	22.03180695

There are obvious problems with this technique, including the following: (1) The arrangement of the equations is not unique; (2) successive substitution may converge slowly or not at all; (3) the values obtained in this calculation may be one solution, but not the one we were actually seeking; and (4) for many nonlinear problems, the analyst may have absolutely no prior knowledge of how the equations behave (what the initial estimates should be). We can underscore these points by noting that the "solution" we obtained previously by successive substitution is actually not a very good one. The reader may want to try a comparison with 1.473, 2.784, and 6.903 for x_1 through x_3, respectively. Successive substitution is related to *tearing systems of equations*, and in recent years, much effort has been expended in the development of tearing algorithms; some of these have been incorporated into commercial simulation software packages. Tearing is particularly useful in the analysis of process flow diagrams where material balances on process units may involve one or more unknown streams. In such cases, it is not possible to proceed sequentially through the process units because of the unknown streams (possibly including recycle streams). The reader interested in chemical engineering applications of tearing may want to consult Ramirez (1997); a more general discussion of issues in tearing has been provided by Elmqvist and Otter (1994).

Fortunately, we have options. For the first of these, let us consider applying Newton's method of tangents to problems of this type. We will need the Jacobian (the matrix of partial derivatives) for the system. Let eq. (2.29) through eq. (2.31) be rewritten as follows:

$$f_1 = 3x_1^2 x_2 + x_3 - 25.032 \qquad (2.35)$$

$$f_2 = x_1 x_3 / x_2^{1.5} - 2.1896 \qquad (2.36)$$

$$f_3 = \sqrt{x_3}(x_1^3 + x_2) - 15.7171. \qquad (2.37)$$

Therefore,

$$\frac{\partial f_1}{\partial x_1} = 6x_1 x_2, \qquad \frac{\partial f_1}{\partial x_2} = 3x_1^2, \qquad \frac{\partial f_1}{\partial x_3} = 1$$

$$(2.38\ a,b,c)$$

$$\frac{\partial f_2}{\partial x_1} = x_3 / x_2^{3/2}, \quad \frac{\partial f_2}{\partial x_2} = -\frac{3}{2}\frac{x_1 x_3}{x_2^{5/2}}, \quad \frac{\partial f_2}{\partial x_3} = \frac{x_1}{x_2^{3/2}}$$

$$(2.39\ a,b,c)$$

$$\frac{\partial f_3}{\partial x_1} = 3\sqrt{x_3}\,x_1^2, \quad \frac{\partial f_3}{\partial x_2} = \sqrt{x_3}, \quad \frac{\partial f_3}{\partial x_3} = \frac{1}{2}\frac{x_1^3 + x_2}{\sqrt{x_3}}.$$

$$(2.40\ a,b,c)$$

We let the set of equations, f_1, f_2, and so on, be represented as $F(x)$. The algorithm we want to use is simply

$$x_{new} = x - J^{-1}(x)F(x), \qquad (2.41)$$

which you will recognize as the equivalent to eq. (2.11), which we used previously for the solution of a single equation. As Faucett (2002) points out, it is generally not practical to evaluate the inverse of the Jacobian, so as an alternative, we solve the system of linear equations to determine the corrections for x:

$$J(x)y = -F(x). \qquad (2.42)$$

The values for y are used to improve x: $x_{new} = x + y$. We choose our initial estimates for x_1 through x_3 to be 2, 2, and 2, and then get the nine initial values for the Jacobian:

24	12	1
0.7071	−1.0607	0.7071
16.9706	1.4142	3.5355

This yields three simultaneous equations for the correction values (the ys):

$$24y_1 + 12y_2 + y_3 = 0.968 \qquad (2.43)$$

$$0.7071y_1 - 1.0607y_2 + 0.7071y_3 = -0.7754 \qquad (2.44)$$

$$16.971y_1 + 1.4142y_2 + 3.5355y_3 = -1.575. \qquad (2.45)$$

The reader may want to verify that the solution for these three equations is 0.6407, −0.9382, and −3.144 for y_1, y_2, and y_3, respectively. Therefore, our revised estimates for the three unknown variables (the x_is) are 2.641, 1.0618, and −1.1444. Of course, this changes the values in the Jacobian, so it must be recalculated then used to get improved values for the corrections (the y_is). The entire process is repeated until the change in the x_is is sufficiently small to indicate that a solution has been identified.

Often we can expect this to require four to six trials—and possibly more.

Pattern Search for Solution of Nonlinear Algebraic Equations

Sequential Simplex and the Rosenbrock Method We noted previously that search techniques (optimization algorithms) can be used to solve many types of algebraic equations and this author has found sequential simplex to be particularly useful for the exploration of nonlinear systems. The optimization method we first consider was devised by Spendley et al. (1962) and refined by Nelder and Mead (1964); it is most readily visualized in two dimensions. Suppose we are interested in the system of equations:

$$x_1 x_2^2 + \frac{x_1}{x_2} = 61.4167 \qquad (2.46)$$

and

$$75 \frac{x_2}{x_1^3} - \frac{x_1 x_2}{50} = 12.230. \qquad (2.47)$$

We disregard the fact that x_1 can simply be factored out of the first equation, facilitating easy solution. Furthermore, we know that the values we seek for the unknowns lie between 0 and 10. We modify the problem to search for the *minimum* of the function, F:

$$F = \left[x_1 x_2^2 + \frac{x_1}{x_2} - 61.4167 \right]^2 + \left[75 \frac{x_2}{x_1^3} - \frac{x_1 x_2}{50} - 12.230 \right]^2. \qquad (2.48)$$

It is convenient—and conceptually useful—to think of F in terms of a contour plot in the $x_1 - x_2$ plane.

Our objective here is to use the simplex pattern search to identify the "valley" that is prominent in the left-center of Figure 2.3; that is where the minimum is located. We will do this in the following way: Place a small equilateral triangle on Figure 2.3 and calculate the function F at each of the three vertices. We reject the "worst-valued" vertex and project the simplex away from it; that is, we form a new triangle using two of the original vertices with the third on

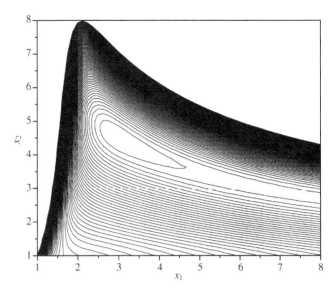

FIGURE 2.3. Contour plot for the function, F. The actual solution for this problem is $x_1 = 3$ and $x_2 = 4.5$.

the other side. We evaluate F at the new vertex and repeat the process.

There are several observations we should make about this scheme. Obviously, if we use a fixed-size triangle that is not small, we will not get a very good estimate for the solution location, regardless of the number of cycles employed. At the same time, if we make the equilateral triangle very small, it may take a very long time to follow an elongated valley such as the one seen in Figure 2.3. Clearly, we must change the size and shape of the triangle dynamically through the course of the search. This is where the improvements offered by Nelder and Mead (1964) come into play. For example, if we identify a new vertex, and the function F is improved in that direction, then why not accelerate the movement by elongating (stretching) the triangle? In this two-dimensional case, there is no reason why we must maintain the equilateral form. If we can both stretch and contract the simplex, as needed, we should be able to follow just about any ridge or valley and do so efficiently.

The Nelder–Mead modifications to sequential simplex produced a powerful tool to find minima of functions. Although Nelder–Mead codes are a bit more complicated than the search examples we have used previously in this chapter, a very effective version appears in detail in chapter 6 in Shoup and Mistree (1987). The critical steps in the logic are as follows:

- The merit function F is constructed and the number of unknowns (dimensions) is specified.
- The initial length of a simplex side is specified and the estimates for the unknowns are provided.
- The vertices of the initial simplex are determined.

- The merit function, F, is evaluated at each vertex.
- The worst and best functional values are determined.
- The centroid of the simplex is evaluated and the position of the reflected point is determined (projected away from the worst F, of course).
- F is determined at the new point; if it is better, expansion is employed (factor of 2).
- If the reflected F is worse than the other vertices, contraction is implemented (factor of ½).
- A check for convergence (based on the change in F) is made by comparison to a set value, say, 1×10^{-6}.
- Results are directed to output if convergence is attained.

Let us see exactly how this works by applying the technique to the two-dimensional problem we introduced earlier. Some of the pertinent information along with results and the final value of the function being minimized are presented in the following table. Note that the final value of the merit function F has been *multiplied by* 1×10^7.

Initial Length for Simplex	Starting Values		Final F ($\times 10^7$)	Final Values	
	x_1	x_2		x_1	x_2
0.05	0.5	0.5	13.575	0.315976	0.0051447
0.10	1.0	1.0	2.574	0.315987	0.0051449
0.25	2.0	2.0	0.142	2.999995	4.500002
0.50	5.0	5.0	3.4615	2.99996	4.500026
0.75	7.0	1.0	4.7736	2.999953	4.500032
0.05	2.0	5.0	0.590	3.00001	4.499987
0.10	2.0	7.0	1.585	2.999982	4.500025
0.10	8.0	6.0	1.813	3.000005	4.499982
0.10	0.5	9.0	2.971	3.000031	4.499991

There are a couple of important characteristics of this technique that are revealed by this example: We must bear in mind that the starting point matters and that the initial size of the simplex leg can impact the results. Moreover, the elongated valley apparent in Figure 2.3 is typical of many engineering problems. Although the Nelder–Mead version of sequential simplex can usually adapt to such circumstances, there are specialized search methods *designed specifically* to follow ridges and valleys. One example is the *Rosenbrock* method in which the axis of the search is aligned with the orientation of the ridge, allowing rapid progress toward the location of the optimum. The basic idea is easy to visualize in two dimensions; suppose we have an optimization problem involving two variables, x_1 and x_2. We begin the search with trials in the x_1-direction, identifying successes and failures. Assume we obtain a success. Then we search the x_2-direction, and if a success is achieved, we rotate the coordinate system such that one new axis is the vector sum of the two successes (and the other orthogonal

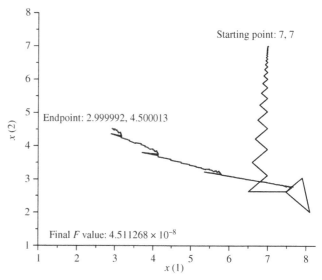

FIGURE 2.4. Progress of the Rosenbrock search method applied to the problem illustrated previously in Figure 2.3. Because of the very poor choices made for scaling factors and the starting point, 535 steps were required to identify the endpoint, 2.999992 and 4.500013. Nevertheless, the final value of the objective function was 4.511268×10^{-8}. Notice how the search direction has changed to follow the valley shown in Figure 2.3.

to it, of course). When successes are achieved, we increase the step size in that direction in the next cycle (Rosenbrock recommended a factor of 3). For a failure, the step size is contracted (and reversed) using 1/2. The appeal of Rosenbrock's approach is that the analyst obtains acceleration in *both distance and direction*. We will now use the previous example, illustrated by Figure 2.3, to reveal how Rosenbrock works in two dimensions. Our code is one that has been adapted from Shoup and Mistree (1987), and it requires specification of

ε (test value for minimum step size, 1×10^{-5})

α (scaling factor for step-size increase, 1.25)

β (scaling factor for step-size reduction, 0.85)

ST (initial step size, 0.01).

Please note that we have selected suboptimal values for both α and β; this has been done to better reveal the Rosenbrock scheme's progress toward finding the valley revealed in Figure 2.3. We are also going to start the search from the initial point, (7, 7), and attempt to minimize the function given by eq. (2.48); we will illustrate the progress in Figure 2.4.

A problem that can be encountered with the Rosenbrock method and one that has been mentioned by Beveridge and Schechter (1970) is that it can be slow to start particularly if the initial step size must be reduced repeatedly. It is, however, a very powerful tool that can be used to advantage for certain problems involving nonlinear, algebraic equations, and more importantly, examples of programming logic for the Rosenbrock search can be found across the Web and throughout the literature.

An Example of a Pattern Search Application Now let us look at a problem type that is quite different (in both intent and form) from the previous examples; we will use the sequential simplex method. We still wish to minimize a merit function developed from a set of simultaneous nonlinear algebraic equations, but our purpose is unlike previous examples in this chapter. Imagine we are seeking an approximate relationship (a correlation) that could be used to predict the likely attainable speed (S) achieved by a conventional ship hull. It is apparent that some of the important factors are displacement, hull length (L), beam (B), available horsepower (P), and hull shape. Of course, we know that certain dimensionless ratios such as Reynolds number, Froude number, and Euler number are important, but we are going to try a minimalist approach and not worry overtly about hydrodynamic forces. We propose that the ship's likely speed in knots can be determined in the following way:

$$S = aL^b B^c P^d. \tag{2.49}$$

We intend to attempt this by using some available data:

	Length	Beam	P (hp)	S (knots)
Titanic	883	92	46,000	21
Bismarck	793	118	150,170	30
Missouri	887	108	212,000	33
Nimitz	1,040	134	260,000	31.5
Edmund Fitzgerald	711	75	7,500	14
Elco PT boat	80	21	4,500	41
Monitor	172	41	320	7–8

Keep in mind that we are using data from ships designed in very different eras for very different purposes, so we expect this may not produce the cleanest result we could imagine. The minimum merit function (F) value was found to be 4.9339 (using all of the data included in the accompanying table) and the correlation that emerged was

$$S = 34.9084 L^{-0.28125} B^{-0.5764} P^{0.3738}. \tag{2.50}$$

For the *Titanic*, the correlation yields 21.15 knots; for the *Elco PT* boat, 40.84 knots; and for the *Monitor*, 8.34 knots. (An aside: There are indications that Ericsson's original plans called for the installation of two engines. However, one engine has been recovered from the site of the *Monitor*'s wreck and is now on display at the Mariners' Museum in Newport News, Virginia. In trials in 1862, the *Monitor* was able to achieve only 6.25 knots.) One of the poorer results is the *Edmund Fitzgerald* for which the correlation yields 12.83 knots (instead of 14). For the battleship *Missouri* (BB-63), our equation indicates 34.1 knots, and for the *Nimitz* (CVN-68), it yields 31.1 knots. This is encouraging, and the correlation may be good enough to make some rough projections. Suppose, for example, we wanted a high-speed (50-knot) hull that was 300 ft long; we choose a beam of 40 ft (though such a design will be prone to roll) because the importance of width to speed is clear. Our eq. (2.50) indicates that about 56,445 hp will be required for this hull to make a speed of 50 knots; the fastest ship in the US Navy, the 45+-knot LCS-1 *Freedom*, is 324 ft long with a beam of 43 ft and she has 96,000 hp. The correlation (2.50) suggests that the *Freedom* should be capable of about 57 knots; however, the *Freedom* uses water jet propulsion, which is known to be somewhat less efficient than a propeller. The Nelder–Mead version of sequential simplex is extremely useful for solving this and related types of nonlinear problems.

ALGEBRAIC EQUATIONS WITH CONSTRAINTS

Suppose we have the algebraic equation

$$2x_1^2 + x_2^2 - 8x_1 + x_2 + 1 = 0 \tag{2.51}$$

accompanied by the constraint $2x_1 - x_2 = 0$. We wish to find the stationary point (values of the independent variables that cause the derivatives to be zero) and we will do so by using two very well-known methods. First, we will make use of the *Lagrange multiplier*, λ. We write a combination of the equation and the constraint as follows:

$$F = 2x_1^2 + x_2^2 - 8x_1 + x_2 + 1 + \lambda(2x_1 - x_2). \tag{2.52}$$

We now differentiate F with respect to all *three* variables:

$$\frac{\partial F}{\partial x_1} = 4x_1 - 8 + 2\lambda = 0$$

$$\frac{\partial F}{\partial x_2} = 2x_2 + 1 - \lambda = 0 \tag{2.53a,b,c}$$

$$\frac{\partial F}{\partial \lambda} = 2x_1 - x_2 = 0.$$

These three equations are solved yielding $x_1 = 1/2$, $x_2 = 1$, and $\lambda = 3$ (the reader may wish to verify this solution). Next, to provide a contrast, we will apply *Courant's penalty method* to the very same problem; in Courant's scheme, we seek a minimum for the *augmented functional*:

$$F = 2x_1^2 + x_2^2 - 8x_1 + x_2 + 1 + \frac{\varepsilon}{2}(2x_1 - x_2)^2. \quad (2.54)$$

Note that *no additional variable* has been introduced, only the penalty parameter, ε. Of course, this means that the penalty method is computationally simpler than the Lagrange multiplier used earlier. This time, we have two derivatives to set equal to zero:

$$\frac{\partial F}{\partial x_1} = 4x_1 - 8 + 2\varepsilon(2x_1 - x_2) = 0 \quad (2.55a)$$

and

$$\frac{\partial F}{\partial x_2} = 2x_2 + 1 - \varepsilon(2x_1 - x_2) = 0. \quad (2.55b)$$

With a bit of effort, one can show that

$$x_2 = \frac{-4 + 12\varepsilon}{8 + 12\varepsilon} = \frac{-1 + 3\varepsilon}{2 + 3\varepsilon},$$

and this result can be used to find $x_1 = f(\varepsilon)$. As the penalty parameter, ε, becomes large, these expressions should converge to the correct values for the two variables. We will explore this process with the following results:

ε	$x_1(\varepsilon)$	$x_2(\varepsilon)$
1,	1.10000002384186	.400000005960464
10,	.59375	.90625
100,	.509933769702911	.990066230297089
1,000,	.500999331474304	.999000668525696
10,000,	.500100016593933	.999899983406067
100,000,	.500010013580322	.999989986419678

We see that Courant's penalty method will approach the correct values for our two variables if the penalty parameter, ε, is sufficiently large.

CONCLUSION

One of the interesting aspects of the solution of algebraic equations is the spectrum of options available. In many cases, the analyst can choose from several solution techniques, all capable of producing the desired result. At the same time, for some cases, no method will yield an appropriate result. For students in the applied sciences, balancing the focus between procedure and result may improve the chance of success and lessen the chance of serious error. Obviously, for "real" problems, we may have little or no information available in advance and when these problems involve *nonlinear* algebraic equations, we may not have much insight into where the solution we are seeking may be found.

Although we will encounter cases where, for example, we will know that the volume cannot be negative, or that the summation of the mole fractions must be 1, situations arise where we have no idea of the appropriate magnitude of the unknowns. Such instances are disconcerting since it is difficult to know whether or not the solution procedure is working properly. These comments are not meant to alarm the student; they are only intended to point out that circumspection is critical. For nonlinear algebraic equations, the analyst should consider the physical situation very carefully *before* initiating an attempt to find a solution.

We can underscore the preceding observations with a final example for which we will use a pair of widely available tools. Consider the set of nonlinear algebraic equations:

$$x^2 + \frac{y^3}{x} = 1.84401 \quad \text{and} \quad x + xy + x^2y^2 = 2.80076. \quad (2.56)$$

Assume we know that both x and y are positive for the solution we seek (in fact, $x = 1.275$ and $y = 0.653$). We first use a scientific calculator from the TI-89, TI-92 family (TI is the abbreviation for Texas Instruments). From the *MATH/Algebra* menu, we select *solve( )*, and enter our two equations as follows:

$$solve(x \wedge 2 + y \wedge 3 / x = 1.84401 \quad \text{and}$$
$$x + x \cdot y + x \wedge 2 \cdot y \wedge 2 = 2.80076, \{x = 2, y = 0.8\})$$

and press *ENTER*. Using the guesses, $x = 2$ and $y = 0.8$, the TI-89 produces $x = 1.275$ and $y = 0.653003$. However, if we start with $x = 1$ and $y = 1$, we obtain $x = 0.985357$ and $y = 0.951078$. Naturally, a software package like Mathcad also has capabilities for problems of this type, using the *Solve Block*:

$$x := 2$$
$$y := 2$$

Given

$$x^2 + \frac{y^3}{x} = 1.8440$$
$$x + xy + x^2y^2 = 2.8007$$
$$\binom{xval}{yval} := Find(x, y)$$
$$xval = 1.275$$
$$yval = 0.653$$

Note that the initial estimates used at the top were $x = 2$ and $y = 2$. Once again, however, if we select 1 and 1 we obtain $x = 0.985$ and $y = 0.951$; when we insert these values into the pair of eq. (2.56), we find 1.84341 (instead of

1.84401) and 2.79921 (instead of 2.80076). Even a very simple nonlinear algebraic problem can produce undesired or incorrect results if a solution tool is used incautiously.

PROBLEMS

2.1. Use the Newton–Raphson method to find a solution for the equation

$$x^5 - 2x^3 + 6x - 0.851372 = 0.$$

It is known that the solution we are seeking is between 0 and 1.

2.2. Solve the set of simultaneous, linear algebraic equations represented by the augmented coefficient matrix:

5	1	1	−1	−1	0.899171
1	2	1	−4	1	0.223846
2	−1	−1	−1	6	0.49982
3	3	−2	−3	−3	0.072511
−1	−1	−1	7	7	0.979619

2.3. Consider a steel pipe with a nominal diameter of 3" (actually 3.068" ID). Water flows through the pipe (which is 980 ft long) at an average velocity of 8.75 ft/s. The Reynolds number associated with this flow is

$$\text{Re} = \frac{d<V>\rho}{\mu} = \frac{(3.068/12)(8.75)(62.4)}{(1)(6.72\times10^{-4})} = 207{,}729.$$

The friction factor for this flow is approximately given by $f = 0.0791/(\text{Re})^{1/4}$. Therefore, the value of f for the original flow is about 0.0037. In an effort to increase the flow rate, a parallel (also nominal 3") pipe is installed *over the last* 600 ft. Find the (increased) flow rate through the new arrangement. The design equation for flow through a horizontal pipe is

$$\Delta \tfrac{1}{2}\langle V\rangle^2 + \frac{p_2 - p_1}{\rho} + \sum \tfrac{1}{2}\langle V\rangle^2 \frac{L}{R_h} f = 0.$$

This allows us to calculate the pressure drop through the original 3" pipe; initially, there is no change in kinetic energy, so

$$\frac{p_1 - p_2}{\rho} = (1/2)(8.75)^2 \left(\frac{980}{0.0639}\right)(0.0037) = 2172.3 \text{ ft}^2/\text{s}^2.$$

We now multiply by the fluid's density and take care of the mass–force conversion problem using g_c. The result is 29.26 psi ($p_1 - p_2$). In the revised installation, the *overall*

pressure drop is exactly the same, but we have an additional 3" pipe over the last 600 ft. Find the new flow rate by writing the design equation three times, for the initial 380 ft of 3" pipe and then once for each of the two parallel legs extending from 380 to 980 ft.

2.4. We will consider a problem in chemical equilibria in which a compound is put into water and dissociation occurs. Let us preface this problem with a little review.

Compounds like silver chloride (AgCl) are sparingly soluble in water; at 10°C, about 0.00089 g AgCl will dissolve in 1 L of water. An equilibrium will be established among the three species:

$$K = \frac{[\text{Ag}^+][\text{Cl}^-]}{[\text{AgCl}]},$$

where K is the equilibrium constant. Water itself is a weak electrolyte for which

$$\text{H}_2\text{O} \Leftrightarrow \text{H}^+ + \text{OH}^-.$$

At 25°C, $K_w = [\text{H}^+][\text{OH}^-] = 1\times10^{-14}$. In contrast, sodium hydroxide (NaOH) is a strong electrolyte that will almost completely dissociate in water. Therefore, if we add 0.005 mol NaOH (about 0.2 g) to a liter of water,

$$1\times10^{-14} = [\text{H}^+][\text{OH}^-],$$

but with complete dissociation, $[\text{OH}^-] = 5 \times 10^{-3}$, so

$$[\text{H}^+] = \frac{1\times10^{-14}}{5\times10^{-3}} = 2\times10^{-12}.$$

Since pH $= -\log_{10}[\text{H}^+]$, pH $= 11.7$.

Now, suppose we place 0.0002 mol HCl in a liter of water at 25°C. We want to find the concentrations of all four species, as well as the pH of the resulting solution. Therefore, we need

$$[\text{H}^+] \quad [\text{Cl}^-] \quad [\text{HCl}] \quad \text{and} \quad [\text{OH}^-].$$

The following relationships are available to us:

$$[\text{H}^+][\text{OH}^-] = K_w = 1\times10^{-14}, \quad \frac{[\text{H}^+][\text{Cl}^-]}{[\text{HCl}]} = K = 1\times10^3$$

$$[\text{HCl}] + [\text{Cl}^-] = 2\times10^{-4} \quad (\text{total concentration})$$

$$[\text{H}^+] = [\text{Cl}^-] + [\text{OH}^-] \quad (\text{electroneutrality}).$$

2.5. Use the technique of your choice to find a solution for the following set of equations, given that all of the unknowns lie between zero and one:

$$x_1{}^2 \exp(-x_2) + x_3 = 0.385028$$

$$\frac{x_1}{x_2{}^2 + x_3{}^2} = 2.20022$$

$$x_1 + x_2 - x_3{}^3 = 0.710317.$$

2.6. The equation, $0.5 + \exp(-0.085t)\sin(2t)$, has a finite number of zeros for positive values of t between 0 and 11. Devise a method that will locate and count all of them.

2.7. Most airline travelers would prefer to spend less time in the air. Clearly, one way this could be accomplished is to increase the speed of aircraft. Presently, large passenger airplanes cruise at high-subsonic speeds, and at altitude, this means about 500–600 mph. If longer flights could cruise at Mach 2 or 3 (or more), travel times would be reduced significantly. The problem is twofold: We have the rise in drag accompanying supersonic flight (which dramatically increases fuel consumption) and we have aerodynamic heating. For example, if we look at the Concorde (now retired from service), we find fuel consumption at full power of 6180 gal (A1 jet fuel)/h; the Concorde was able to carry one passenger about 17 mi on a gallon of fuel. We can contrast this with a Boeing 747; with 500 passengers, a 747 can carry one passenger between 60 and 100 mi on 1 gal of fuel. Of course, it will take two and a half times as long to reach the destination! So this raises an interesting question: What must the design characteristics of an advanced aircraft resemble in order for the craft to carry a hundred or more passengers at a cruising speed of Mach 3 or 4? Here are some approximate data from open sources:

Aircraft	Span	Gross Weight	Thrust	Cap	Max Speed
LH C-140	54	38,940	6,000	12	512
DH Comet	115	162,000	42,000	101	525
B 737-700	117	150,000	52,600	149	540
H FGA.9	34	18,000	10,050	1	627
NA F-100D	39	34,832	16,950	1	864
MD F-101B	40	52,400	33,800	2	1,130
Concorde	84	408,000	152,000	128	1,330
NA A3J-1	53	49,500	32,300	2	1,385
Con B-58A	57	160,000	62,500	3	1,385
LH F-104G	22	27,000	16,150	1	1,450
Con F-106A	38	35,000	24,500	1	1,525
MiG 25	46	80,950	49,400	1	1,900
LH SR-71	56	170,000	68,000	2	2,090
X-15	22	33,000	70,400	1	4,519

Some observations: Note the relatively high take-off weights for the Concorde and the SR-71; this is indicative of the fuel load required if one wants to cruise at high speeds. Even so, the SR-71 had to be refueled routinely in

flight to accomplish its long-range surveillance missions. The very low weight of the X-15 was possible because it was air-launched (dropped from underneath a B-52). One factor we referred to previously that is not reflected here is the trouble caused by aerodynamic heating. For many supersonic aircraft, this was/is a serious limitation on the performance envelope. The MiG 25, for example, could only be flown at high Mach for a few minutes before the structural components would attain unsafe temperatures. The X-15 suffered serious thermal damage during its October 1967 flight in which it achieved Mach 6.7. The tabulated data make it pretty obvious that to carry 100 passengers at the desired speeds, we will need a wing span of 90–100 ft and a take-off weight of least 200,000–300,000 lb.

2.8. Find a solution for the following set of nonlinear algebraic equations:

$$x_1{}^5 \exp(-2x_2) + 10/x_3 = 2.34284$$

$$x_1 x_2{}^2 + \sqrt[3]{x_3} = 25.5966$$

$$\frac{x_1 + x_3{}^3}{\sqrt{x_2}} = 48.1033.$$

2.9. Müller's method, a root finding technique in which a parabola is fit to three points, can be modified to find complex roots. Consider the fourth-degree polynomial:

$$x^4 - \frac{17}{2}x^3 + \frac{353}{18}x^2 - \frac{17}{2}x - \frac{185}{18} = 0.$$

Search for real roots using the method of your choice and then use polynomial deflation to locate the complex conjugate pair. Confirm your findings using a calculator (like a TI-89) or a suitable commercial software (you should get $-1/2$, $+5$, and $+2 \pm i/3$). Then, repeat the analysis by consulting the literature (e.g., section 7.4 in Chapra and Canale, 2002) and developing a routine to use Müller's method for the same purpose. The following discussion should help you get started.

We assume that the function can be approximated with the polynomial:

$$f(x) = a(x - x_0)^2 + b(x - x_0) + c.$$

This choice ensures that $f(x_0) = c$ and thereby lessens our workload. We arbitrarily select 4, 6, and 8 for our three trial points; we find $f(4) = -18.5$, $f(6) = 104.714$, and $f(8) = 920.783$. We choose $x_0 = 8$ for our reference point, so $c = 920.783$. Our function values at $x = 4$ and $x = 6$ give us two simultaneous equations for a and b: $a = 86.607$ and

$b = 581.249$. Of course, we want to identify the value for x where $f = 0$. Thus (after dividing by 86.607),

$$0 = (x-8)^2 + 6.7113(x-8) + 10.632.$$

Letting $z = (x - 8)$, we have the quadratic: $z^2 + 6.7113z + 10.632$, with solutions at -4.14835 and -2.56295. Remembering to add 8, we get our improved estimates and the closest one is $+5.43705$. One advantage of Müller's technique is the use of the quadratic equation—we will be able to locate complex roots should they arise. We can also minimize our workload by reusing the trial points closest to the root estimate. In this case, we would discard $x = 8$, retaining 4, 5.43705, and 6 for our next cycle.

2.10. The cargo ship *MSC Fabiola* is one of the largest container transports ever constructed. She is 366 m (1200 ft) long with a beam of 48 m; the gross tonnage is 140,259. This class of freighter will not fit through the Panama Canal (which is one of the reasons why the canal is being enlarged). Estimate the maximum speed of the *MSC Fabiola* given that her turbines produce 98,000 hp (single screw). It has been reported that she is capable of 19.5 knots. Is that speed realistic? Does it appear that the correlation given by eq. (2.50) is appropriate for the *Fabiola*?

2.11. Solve the set of simultaneous algebraic equations:

$$(X_1 - X_2)^2 + X_1 X_2 X_3 = 61.0229$$

$$X_1 + X_2 + X_3 = 15.8225$$

$$\frac{X_2}{X_3^2} + \frac{X_1}{X_2} = 11.366.$$

We know that for all three variables, $0 \le X_n \le 10$. Use the method of your choice.

2.12. The Lagrange multiplier (λ) is often used for constrained optimization problems. Consider the equation

$$2x_1^2 - 3x_2^2 - 5x_1 + x_2 = -5.1207,$$

accompanied by the constraint, $2x_1 - 3.066202x_2 = 0$. Use the Lagrange multiplier to solve this problem; that is, find the stationary point subject to the constraint.

2.13. Use your choice of methods to seek a solution for the two simultaneous equations:

$$\frac{x_1 \exp(-x_1^2)}{x_2} + x_2^3 = 1.42957$$

and

$$x_1^2 x_2 + \frac{x_1 \tan(x_1)}{x_2^2} = 77.4525.$$

We know that both unknowns are between 0 and 2.

2.14. The first seven *Chebyshev* (or Tschebysheff) polynomials of the *first kind* are:

$$T_0(x) = 1, \qquad\qquad T_1(x) = x$$

$$T_2(x) = 2x^2 - 1, \qquad\qquad T_3(x) = 4x^3 - 3x$$

$$T_4(x) = 8x^4 - 8x^2 + 1, \quad T_5(x) = 16x^5 - 20x^3 + 5x$$

$$T_6(x) = 32x^6 - 48x^4 + 18x^2 - 1.$$

These polynomials are solutions for the Chebyshev differential equation (for Mathcad, see *Tcheb(n, x)*). Suppose we need to identify all of the zeros for $T_6(x)$ on the interval, $-1 \le x \le +1$. Prepare a plot of this polynomial and then use the method of your choice to find accurate values for the roots.

2.15. A liquid-phase reaction was carried out in a batch reactor to study $A + B \rightarrow C + D$. The experiment was conducted for four different trials and stopped at different elapsed times. The data obtained consisted of the duration of the experiment and the concentration of species D:

Time (s)	780	2040	3540	7200
[D] (mol/L)	0.0112	0.0257	0.0367	0.0552

Two different rate expressions have been proposed for this reaction:

$$\frac{d[A]}{dt} = -k_1[A] \quad \text{and} \quad \frac{d[A]}{dt} = -k_2[A][B] = -k_2[A]^2,$$

which, when integrated, yield $\ln[A] = -k_1 t + C_1$ and $-(1/[A]) = -k_2 t + C_2$, respectively. Which rate expression is more nearly correct? There is no D present in the reactor initially, and the initial concentrations for A and B are both 0.1.

REFERENCES

Beveridge, G. S. G. and R. S. Schechter. *Optimization: Theory and Practice*, McGraw-Hill, New York (1970).

Chapra, S. C. and R. P. Canale. *Numerical Methods for Engineers*, 4th edition, McGraw-Hill, New York (2002).

Elmqvist, H. and M. Otter. Methods for Tearing Systems of Equations in Object-Oriented Modeling. *Proceedings of the European Simulation Multiconference*, Barcelona (1994).

Faucett, L. *Numerical Methods Using Mathcad*, Prentice Hall, Upper Saddle River, NJ (2002).

Hamming, R. W. *Introduction to Applied Numerical Analysis*, McGraw-Hill, New York (1971).

Heath, T. L. *Diophantus of Alexandria: A Study in the History of Greek Algebra*, 2nd edition, Dover Publications, New York (1964). reprinted from Cambridge University Press, 1910 Edition.

James, M. L., Smith G. M., and J. C. Wolford. *Applied Numerical Methods for Digital Computation with FORTRAN and CSMP*, 2nd edition, Harper & Row, New York (1977).

Nelder, J. A. and R. Mead. A Simplex Method for Function Minimization. *Computer Journal*, 7:308 (1964).

Ramirez, W. F. *Computational Methods for Process Simulation*, 2nd edition, Butterworth-Heinemann, Oxford (1997).

Shoup, T. E. and F. Mistree. *Optimization Methods with Applications for Personal Computers*, Prentice-Hall, Englewood Cliffs, NJ (1987).

Spendley, W. G., Hext G. R., and F. R. Himsworth. Sequential Applications of Simplex Designs in Optimization and Evolutionary Operation. *Technometrics*, 4:441 (1962).

3

VECTORS AND TENSORS

INTRODUCTION

A significant area of interest in continuum mechanics is the application of forces to material objects and the response of those objects to the applied forces. Most of us tend to think of a force simply as a directed line segment (possessing magnitude and direction), but a little caution is in order here: We must remember that the *effect* of a force applied to a body depends on both its line of action and the point at which it is applied. For example, if one were using a lever to move a boulder, the distance between the point of application and the fulcrum would have enormous impact on the effectiveness of the action. We also need to emphasize that our discussions here are concerned with continuum mechanics in *three-dimensional Euclidean space*. Thus, when we speak of tensors, for example, we mean *Cartesian tensors*. Tensors do figure prominently in non-Euclidean spaces, but those applications are not relevant to our principal objectives.

We will begin by reviewing what we mean when we refer to scalars, vectors, and tensors. A scalar is a quantity that has magnitude only; for example, we might find that an enclosure has a volume of 1.2 m^3 or that the fluid contained within has a temperature of 215°F (101.67°C). We also observe that a scalar is a zero-order tensor (we will use *order and rank* synonymously). In contrast, a vector has both magnitude and direction, and we can think of force and velocity as examples. A vector with three components is a first-order tensor. By tensor we merely mean an ordered set of numbers; second-order Cartesian tensors (with nine components) are very important in the mechanics of materials and in hydrodynamics. The third-order alternating tensor, ε_{ijk}, has 27 components and is equal to $+1$ if the subscripts are in cyclic order (e.g., 1, 2, 3 or 2, 3, 1), -1 for anticyclic order, and zero otherwise. The alternating tensor is particularly useful for the cross product of certain vectors. You probably have detected a pattern:

Tensor Order	Number of Components
0	1
1	3
2	9
3	27

To illustrate how second-order tensors come about, let us think about a force acting on a surface such that the dimensions are F/L^2 or $M/(Lt^2)$. Of course, pressure and shear stress are prime examples and we recognize that two directions are important for such quantities. Consider a fluid flowing in the x-direction past a fixed solid surface located at $y = 0$; because of the Newtonian no-slip condition (Newton's law of viscosity), a shear stress will be created by the surface as shown in Figure 3.1.

We will interpret this shear stress, τ_{yx}, as a force acting on the y-plane (the fixed surface is located at $y = 0$) due to fluid motion in the x-direction. Since we have three principal directions, it is clear that each of the two indices on τ can assume one of three values, corresponding to x, y, and z in

Applied Mathematics for Science and Engineering, First Edition. Larry A. Glasgow.
© 2014 John Wiley & Sons, Inc. Published 2014 by John Wiley & Sons, Inc.

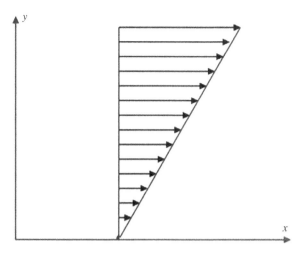

FIGURE 3.1. Shear stress created by fluid motion past a solid surface. For this illustration, $\nu_x = my$ and $\tau_{yx} = -\mu(d\nu_x/dy) = -\mu m$. Momentum is transferred to the stationary surface (in the negative y-direction).

rectangular coordinates. Thus, the stress tensor consists of the set of components:

$$
\begin{array}{ccc}
\tau_{xx} & \tau_{xy} & \tau_{xz} \\
\tau_{yx} & \tau_{yy} & \tau_{yz} \\
\tau_{zx} & \tau_{zy} & \tau_{zz}
\end{array}
\qquad (3.1)
$$

You will see immediately that the subscripts are repeated on the diagonal; these are called *normal* stresses and their sum is called the *trace* of the tensor. The off-diagonal components are shear stresses, and we should recognize that the corresponding off-diagonal stresses must be equal; that is, $\tau_{yx} = \tau_{xy}$. If this were not the case, then an infinitely small element of fluid could experience very large (infinite!) angular acceleration. This requirement means that the stress tensor is *symmetric* and that it contains just six independent quantities.

It is common practice to write second-order tensors using the Cartesian summation convention. For example, we might write S_{ij} where the indices i and j can each assume the values 1, 2, and 3. Consequently, if we write S_{ii} (with repeated subscripts), we mean

$$
S_{ii} = S_{11} + S_{22} + S_{33}, \qquad (3.2)
$$

which of course is the trace of the tensor as we observed previously.

MANIPULATION OF VECTORS

Let us review some vector algebra, noting that **A**, **B**, and **C** are vectors and that a and b are scalars:

$$\mathbf{A} + \mathbf{B} = \mathbf{B} + \mathbf{A} \qquad \text{commutative} \qquad (3.3)$$

$$\mathbf{A} + (\mathbf{B} + \mathbf{C}) = (\mathbf{A} + \mathbf{B}) + \mathbf{C} \qquad \text{associative (addition)} \qquad (3.4)$$

$$a(b\mathbf{A}) = (ab)\mathbf{A} = b(a\mathbf{A}) \qquad \text{associative (multiplication)} \qquad (3.5)$$

$$(a + b)\mathbf{A} = a\mathbf{A} + b\mathbf{A} \qquad \text{distributive} \qquad (3.6)$$

$$a(\mathbf{A} + \mathbf{B}) = a\mathbf{A} + a\mathbf{B} \qquad \text{distributive.} \qquad (3.7)$$

When we refer to *unit vectors*, we mean unit length. Therefore, for vector **A** with length a, unit length is otained simply by $\mathbf{A}/a$. In rectangular coordinates, it is common practice to write the unit vectors as i, j, k, corresponding to the x-, y-, z-directions.

We will review these basic operations by looking at a few examples. Let the vectors **A** and **B** be given by

$$A = 4i - 3j + 2k \qquad (3.8)$$

$$B = 5i + 5j + 3k. \qquad (3.9)$$

First, we look at the sum:

$$A + B = (4 + 5)i + (-3 + 5)j + (2 + 3)k = 9i + 2j + 5k. \qquad (3.10)$$

The magnitude of A is

$$|A| = \left[(4)^2 + (-3)^2 + (2)^2\right]^{1/2} = \sqrt{29} = 5.38516. \qquad (3.11)$$

If we multiply B by a scalar, b, letting $b = 3$,

$$bB = 3(5i + 5j + 3k) = 15i + 15j + 9k. \qquad (3.12)$$

Now suppose a body, M, is being acted on by three different forces, F_1, F_2, and F_3. Let

$$F_1 = 5i + 5j - 1k \qquad (3.13)$$

$$F_2 = 6i - \frac{1}{2}j + 1k \qquad (3.14)$$

$$F_3 = 3i - 3j + 2k. \qquad (3.15)$$

The resultant force is $F_1 + F_2 + F_3$:

$$R = F_1 + F_2 + F_3 = 14i + \frac{3}{2}j + 2k. \qquad (3.16)$$

If we wished to prevent movement of the body, M, we could apply a force corresponding to $-R$: $-R = -14i - \frac{3}{2}j - 2k$. This force would produce equilibrium and the body, M, would naturally remain at rest if it were initially stationary. Let us now assume that we would like to identify the unit vector that is parallel to the resultant of two forces given by

$$F_1 = 3i + 2j + 2k \qquad (3.17)$$

and

$$F_2 = 2i - 1j + 1k. \qquad (3.18)$$

The resultant is

$$R = F_1 + F_2 = (3i + 2j + 2k) + (2i - 1j + 1k) = 5i + 1j + 3k. \qquad (3.19)$$

The magnitude of the resultant is

$$|R| = \left[(5)^2 + (1)^2 + (3)^2 \right]^{1/2} = \sqrt{35} = 5.91608. \qquad (3.20)$$

So the unit vector we are seeking is

$$\frac{R}{|R|} = \frac{5i + 1j + 3k}{5.91608} = 0.84515i + 0.16903j + 0.50709k. \qquad (3.21)$$

Often we are interested in the angle between a given vector and the coordinate axes. For transparency, we will start with a vector in the x-y plane and determine the angle (θ) between this vector and the x-axis. Let

$$F_1 = 6i + 3j. \qquad (3.22)$$

It is apparent that $\tan\theta = 3/6$ such that $\theta = 0.4636$ rad or 26.57°. More generally, if a vector in three-space is given by

$$F_1 = 6i + 3j + 2k, \qquad (3.23)$$

then the angles between this vector and the coordinate axes are

$$\theta_1 = \cos^{-1}\frac{x}{|F_1|}, \qquad (3.24)$$

$$\theta_2 = \cos^{-1}\frac{y}{|F_1|}, \qquad (3.25)$$

and

$$\theta_3 = \cos^{-1}\frac{z}{|F_1|}. \qquad (3.26)$$

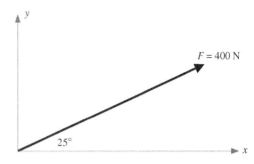

FIGURE 3.2. A force of 400 N applied in the x-y plane at an angle of 25°.

Since $|F_1| = \sqrt{36 + 9 + 4} = 7$, the three angles we are seeking are 0.5411, 1.1279, and 1.2810 rad, respectively. Please note that

$$\cos^2\theta_1 + \cos^2\theta_2 + \cos^2\theta_3 = 1. \qquad (3.27)$$

Now we will consider a force, F, applied at the origin in two-space (the x-y plane); we will illustrate this in Figure 3.2. We can resolve F into x- and y-components if we know the angle, θ. Suppose, for example, that $F = 400\,\text{N}$ and $\theta = 25°$; then

$$F_x = F\cos(\theta) = (400)(0.9063) = 362.5\,\text{N} \qquad (3.28)$$

and

$$F_y = F\sin(\theta) = (400)(0.4226) = 169\,\text{N}. \qquad (3.29)$$

Of course, the resultant of F_x and F_y can be found by rectangular resolution; that is,

$$F = \sqrt{F_x^2 + F_y^2} = 399.96\,\text{N}. \qquad (3.30)$$

The discrepancy is merely the result of roundoff error.

An illustration: Suppose a light aircraft flies due east with a ground speed of 120 mph; it is subjected to a constant wind from the south blowing at 14 mph, as shown in Figure 3.3. What is the effective ground speed of the airplane, and where will it be after 1 hour (if the pilot makes no corrections)?

In this case, $\tan(\theta) = 14/120$, so $\theta = 6.654°$. After 1 hour, the aircraft will have traveled 120.81 mi and it will, of course, be 14 mi north of its intended destination. *This elementary example underscores one of the difficulties of navigation prior to GPS. On a long flight, a small crosswind could result in a catastrophic error if not caught by the navigator. It is thus easy to understand how Amelia Earhart and Fred Noonan missed Howland Island in 1937 on their 2500-mile flight from Lae (to Howland) when their radio direction finder (RDF) did not function properly.*

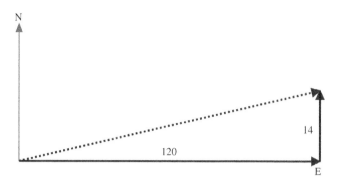

FIGURE 3.3. Effect of a crosswind on a light aircraft flying due east at 120 mph.

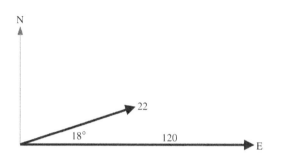

FIGURE 3.4. Effect of a quartering wind on a light aircraft flying east at 120 mph. The wind is blowing to the ENE at 22 mph.

An important variation of such problems arises when the second vector is not perpendicular to the first. Now assume that the same aircraft, flying east at 120 mph, experiences a quartering wind blowing 22 mph at an angle 18° north of east as shown in Figure 3.4.

In this case, the plane's speed to the east will be

$$120 + 22\cos(18°) = 140.92 \text{ mph.} \qquad (3.31)$$

However, its velocity to the north will be

$$22\sin(18°) = 6.80 \text{ mph.} \qquad (3.32)$$

After 1 hour, the craft will have flown a distance of about 141.08 mi and it will be 6.8 mi north of the intended point.

Force Equilibrium

Let us review a familiar type of problem in which we utilize equilibrium at a point. Imagine, for example, that during the erection of a building, a structural member (a beam 30 ft long) weighing 800 lb is temporarily leaned against a vertical wall as illustrated in Figure 3.5 (the angle between the ground and the beam is 50°). At the upper end, the wall exerts a horizontal force on the beam indicated by F_1. The

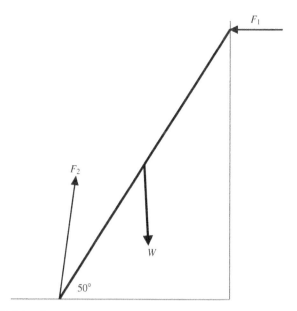

FIGURE 3.5. A structural member (or beam) lying against a vertical wall. The beam is 30 ft long and weighs 800 lb$_f$; the angle of inclination with respect to the ground is 50°.

ground exerts a force on the bottom end of the beam indicated by F_2. We recognize that under the equilibrium condition, the summation of forces in the horizontal (x-) direction must be zero. Similarly, the summation of forces acting in the vertical (y-) direction must also be zero; therefore,

$$F_2\cos(\theta) - F_1 = 0 \quad \text{and} \quad F_2\sin(\theta) - W = 0. \qquad (3.33)$$

For us to solve these equations, we must determine the angle, θ: We do this by using two right triangles and the first of these is formed using the lower half of the beam such that the hypotenuse is 15 ft; the vertical leg is the distance from the midpoint of the beam to the ground. Therefore, $x = 15\cos(50°) = 9.642$ ft. Now we can determine θ (the angle between the line of action of F_2 and the ground): $\tan(\theta) = 22.98/9.642$, so $\theta = 67.24°$. Our horizontal and vertical force summations are now used to compute $F_2 = 867.56$ lb and $F_1 = 335.64$ lb. Of course, the very same problem can be worked more easily by noting that at equilibrium, the summation of *moments* about the z-axis must be zero.

Equating Moments

Recall that the moment of a force is the measure of the torque introduced about a chosen point; the moment is the product of the magnitude of the force and the perpendicular distance between the selected origin and the line of action of the force. Furthermore, according to Varignon's theorem, the moment of the resultant of two forces equals the sum of

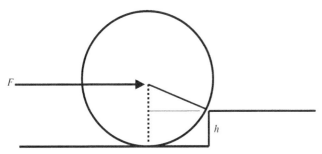

FIGURE 3.6. A cylindrical object of radius, R, and weight, W, has rolled up against a step of height, h. A force is to be applied at the center with the intent of moving the cylinder to the top of the step.

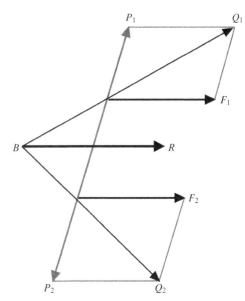

FIGURE 3.7. Coplanar forces acting on a rigid body at different locations.

the two moments of the two forces. Therefore, if an object is acted on by multiple forces—yet remains at equilibrium—then the sum of the moments must be zero. Let us apply this result to an example.

We imagine a cylindrical object with radius, R, that has rolled up against a curb (or step) of height, h. We wish to apply a horizontal force, F, to get the cylinder to roll up over the curb. The force will be applied at the center of the cylinder, although it is obvious that this is not optimal. The cylindrical object has weight, W, and the arrangement is illustrated in Figure 3.6.

The moment arm for the applied force, F, is $R - h$. The moment arm for the weight (due to gravity) is x, the horizontal leg of the triangle. Therefore,

$$F(R - h) = Wx, \qquad (3.34)$$

with x obtained as follows:

$$\theta = \sin^{-1} \frac{R - h}{R},$$

and $x = R\cos\theta$. We can examine this result quantitatively by setting $R = 10$ and $h = 2$ (a fairly small step). For this case, $\theta = 0.9273$ rad (53.1°) and $x = 6$. Therefore, $F = 0.75\ W$. This will, of course, change dramatically if the step height is greater; we can repeat the problem but with $h = 7$. The reader may wish to verify that $F = 3.18\ W$ in this case. We observe that the point of the step exerts a force on the cylinder; at equilibrium, the x- (horizontal) component of that force must counter F and the y-component must balance W. For the case in which $h = 7$, this force will total $3.33\ W$ and the angle between the line of action and the upper horizontal surface will be 17.46°.

A different situation can arise when coplanar forces act on a body at different locations. To illustrate, let us take the case where forces F_1 and F_2 are acting on a rigid body as depicted in Figure 3.7.

We begin by constructing auxiliary forces P_1 and P_2, which are equal in magnitude but opposite in direction. Then the lines of action for Q_1 (the resultant of F_1 and P_1) and Q_2 (the resultant of F_2 and P_2) are constructed and extended back to the point of intersection, B. The force transmitted to the solid body is R, which is the resultant of the pair, Q_1 and Q_2. Naturally, the magnitude of R is just the sum of F_1 and F_2; the line of action of the resultant is obtained by noting that we have similar triangles (created by extending Q_1 and Q_2 back to the point of intersection, B). The case of coplanar, nonparallel forces (applied to a rigid body) is simpler—it is only necessary to extend the lines of action to the point of intersection. Of course, it is entirely possible that a pair of parallel, coplanar forces act in opposite directions. If they have the same magnitude, the result will be a couple, and a couple acting on a rigid body will produce rotation.

Projectile Motion

Imagine a situation in which a projectile is launched in the x-y plane at an angle (relative to level ground) of θ with an initial velocity of V_0. If we neglect drag (a potentially serious omission), then the initial horizontal and vertical velocity vector components are

$$V_{x0} = V_0 \cos\theta \quad \text{and} \quad V_{y0} = V_0 \sin\theta. \qquad (3.35)$$

Since gravity acts in the negative y-direction, we know

$$V_y = V_0 \sin\theta - gt; \qquad (3.36)$$

the projectile will reach its apogee when $V_y = 0$, or $V_0 \sin \theta = gt$. During flight, the angle between the instantaneous trajectory and the ground is $\theta = \tan^{-1}(V_y/V_x)$. We can look at a simple example for illustration: Let $V_0 = 300$ ft/s and $\theta = 35°$. The initial horizontal velocity is 245.7 ft/s and the initial vertical velocity is 172.2 ft/s. The maximum vertical height above ground will occur at

$$t = \frac{(300)(0.574)}{(32.17)} = 5.35 \text{ s}, \qquad (3.37)$$

and that maximum height will be

$$y_{max} = \left[(V_0 \sin \theta)t - \frac{1}{2}gt^2 \right]_{t=5.35} = 460.9 \text{ ft}, \qquad (3.38)$$

a number that is actually quite unrealistic. After 8 seconds, the angle between the ground and the projectile's trajectory would be

$$\theta = \tan^{-1}\left(\frac{(300)(0.574) - (32.17)(8)}{(245.7)} \right) = -19°; \qquad (3.39)$$

that is, 19° below horizontal. As we implied earlier, the effect of drag on the motion of projectiles is profound. For example, returning to the previous illustration, if the projectile were a sphere launched in air (with $R = 1$ in. and $m = 1$ lb$_m$) the maximum height would be about 394 ft and that would be attained in about 4.83 seconds. A larger object of the same mass would experience greater drag, of course. If we let $R = 2$ in., then $y_{max} = 288$ ft and that height is achieved at $t = 3.93$ seconds.

Dot and Cross Products

In physics, we often think of the dot product in the context of work: Work is performed when a force is applied to a body, the body moves, and a component of the force vector is acting in the direction of the motion. Thus, $W = \int_1^2 F \cdot ds$. The dot product of vectors **A** and **B** is written **A·B** and it is a scalar. For example, given the situation depicted in Figure 3.8, then $\mathbf{A·B} = (100)(50)\cos(\theta)$. If the angle $\theta = 30°$, then $\mathbf{A} \cdot \mathbf{B} = 4330.13$, but if θ were changed from 30° to 85°, then $\mathbf{A} \cdot \mathbf{B} = 435.78$.

More generally, given two vectors A and B:

$$A = A_1 i + A_2 j + A_3 k \quad \text{and} \quad B = B_1 i + B_2 j + B_3 k, \qquad (3.40)$$

then

$$A \cdot B = A_1 B_1 + A_2 B_2 + A_3 B_3. \qquad (3.41)$$

Similarly, $A \cdot A = A_1^2 + A_2^2 + A_3^2$. If $A \cdot B = 0$, and neither is null, then A and B *must be perpendicular*. For example, suppose $A = 2i + 3j$, and further, that A is perpendicular to vector B (and both in the x-y plane). We have $A \cdot B = |A||B|\cos \theta$ and since $\theta = 90°$, we must have $(2i + 3j) \cdot (b_1 i + b_2 j) = 0$, or $2b_1 + 3b_2 = 0$. Thus, $b_1/b_2 = -(3/2)$. We also make note of the following properties of dot products:

$$A \cdot B = B \cdot A \quad \text{commutative} \qquad (3.42)$$

$$A \cdot (B + C) = A \cdot B + A \cdot C \quad \text{distributive} \qquad (3.43)$$

$$a(A \cdot B) = (aA) \cdot B = A \cdot (aB). \qquad (3.44)$$

The cross product of two vectors, indicated by $\mathbf{A} \times \mathbf{B}$, is also a vector, **C**. Thus,

$$A \times B = C, \qquad (3.45)$$

where **C** is perpendicular to the plane of A and B. Furthermore, **C** follows the right-hand screw rule, so for the case in which **A** and **B** correspond to the x- and y-axes in the plane of the page, force **C** is directed out of the page toward the reader. We also have the following relations for the cross (or vector) product:

$$A \times B = -B \times A \quad not \text{ commutative} \qquad (3.46)$$

$$Ax(B + C) = A \times B + A \times C \quad \text{distributive} \qquad (3.47)$$

$$a(A \times B) = (aA) \times B = Ax(aB) \qquad (3.48)$$

There are applications requiring triple products of the vectors, A, B, and C; the *scalar triple product* is

$$A \cdot (B \times C) = B \cdot (C \times A) = C \cdot (A \times B). \qquad (3.49)$$

This relation gives us the volume of a parallelepiped defined by the three vectors. Given $A = 2i + 4j + k$, $B = 6i - j + k$, and $C = 0 + j + 3k$, for example, we have

$$\begin{matrix} 2 & 4 & 1 \\ 6 & -1 & 1 \\ 0 & 1 & 3 \end{matrix} = -6 + 0 + 6 - 0 - 2 - 72 = -74.$$

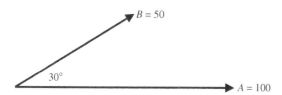

FIGURE 3.8. Illustration of coplanar vectors **A** and **B** forming an angle of 30°.

The vector triple product is

$$Ax(B \times C) = B(A \cdot C) - C(A \cdot B). \qquad (3.50)$$

Equation (3.50) is often referred to as the "*BAC – CAB*" rule—a useful mnemonic device as long as one remembers the minus (−) sign.

Differentiation of Vectors

Let us suppose that a vector is a function of a scalar like time, t. In particular, let $A = A_1 i + A_2 j + A_3 k$, where A_1, A_2, and A_3 depend on t. Then

$$\frac{dA}{dt} = \frac{dA_1}{dt} i + \frac{dA_2}{dt} j + \frac{dA_3}{dt} k. \qquad (3.51)$$

Naturally, $dA/dt = 0$ if A is a *constant* vector. We also want to consider the case in which $B = B_1(x,y,z)i + B_2(x,y,z)j + B_3(x,y,z)k$, and spatial derivatives are required. Let $B = 3xy^2 i - 4xyz j + 7y^2 z^2 k$, then

$$\frac{\partial B}{\partial x} = 3y^2 i - 4yz j, \qquad (3.52a)$$

$$\frac{\partial B}{\partial y} = 6xy i - 4xz j + 14yz^2 k, \qquad (3.52b)$$

and

$$\frac{\partial B}{\partial z} = -4xy j + 14y^2 z k. \qquad (3.52c)$$

Gradient, Divergence, and Curl

We define the *del* operator (∇) in rectangular coordinates in the following way:

$$\nabla = i \frac{\partial}{\partial x} + j \frac{\partial}{\partial y} + k \frac{\partial}{\partial z}. \qquad (3.53)$$

Now let us suppose that T is a scalar (perhaps temperature), then

$$\nabla T = i \frac{\partial T}{\partial x} + j \frac{\partial T}{\partial y} + k \frac{\partial T}{\partial z}. \qquad (3.54)$$

This is called the *gradient* of T and it is a vector. When we speak of the *divergence* of a vector (say, **A**), we mean

$$\nabla \cdot A = \frac{\partial A_1}{\partial x} + \frac{\partial A_2}{\partial y} + \frac{\partial A_3}{\partial z}, \qquad (3.55)$$

and this quantity is a scalar. Therefore, in the case of the velocity vector, V, the divergence would be written as

$$\nabla \cdot V = \frac{\partial v_x}{\partial x} + \frac{\partial v_y}{\partial y} + \frac{\partial v_z}{\partial z}. \qquad (3.56)$$

For an incompressible fluid, the divergence of the velocity vector is a statement of conservation of mass, and since the fluid density (ρ) is constant, $\nabla \cdot V = 0$. A vector field for which the divergence is zero is said to be *solenoidal*. In cylindrical coordinates,

$$\nabla \cdot V = \frac{1}{r} \frac{\partial}{\partial r}(r v_r) + \frac{1}{r} \frac{\partial v_\theta}{\partial \theta} + \frac{\partial v_z}{\partial z}. \qquad (3.57)$$

The *curl* of a vector, A, is written as $\nabla x A$, and it is a vector. In much of the older (particularly German) literature, the curl of A is written $rot(A)$. It is convenient for us to construct a determinant to help with our interpretation:

$$\begin{vmatrix} i & j & k \\ \dfrac{\partial}{\partial x} & \dfrac{\partial}{\partial y} & \dfrac{\partial}{\partial z} \\ A_1 & A_2 & A_3 \end{vmatrix} \qquad (3.58)$$

$$= \left(\frac{\partial A_3}{\partial y} - \frac{\partial A_2}{\partial z} \right) i + \left(\frac{\partial A_1}{\partial z} - \frac{\partial A_3}{\partial x} \right) j + \left(\frac{\partial A_2}{\partial x} - \frac{\partial A_1}{\partial y} \right) k.$$

We also often need the curl of a vector in cylindrical coordinates as well so it is worthwhile to provide it here:

$$\left(\frac{1}{r} \frac{\partial A_3}{\partial \theta} - \frac{\partial A_2}{\partial z} \right) i + \left(\frac{\partial A_1}{\partial z} - \frac{\partial A_3}{\partial r} \right) j + \left(\frac{1}{r} \frac{\partial}{\partial r}(r A_2) - \frac{1}{r} \frac{\partial A_1}{\partial \theta} \right).$$

$$(3.59)$$

In fluid mechanics, $\nabla x V$ is particularly significant—it is the *vorticity* vector, ω_i.

$$\left(\frac{\partial v_z}{\partial y} - \frac{\partial v_y}{\partial z} \right) i = \omega_x \qquad (3.60a)$$

$$\left(\frac{\partial v_x}{\partial z} - \frac{\partial v_z}{\partial x} \right) j = \omega_y \qquad (3.60b)$$

$$\left(\frac{\partial v_y}{\partial x} - \frac{\partial v_x}{\partial y} \right) k = \omega_z. \qquad (3.60c)$$

The three components of the vorticity vector are directly related to the angular velocity of the fluid (rotation) about the three axes (in fact, each is exactly twice the rate of rotation about that particular axis). A vector field is said to be *irrotational* if $\nabla x V = 0$. Let us suppose, for example, that an incompressible fluid undergoes two-dimensional motion which occurs in the x-y plane such that

$$v_x = 3xy^2 + 2y \quad \text{and} \quad v_y = -y^3. \tag{3.61}$$

We observe immediately that $\nabla \cdot V = 3y^2 - 3y^2 = 0$ and that

$$\frac{\partial v_y}{\partial x} - \frac{\partial v_x}{\partial y} = 0 - 6xy - 2 = \omega_z. \tag{3.62}$$

Thus, there will be rotation about the z-axis unless the product $xy = -1/3$. Now, suppose that an arbitrary vector field is described by

$$A = x^2 y i + xz^2 j + xyz k. \tag{3.63}$$

We begin by finding the divergence, $\nabla \cdot A$:

$$\frac{\partial}{\partial x}(x^2 y) + \frac{\partial}{\partial y}(xz^2) + \frac{\partial}{\partial z}(xyz) = 2xy + xy. \tag{3.64}$$

Therefore, at the point $(1, 1, 1)$, we find $\nabla \cdot A = 2 + 1 = 3$. We can also evaluate the curl of this vector field:

$$\left[\frac{\partial}{\partial y}(xyz) - \frac{\partial}{\partial z}(xz^2)\right]i + \left[\frac{\partial}{\partial z}(x^2 y) - \frac{\partial}{\partial x}(xyz)\right]j$$
$$+ \left[\frac{\partial}{\partial x}(xz^2) - \frac{\partial}{\partial y}(x^2 y)\right]k \tag{3.65}$$
$$= (xz - 2xz)i + (0 - yz)j + (z^2 - x^2)k.$$

At the point $(1, 1, 1)$ this yields $-1i - 1j$.

Let us summarize some of the common operations involving the del operator here:

$$\nabla(a + b) = \nabla a + \nabla b \tag{3.66}$$

$$\nabla^2 a = \nabla \cdot \nabla a \tag{3.67}$$

$$\nabla^2 A = (\nabla \cdot \nabla)A \tag{3.68}$$

$$\nabla \cdot (A + B) = \nabla \cdot A + \nabla \cdot B \tag{3.69}$$

$$\nabla \times \nabla a = 0 \tag{3.70}$$

$$\nabla \cdot \nabla \times A = 0 \tag{3.71}$$

$$\nabla x(A + B) = \nabla \times A + \nabla \times B \tag{3.72}$$

$$\nabla \cdot (aA) = \nabla a \cdot A + a\nabla \cdot A \tag{3.73}$$

$$\nabla x(aA) = \nabla a \times A + a(\nabla \times A) \tag{3.74}$$

$$\nabla \cdot (A \times B) = B \cdot (\nabla \times A) - A \cdot (\nabla \times B) \tag{3.75}$$

$$\nabla(A \cdot B) = (B \cdot \nabla)A + (A \cdot \nabla)B + Bx(\nabla \times A) + Ax(\nabla \times B) \tag{3.76}$$

$$\nabla x(A \times B) = (B \cdot \nabla)A - B(\nabla \cdot A) - (A \cdot \nabla)B + A(\nabla \cdot B) \tag{3.77}$$

$$\nabla \cdot (\nabla a) = \nabla^2 a$$
$$= \frac{\partial^2 a}{\partial x^2} + \frac{\partial^2 a}{\partial y^2} + \frac{\partial^2 a}{\partial z^2} \quad \text{(the Laplacian operator)} \tag{3.78}$$

$$\nabla x(\nabla \times A) = \nabla(\nabla \cdot A) - \nabla^2 A. \tag{3.79}$$

Recall that we said previously that there are two types of vector fields: *solenoidal*, for which $\nabla \cdot A = 0$, and *irrotational*, for which $\nabla \times A = 0$. In the case of the latter, if a vector field is irrotational, then a scalar function (ϕ) exists such that $A = \nabla\phi$. We refer to ϕ as a *potential* function; if we consider the x-component of the velocity vector in a potential flow problem, $v_x = \partial\phi/\partial x$. In an electric field, the component of (negative) electric intensity in a given direction is equal to the derivative of the potential in that direction. We also know by Stokes' theorem that if $\nabla \times A = 0$, then

$$\oint_C A \cdot dr = \iint_S (\nabla \times A) \cdot n dS = 0. \tag{3.80}$$

Let us spend a little time contemplating this statement prior to discussing Green's and Stokes' theorems. In physics, when we consider work (W), we mean that a force has been applied to a moving body in such a way that there is a non-zero component of F in the direction of the motion. The work done in moving an object along the x-axis from position x_1 to x_2 would be written as

$$W = \int_{x_1}^{x_2} F_x dx. \tag{3.81}$$

More generally, we can write a *line integral* of the vector, F, from point 1 to point 2 (along a curve, C) as

$$\int_{P_1}^{P_2} F \cdot dr = \int_C F \cdot dr = \int_C F_1 dx + F_2 dy + F_3 dz. \tag{3.82}$$

To illustrate, suppose we have a vector (force) in the x-y plane: $F = -3x^2 i + 5xy j$. This force is to be applied to an object moving along a path described by $y = 2x^2$. Therefore,

$$\int_C (-3x^2 dx + 5xy dy). \qquad (3.83)$$

Since $y = 2x^2$, $dy = 4xdx$; consequently, the integral of interest is

$$\int_{x=0}^{x=1} \left[-3x^2 dx + 5x(2x^2)4xdx \right] = \int_0^1 (-3x^2 + 40x^4)dx \qquad (3.84)$$
$$= -1 + 8 - 0 = 7.$$

A line integral that is particularly important in fluid mechanics is the integral of the velocity vector around a closed path; it is called *circulation*, $\Gamma = \oint V \cdot dr$. Of course, it is this circulation that is necessary for a wing (airfoil) to generate lift.

When we say that a force field is *conservative*, we mean that the work required to move an object, say, from point 1 to point 2, is *independent of path*. If the path is closed, that is, if we start from point 1 and return to 1, then the work is zero. This is where our result earlier, eq. (3.80), comes into play. Obviously, based on this discussion, the curl of a conservative force field must be zero—a conservative force field is *always irrotational*. We note that gravity is an example of a conservative force but friction is not. We will explore this a little further; suppose a force field is given by

$$F = (2xz^3 + 6y)i + (8x - 2yz)j + (3x^2z^2 - y^2)k. \qquad (3.85)$$

We would like to know if it is conservative:

$$\nabla \times F = \begin{array}{ccc} i & j & k \\ \dfrac{\partial}{\partial x} & \dfrac{\partial}{\partial y} & \dfrac{\partial}{\partial z} \\ 2xz^3 + 6y & 8x - 2yz & 3x^2z^2 - y^2. \end{array} \qquad (3.86)$$

Our result is $(-2y + 2y)i + (6xz^2 - 6xz^2)j + (8 - 6)k \neq 0$; this force field is *not* conservative (but nearly so—it would be if $(8x - 2yz)j$ were changed to $(6x - 2yz)j$).

GREEN'S THEOREM

Green's theorem plays an important role in the solution of many problems in mathematical physics, and examples are found in areas such as electrostatics and hydrodynamics. Green's theorem provides a relationship between line integrals on closed paths (denoted by C) and double integrals over the region enclosed by C (which we denote by D). We stipulate that the path C has positive orientation, which means that the enclosed region, D, is always on the left-hand

side as we traverse the path. We also require that the functions P and Q appearing in the theorem have continuous first-order partial derivatives over D:

$$\oint_C P dx + Q dy = \iint_D \left(\frac{\partial Q}{\partial x} - \frac{\partial P}{\partial y} \right) dA. \qquad (3.87)$$

One of the difficulties posed by eq. (3.87) to students of engineering and the applied sciences is its apparent lack of connection to the physical world. However, there was a mechanical device constructed in the nineteenth century by Jakob Amsler-Laffon that makes the significance of eq. (3.87) very clear: The *planimeter* was used to compute the area of a region by tracing the enclosing curve (C). At one time, just about every draftsman had a planimeter, and many used the instrument routinely.

Let us now illustrate Green's theorem with an example. Suppose the (enclosed) region under consideration is a triangle, and we wish to evaluate

$$\int_C xy dx + x^2 y^3 dy. \qquad (3.88)$$

Let the triangular region be bounded by the x-axis from 0 to 1, a vertical line extending from $(1, 0)$ to $(1, 2)$, and the hypotenuse, which returns from $(1, 2)$ to the origin $(0, 0)$. In this case, the hypotenuse is represented very simply as $y = 2x$ and the x, y region of interest is enclosed by $0 \leq x \leq 1$ and $0 \leq y \leq 2x$. We can see by inspection of eq. (3.87) and eq. (3.88) that $P = xy$ and $Q = x^2 y^3$. Therefore,

$$\frac{\partial Q}{\partial x} = 2xy^3 \quad \text{and} \quad \frac{\partial P}{\partial y} = x. \qquad (3.89)$$

By virtue of Green's theorem, we write

$$\iint_D (2xy^3 - x)dA = \int_0^1 \int_0^{2x} (2xy^3 - x)dy dx. \qquad (3.90)$$

It is left to the reader to show that eq. (3.90) is equal to 2/3. Aside from merely allowing us to rewrite some complicated integrals in more tractable form, Green's theorem can be thought of as providing us with the total flow (of, say, mass, momentum, or heat) out of region D.

Now suppose we have an annular region formed by two concentric circles located at the origin. We let the radius of the inner circle be 1 and the radius of the outer circle be 2. Since Green's theorem does not apply to regions with holes, we must take a different approach and there are two possibilities: We can take the annular region and break it into two halves by drawing a horizontal line just through the middle. We choose the orientation such that as we travel on the path(s) the two regions always lie to the left. But we could also look at the difference between the outer and inner circles; we will take this approach for illustration. By inspection, we see that

$$P = y^3 \quad \text{and} \quad Q = -x^3; \qquad (3.91)$$

therefore,

$$\frac{\partial P}{\partial y} = 3y^2 \quad \text{and} \quad \frac{\partial Q}{\partial x} = -3x^2. \qquad (3.92)$$

Thus, we have

$$\iint_D (-3x^2 - 3y^2)dA \quad \text{with} \quad dA = rdrd\theta. \qquad (3.93)$$

Since $x^2 + y^2 = r^2$, we have

$$-3\int_0^{2\pi}\int_0^2 r^3 drd\theta = -12\theta\big|_0^{2\pi} = -24\pi. \qquad (3.94)$$

Now we apply the same technique to the inner circle (but with a radius of 1 rather than 2):

$$-3\int_0^{2\pi}\int_0^1 r^3 drd\theta = -\frac{3}{2}\pi. \qquad (3.95)$$

So for the integral of interest to us, $-24\pi - (-1.5\pi) = -22.5\pi$.

STOKES' THEOREM

Stokes' theorem applies to a closed curve, C, in three-dimensional space that bounds some surface, S. It reduces to Green's theorem when the enclosing curve lies in a plane (is two-dimensional). It has been pointed out repeatedly that Stokes' theorem was not actually developed by Sir G. G. Stokes; Lord Kelvin (William Thomson) is credited with having done so after he discovered George Green's 1828 essay. Neeley (2008) states that it became known as Stokes' theorem due to the frequency with which Stokes placed it on the Cambridge prize examinations.

Given a vector field, F, where S is any surface bounded in three-dimensional space by closed curve, C,

$$\oint F \cdot dr = \iint_S (\nabla \times F) \cdot nds. \qquad (3.96)$$

This is a powerful relationship that finds application in the connection between electric (E) and magnetic (B) fields.

For example, you may have seen the Maxwell–Faraday equation previously:

$$\oint_C E \cdot dl = -\int_S \frac{\partial B}{\partial t} \cdot dS. \qquad (3.97)$$

A verbal statement will help our comprehension here and we cite the one provided by Sears and Zemansky (1964), "the line integral of an induced electric field around a closed path, or the electromotive force (ε) in the path, equals the time rate of change of magnetic flux across the area bounded by the path."

Let us look at some examples to better reveal the utility of Stokes' theorem. To begin, we assume we have a vector field, $F = -yi + xj + x^2k$, with a surface corresponding to a right circular cylinder of height, h, and radius, R (with its bottom placed at the origin). For the right-hand side of eq. (3.96), we need the curl of F:

$$\begin{vmatrix} i & j & k \\ \dfrac{\partial}{\partial x} & \dfrac{\partial}{\partial y} & \dfrac{\partial}{\partial z} \\ -y & x & x^2 \end{vmatrix} \qquad (3.98)$$

Therefore, $\nabla xF = -2xj + 2k$. The surface integral will take into account the top (upper end) and the side of course. For example, for the top we have

$$\iint_S (\nabla \times F) \cdot nds = \int_0^{2\pi}\int_0^R (-2xj + 2k) \cdot krdrd\theta = 2\pi R^2. \qquad (3.99)$$

For the side of the cylinder, $ds = Rd\theta dz$, so in this case, we note $x = R\cos\theta$ and $y = R\sin\theta$, to obtain

$$\iint_S (\nabla \times F) \cdot nds = -2R\int_0^h\int_0^{2\pi} \sin\theta\cos\theta Rd\theta dz = 0. \qquad (3.100)$$

The two parts combined yield $2\pi R^2$.

Let us work through an example adapted from Spiegel (1971); in this case, the vector field, F, is $3yi - xzj + yz^2k$. The closed curve is a circle located at $z = 2$ and the surface is a paraboloid, given by $2z = x^2 + y^2$. We can determine the curl:

$$\nabla \times F = \begin{vmatrix} i & j & k \\ \dfrac{\partial}{\partial x} & \dfrac{\partial}{\partial y} & \dfrac{\partial}{\partial z} \\ 3y & -xz & yz^2 \end{vmatrix} = (z^2 + x)i - (z+3)k \qquad (3.101)$$

We could set about computing $\iint_S (\nabla \times F) \cdot n \, ds$, but in this case, we note that the enclosing curve is a circle in the plane corresponding to $z = 2$; accordingly, $4 = x^2 + y^2$. Since $x = R \cos \theta$ and $y = R \sin \theta$, where $R = 2$, we write

$$\oint_C 3y \, dx - xz \, dy + yz^2 \, dz. \qquad (3.102)$$

We substitute for x and y and note that $dz = 0$:

$$\oint_C -3R^2 \sin^2 \theta \, d\theta - 2R^2 \cos^2 \theta \, d\theta. \qquad (3.103)$$

The reader may wish to verify

$$-R^2 \int_{2\pi}^{0} (3\sin^2 \theta + 2\cos^2 \theta) d\theta = -4(-5\pi) = 20\pi. \quad (3.104)$$

It will be left as a student exercise to verify that the integral of the curl of F over the surface yields the same result (20π).

CONCLUSION

We have assumed that everyone who comes to this study of applied mathematics has exposure to basic physics and therefore has some familiarity with vectors and the resolution of forces. Accordingly, our purpose with this chapter is to provide a brief review of a few concepts that the student has seen previously. For the reader who needs more, an excellent treatment of vectors and vector manipulation has been provided by Spiegel et al. (2009). This book is especially useful for self-study because it contains numerous examples that are completely solved.

For the reader interested in the connection between vectors, tensors, and fluid mechanics, it would be difficult to find a better starting point than the book *Vectors, Tensors, and the Basic Equations of Fluid Motion* by Rutherford Aris (1962). This book is particularly important because it consistently tries to reveal the connection between mathematics and the physical reality of fluid flow. For applied scientists and engineers—many of whom are mainly visual and tactile learners—this is critical. G. K. Batchelor pointed out about 45 years ago that educators should always underscore the relation between "analysis and the behavior of real fluids; fluid dynamics is much less interesting if it is treated largely as an exercise in mathematics." This observation applies to effective learning for students in all areas of applied science and engineering.

PROBLEMS

3.1. A commercial aircraft flying straight east with a ground speed of 495 mph encounters a jet stream flowing at 118 mph at 75° (15° north of east). If no corrective action is taken, what is the speed of the aircraft (relative to the ground) and what is the aircraft's modified course?

3.2. A ladder (24 ft long) leans against a house, forming an angle with the ground of 68°. A man stands midway on the ladder, and the ladder with the man weighs 219 lb$_f$. Determine the forces transmitted to the ladder by the vertical wall and the ground. At what angle with respect to the ground does the bottom force act?

3.3. Two vectors, A and B, are given by

$$A = 6i + 3j + 7k \quad \text{and} \quad B = -2i + 5j - 1k.$$

Are these two vectors perpendicular?

3.4. Two vector fields are described by

$$A = xy^2 i + xz^3 j - xyzk \quad \text{and} \quad B = x^2 y^3 i - y^2 z^2 j + y^3 zk.$$

Is either vector field solenoidal? Irrotational? What simple change could be made to B to make it irrotational?

3.5. Consider the three vectors:

$$A = 25i + 3j \quad B = 5i + 12j, \quad \text{and} \quad C = 5k.$$

The volume of the parallelepiped formed by these three vectors is 1636.66. Confirm that this is equivalent to

$$|A \cdot (B \times C)|.$$

3.6. An elementary—but quite interesting—demonstration that is frequently used in physics classrooms results in the collision of two spheres (when the experiment is successful). Sphere 1 is dropped vertically from some height, h, above a flat, planar surface. Sphere 2 is launched, at the exact instant that sphere 1 is released, from the flat surface some horizontal distance away. Sphere 2 is aimed as precisely as possible at the initial vertical position of sphere 1. If gravity is the only significant force acting on the spheres, then they should collide, irrespective of the initial velocity of sphere 2. Now, suppose we wanted to attempt this experiment on a larger scale; in particular, assume that sphere 1 (S1) is raised to an initial position 100 ft above the ground surface. Sphere 2 (S2) is to be launched at S1 from a (horizontal) distance 275 ft away. We want the collision between the spheres to occur *exactly* 16 ft above the ground surface. What are the launch parameters for S2 that *may* achieve the desired impact? In the classroom, drag and buoyancy can usually be neglected. Will drag affect the two spheres differently (the dropped sphere's velocity is low relative to the launched projectile)? Explain the likely impact of drag on the large-scale experiment.

3.7. An archer releases an arrow at an angle (with respect to level ground) of 70°. If the arrow's initial velocity is 260 ft/s, how far will the arrow have traveled *horizontally* when it reaches its apogee? At $t = 10.2$ seconds, what is the angle between the arrow's trajectory and the ground? Neglecting drag, what is the *total distance* traveled by the arrow after its return to level ground?

3.8. A two-dimensional flow field has velocity vector components (in the *x-y* plane) described by

$$v_x = ay\exp(-by) \quad \text{and} \quad v_y = c\sqrt{x}.$$

Evaluate ∇xV at the points (x, y): (1, 1), (2, 2), (3, 3), and (4, 4), if $a = b = c = \frac{1}{2}$.

3.9. Elementary physics texts often cite 45° as the initial angle (of a projectile relative to the earth's surface) that produces maximum range. This is incorrect for spheroidal or cylindrical objects that rotate during their translational motion. Obvious examples include golf balls, tennis balls, and baseballs. In golf, impact between the clubhead and the ball imparts reverse spin on the ball, generating a lift force due to the *Magnus effect*. This has the tendency to steepen the trajectory of the ball during ascent. If a golf ball leaves the clubhead at an initial angle of 16°, but with reverse spin of 900 rpm, what will the steepest angle achieved during ascent be? Assume the initial ball velocity is 145 mph, neglect drag, and assume the rotational motion does not decay. The lift force acts perpendicularly to the ball's trajectory, and we will estimate its magnitude with the following expression (actually valid for a *right circular cylinder* on a per-unit-length basis):

$$\text{Lift} = 2\pi R\rho_{\text{air}}V_\theta V_{\text{trans}}.$$

We will take the density of the air to be 0.0012 g/cm³, the tangential velocity (ball's surface) to be 188.4 cm/s, and the translational velocity to be 6482 cm/s. Assume the effective radius of the ball is 2 cm. Since the lift force is expressed on a per-unit-length basis, take the "length" of the golf ball to be 3 cm.

3.10. Given

$$A = x^2yi - 3yzj + xyzk$$

and

$$B = 2xzi + y^3zj + x^2y^2zk,$$

show that $\nabla x(A \times B) = (B \cdot \nabla)A - B(\nabla \cdot A) - (A \cdot \nabla)B + A(\nabla \cdot B)$.

3.11. Show that $\iint_S (\nabla \times F)\cdot nds = 20\pi$ where $F = 3yi - xzj + yz^2k$. See eq. (3.80) and the accompanying discussion for elaboration.

3.12. On May 22, 1963, Mickey Mantle hit a fastball pitched by Bill Fischer of the Kansas City As that bounced off of the façade at the very top of old Yankee Stadium. In an interview after his retirement, Mantle stated that that homerun was the hardest he had ever struck a baseball. The straight-line distance from home plate to the façade was 370 ft, and the point of impact was 118 ft *above* right field. One website has estimated that this ball would have traveled 734 ft, making it the longest homerun ever hit in Major League baseball! This estimate is flawed for a number of reasons, although some witnesses did report that the ball was *still rising* when it struck the façade. Given that a baseball has a circumference of 9 in. and weighs 5 oz, estimate the distance this ball might have traveled taking drag into account:

$$F_{\text{drag}} = AKf = (\pi R^2)(\tfrac{1}{2}\rho V^2)(0.44).$$

For point of reference, the initial velocity of a very well-struck baseball is on the order of 45–50 m/s. At 47 m/s, the initial Reynolds number is

$$\text{Re} = \frac{d\rho V}{\mu} = \frac{(7.28)(4700)}{(0.151)} = 226,489.$$

In this region, the drag coefficient is approximately constant and therefore we will use 0.44 for *f*. Prove that this baseball *could not have* traveled 734 ft.

3.13. Consider a vector with initial and final points (in terms of *x, y, z* values) of (5, 5, 5) and (3, 2, 2), respectively. Find the magnitude of this vector. Then, given the vector, $A = 2i - 3j + k$, demonstrate that $\nabla \cdot (\nabla \times A) = 0$.

3.14. The distribution of a two-dimensional scalar field in a rectangle with $0 \leq x \leq 10$ and $0 \leq y \leq 5$ is given by

$$\frac{50}{y^2 - 8y + x^2 - 14x + x^2y^2 - 36xy + 389}.$$

Find the maximum gradient at the point $x = 7$ and $y = 3$, and find the corresponding direction of this maximum gradient. Do your answers correspond with estimates made using Figure 3.9?

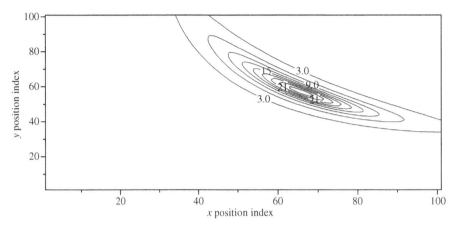

FIGURE 3.9. Scalar field for Problem 3.14.

3.15. A *cardioid* is a figure consisting of the locus of points given by $x^2 + y^2 + Bx = B\sqrt{x^2 + y^2}$. This shape bears some resemblance to a human heart, hence the name. Suppose $B = 6$ and assume we need to determine the area enclosed by the curve. We should be able to find the area of a plane region through the use of a line integral over the boundary according to Green's theorem; specifically,

$$A = \iint\limits_{R} dxdy = \tfrac{1}{2} \int\limits_{C} (xdy - ydx).$$

Prove that the area of the cardioid is $108\pi/2$. *Hint: Try putting the problem in polar form.*

REFERENCES

Aris, R. *Vectors, Tensors, and the Basic Equations of Fluid Mechanics*, Prentice-Hall, Englewood Cliffs, NJ (1962).

Neeley, M. Exploring Stokes' Theorem. University of Tennessee, Knoxville (2008).

Sears, F. W. and M. W. Zemansky. *University Physics*, 3rd edition, Addison-Wesley, Reading, MA (1964).

Spiegel, M. R. *Advanced Mathematics for Engineers and Scientists*, Schaum's Outline Series, McGraw-Hill, New York (1971).

Spiegel, M. R., Lipschutz S., and D. Spellman. *Vector Analysis*, 2nd edition, Schaum's Outline Series, McGraw-Hill, New York (2009).

4

NUMERICAL QUADRATURE

INTRODUCTION

Definite integrals must be evaluated routinely and for cases in which an antiderivative cannot be found, or for cases in which the analytic process is simply too difficult, a numerical scheme (numerical integration, or quadrature) may be our only recourse. Consider the integral

$$\int_0^\infty x^2 \exp(-x^2)dx, \qquad (4.1)$$

which we know to have the value $\sqrt{\pi}/4$. Let us consider a plot of the integrand as a function of x, shown in Figure 4.1.

We will conduct an elementary experiment with the graph of this integrand: First, we will count the smaller rectangular regions under the curve and add their areas together to obtain $(70)(0.25)(0.025) = 0.4375$. Next, we will cut out the region under the curve and weigh it and compare that weight to the average weight per rectangle. For the paper under the curve, we find $W = 0.5754$ g, and for each box, 0.008003 g. Therefore, the area under the curve is approximately 71.89 rectangular boxes or $(71.89)(0.25)(0.025) = 0.4493$. Our first estimate is in error by about 1.3% (under) and our second estimate by about 1.4% (over). For many applications, a rough evaluation like this would be acceptable, but suppose it was not. We might have a critically important problem where accuracy was paramount. Therefore, it will be productive for us to review some techniques that may be capable of producing significantly improved estimates.

TRAPEZOID RULE

One of the simpler schemes for making this evaluation is the trapezoid rule; we use a straight-line approximation for $f(x)$ over a finite interval, Δx, and then multiply the average value for $f(x)$ by the width. For example, let $f_1 = f(x)$ and $f_2 = f(x + \Delta x)$. Our approximation, therefore, is

$$\int_x^{x+\Delta x} f(x)dx = \left(\frac{f_1 + f_2}{2}\right)\Delta x.$$

If Δx is sufficiently small, this may work well. We will apply it first to the previous example; notice in the following code that the upper limit for the integral has been set to 8. We conclude that this *may be reasonable* since $(8)^2 \exp(-8)^2 \cong 1.03 \times 10^{-26}$. This is an example of *limit truncation*; we have changed the upper limit of the integral to 8 from ∞. One might want to assess this by using different upper limits to better gauge the contribution of the right-hand tail. Of course limit truncation may introduce error, but in this particular case, the behavior of the integrand is easily anticipated.

```
#COMPILE EXE
#DIM ALL
  REM *** Numerical integration by trapezoid rule
       GLOBAL x,dx,fx,f1,f2,a,b,fbar,sum,zz AS
            DOUBLE
```

Applied Mathematics for Science and Engineering, First Edition. Larry A. Glasgow.
© 2014 John Wiley & Sons, Inc. Published 2014 by John Wiley & Sons, Inc.

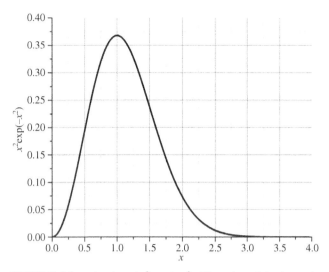

FIGURE 4.1. Behavior of $x^2 \exp(-x^2)$. The value of the integral, eq. (4.1), is known to be $\sqrt{\pi}/4 = 0.443113$.

```
FUNCTION PBMAIN
        a=0:b=8.0:x=a
        dx=(b-a)/1000
            GOSUB 300
            f1=fx
50 REM *** continue
            x=x+dx
                IF x>b THEN 200
            GOSUB 300
            f2=fx
        fbar=(f1+f2)/2
            sum=sum+fbar*dx
            f1=f2
            GOTO 50
200 REM *** continue
            PRINT "Value of definite integral is: ";sum
            INPUT "Shall we continue?";ZZ
            IF ZZ>0 THEN END
300 REM *** function evaluation for integral
            fx=x^2*EXP(-x^2)
            RETURN

END FUNCTION
```

This trapezoid rule code produces a result of 0.4431135 using 1000 intervals between 0 and 8. Despite the very small error produced in this case, one might wonder if a more demanding application could require a more accurate approximation. After all, there is nothing to prevent us from using a *polynomial* to approximate $f(x)$, as opposed to a straight line. Before we attempt that, however, let us examine one other case where the value of the definite integral is known so we can explore the impact of increasing the number of intervals using both single- and double-precision

arithmetic. Again we will use the trapezoid rule to consider

$$\int_0^{\pi} \sin^2(3x)dx = \frac{\pi}{2} = 1.570796327. \qquad (4.2)$$

n	Single Precision, $I(x)$	Double Precision, $I(x)$
125	1.570797	1.5707963268
250	1.570800	1.5707874013
500	1.570789	1.5707952107
1000	1.570781	1.5707961870
2000	1.570817	1.5707963270
4000	1.570853	1.5707963250
8000	1.570871	1.5707963264

Notice that once we exceed $n \approx 2000$ with the single-precision calculations, the result deviates increasingly from the correct value given in eq. (4.2). We will consider this unwelcome development—cumulative roundoff error—more fully later in this chapter.

SIMPSON'S RULE

We saw that we could employ a linear approximation with the trapezoid rule and get satisfactory results for integrals that were not particularly "difficult." An obvious extension that might yield better results could be obtained by merely increasing the degree of the polynomial. For example, we might let $f(x)$ be approximated by

$$y = f(x) = ax^2 + bx + c. \qquad (4.3)$$

If we place the three necessary function evaluations (for y_1, y_2, and y_3) at $x = -\Delta x$, $x = 0$, and $x = +\Delta x$, then we find

$$y_1 = a(\Delta x)^2 - b\Delta x + c \qquad (4.4)$$

$$y_2 = c \qquad (4.5)$$

$$y_3 = a(\Delta x)^2 + b\Delta x + c. \qquad (4.6)$$

Now we will actually integrate the polynomial (eq. 4.3) from $-\Delta x$ to $+\Delta x$:

$$\int_{-\Delta x}^{+\Delta x} (ax^2 + bx + c)dx$$
$$= \left(\frac{a}{3}x^3 + \frac{b}{2}x^2 + cx \right)\Big|_{-\Delta x}^{+\Delta x} = \frac{2a}{3}(\Delta x)^3 + 2c\Delta x. \qquad (4.7)$$

We use the above-mentioned three function evaluations to obtain a, b, and c. The resulting algorithm for *Simpson's rule* is

$$\frac{\Delta x}{3}(f_1 + 4f_2 + f_3) \quad \text{or} \quad \frac{\Delta x}{3}(y_1 + 4y_2 + y_3). \quad (4.8)$$

For an easy comparison, let us apply this method to the integral (eq. 4.2) examined previously, $\int_0^\pi \sin^2(3x)dx = \pi/2$. Using 500 intervals and single precision, we obtain 1.570791, and with 500 intervals and double precision, 1.57079632679; note how these values compare to the appropriate entries in the table provided earlier.

An illustration of logic required for Simpson's rule is shown as follows for the slightly more difficult definite integral:

$$\int_{4.75}^{5} \frac{\exp(2x)}{x}dx = 887.20057$$

(numerical result from the code as it appears here). Mathcad™ produces 887.201 for this integral using its *adaptive* scheme, and you may want to contemplate this integrand to see if you can determine why we might need an adaptive algorithm. For transparency, we will make no effort to be computationally efficient.

```
#COMPILE EXE
#DIM ALL
    REM *** Integration by Simpson's rule
        GLOBAL xi,xf,dx,x,y1,y2,y3,fx,sum,ZZ AS
        DOUBLE
FUNCTION PBMAIN
        xi=4.750:xf=5.0:sum=0
        dx=(xf-xi)/64
        x=xi
50 REM *** continue
        GOSUB 300
        y1=fx
        x=x+dx
        GOSUB 300
        y2=fx
        x=x+dx
        GOSUB 300
        y3=fx
            sum=sum+dx/3*(y1+4*y2+y3)
        IF x&#x003C;xf THEN 50 ELSE 100
100 REM *** continue
        PRINT sum
        INPUT "Shall we continue?";ZZ
        IF ZZ>0 THEN END
```

```
300 REM *** function evaluation
        fx=1/x*EXP(2*x)
        RETURN
END FUNCTION
```

This simple code uses double-precision floating point numbers with about 15 or 16 digits of precision; the actual program output is 887.20056962....

So far, we have examined definite integrals that pose no particularly great challenge, but to see the benefits offered by increasing the order of the polynomial used for the approximation, we need something else. Consider the integral

$$\int_0^{2.99} \frac{dx}{(3-x)^2} = 99.6667. \quad (4.9)$$

We will apply Simpson's rule to this problem, using single precision and starting with a very modest number of intervals, successively doubling n.

n	100	200	400	800	1,600	3,200	6,400	12,800
Definite integral	140.82	109.82	1779.02	429.62	158.65	122.10	101.83	99.65

It is a worthwhile exercise for the reader to test Simpson's rule on the integral in eq. (4.9) using other values for n, perhaps 600. The results we see here suggest that we might encounter cases for which Simpson's rule will not really be adequate. What would happen if we increased the order of the polynomial approximation again?

NEWTON–COTES FORMULAE

We posed a question regarding the impact of increasing the order of the polynomial approximation used for numerical quadrature. After all, we used a straight line with the trapezoid rule, and a quadratic approximation with Simpson's rule. There is nothing to prevent us from adding more function evaluations in an effort to make our curve more accurately reflect the behavior of the integrand. This succession of polynomials is often referred to collectively as the Newton–Cotes formulae. Let us tabulate these expressions for orders 1 through 5:

$$\frac{\Delta x}{2}(y_1 + y_2) \quad \text{trapezoid rule} \quad (4.10)$$

$$\frac{\Delta x}{3}(y_1 + 4y_2 + y_3) \quad \text{Simpson's rule} \quad (4.11)$$

$$\frac{3\Delta x}{8}(y_1 + 3y_2 + 3y_3 + y_4) \quad \text{Simpson's 3/8 rule} \quad (4.12)$$

$$\frac{2\Delta x}{45}(7y_1 + 32y_2 + 12y_3 + 32y_4 + 7y_5) \quad (4.13)$$

$$\frac{5\Delta x}{288}(19y_1 + 75y_2 + 50y_3 + 50y_4 + 75y_5 + 19y_6). \quad (4.14)$$

There is an obvious cost associated with this process: The number of function evaluations required per step for an nth order approximation will be $n + 1$. Will the extra effort required be worthwhile? We can illustrate the effectiveness of this approach by using the fourth-order Newton–Cotes algorithm on the integral,

$$\int_0^{\pi/2} \frac{dx}{(4\sin^2 x + 9\cos^2 x)} = 0.26179939. \quad (4.15)$$

If we specify the number of intervals as 50, 100, and 200, successively, we obtain the results 0.2774816, 0.2617994, and 0.2617994. We have obtained a high-quality result using a limited number of intervals *with single precision*! Of course, this was not a very demanding example, so we may want to ask if the increased complexity of the algorithm will help us deal with the difficulty we encountered with eq. (4.9). The unfortunate answer is a qualified "no." There are cases where merely increasing the order of the polynomial will not materially improve our estimate for the definite integral. We may occasionally need better tools, and in the following material, we will look at several candidates. Before we explore some alternatives, however, we should consider some of the consequences of the types of calculations we have proposed for numerical quadrature.

ROUNDOFF AND TRUNCATION ERRORS

Let us take Simpson's algorithm (in single precision) and apply it to an elementary definite integral. We have an intuitive impression at this point that we can always improve our estimate by reducing the size of the interval (increasing the number of steps employed). But we also know that floating point variables are only represented to the precision of the computing device we use. If the *roundoff* errors associated with that process accumulate, we might get results that are incorrect or unexpected. We will conduct this experiment with

$$\int_0^{\pi/2} \frac{dx}{4\sin^2 x + 9\cos^2 x} = 0.26179939.$$

Let us increase n by a factor of 10 successively to see what happens:

n	$I(x)$
50	0.2617994
500	0.2617993
5,000	0.2619558
50,000	0.2615938
500,000	0.2633153
5,000,000	0.2490056
50,000,000	2.1934250

Our effort to get an extremely accurate value for the definite integral was misguided as we have discovered that roundoff error can corrupt even a very simple calculation. There are two important lessons here: Use double precision whenever possible, and do not blindly assume that decreasing the interval size (Δx) will produce a better result.

We now shift our focus to *truncation* errors that result from terminating an infinite series expression (this will occur anytime a continuous function is represented discretely). Numerical quadrature is susceptible to this problem too, and it would be extremely useful to be able to assess the size of this error. We will begin by using the trapezoid rule to illustrate this phenomenon. Let us contemplate the integral:

$$\int_3^4 (2 + 5x - x^2)dx = 7.16667. \quad (4.16)$$

We set $\Delta x = 1$ and we evaluate the *integrand* at both $x = 3$ and $x = 4$; let those values be represented by $f'(3) = 8$ and $f'(4) = 6$. We also find the *antiderivative* of the integrand and evaluate it at both $x = 3$ and $x = 4$: The values are $f(3) = 19.5$ and $f(4) = 26.6667$, respectively. Of course, if we subtract the former from the latter, we get the exact value provided in eq. (4.16), 7.16667. For the trapezoid rule, we would take $[f'(3) + f'(4)]/2$ and multiply by Δx (which is 1); the result is 7, of course, which means that the absolute error is 0.16667 (or about 2.3%). Now let $f(3)$ be represented by $y(x)$ and $f(4)$ be $y(x + \Delta x)$—we expand the latter in a Taylor series:

$$y(x + \Delta x) = y(x) + \left(\frac{dy}{dx}\right)_x \Delta x + \left(\frac{d^2 y}{dx^2}\right)_x \frac{(\Delta x)^2}{2}$$
$$+ \left(\frac{d^3 y}{dx^3}\right)_x \frac{(\Delta x)^3}{6} + \cdots. \quad (4.17)$$

We subtract $y(x)$ noting that the remainder on the right-hand side is approximately the area we seek:

$$y(x + \Delta x) - y(x) = f'_x \Delta x + f''_x \frac{(\Delta x)^2}{2} + f'''_x \frac{(\Delta x)^3}{6} + \cdots.$$

$$(4.18)$$

For the trapezoid rule, we need both f'_x and $f'_{x+\Delta x}$, so using the Taylor series expansion again,

$$f'_{x+\Delta x} \cong f'_x + f''_x (\Delta x) + f'''_x \frac{(\Delta x)^2}{2} + \cdots. \quad (4.19)$$

Therefore,

$$f''_x \frac{(\Delta x)^2}{2} = \frac{\Delta x}{2} \left[f'_{x+\Delta x} - f'_x - f'''_x \frac{(\Delta x)^2}{2} - \cdots \right]. \quad (4.20)$$

Now we take this result back to eq. (4.18):

$$y(x + \Delta x) - y(x) = \left[\frac{f'_{x+\Delta x} + f'_x}{2} \right] \Delta x - f'''_x \frac{(\Delta x)^3}{12} - \cdots.$$

$$(4.21)$$

Notice that the *exact* area is on the left-hand side, and the trapezoid rule approximation is the bracketed term on the right. Therefore, the error associated with this *one interval* example is on the order of

$$e_1 \cong -f'''_x \frac{(\Delta x)^3}{12}. \quad (4.22)$$

Remember, f'''_x is the *second derivative* of the integrand, and it is constant in this case:

$$f'_x = 2 + 5x - x^2, \quad (4.23)$$

$$f''_x = 5 - 2x, \quad (4.24)$$

and

$$f'''_x = -2. \quad (4.25)$$

Therefore, we have

$$f'''_x \frac{(\Delta x)^3}{12} = \frac{(2)(1)^3}{(12)} = \frac{1}{6}. \quad (4.26)$$

In this instance, the estimate of error is borne out exactly by our previous result. Generally, we will not be that fortunate and f'''_x will vary over the range of the independent variable, x. We will rewrite the error so that

$$-f'''_x \frac{(\Delta x)^3}{12} \approx -f'''_\chi \frac{(\Delta x)^3}{12}, \quad (4.27)$$

where $x \leq \chi \leq x + \Delta x$ (think of the mean value theorem from calculus). If we wish to employ our estimate of error for an arbitrary case with n-intervals,

$$\sum e_n = -\frac{(\Delta x)^3}{12} [f'''(\chi_1) + f'''(\chi_2) + f'''(\chi_3) + \cdots f'''(\chi_n)]. \quad (4.28)$$

How useful this expression is depends entirely on how difficult it is to evaluate the second derivative of the integrand, f'''. If one were integrating tabular data, for example, both f'' and f''' would be unknown. There is a great deal of information available related to error estimation in numerical quadrature and the interested reader might begin by reviewing chapter 5 in James et al. (1977). We will also make error analysis the focus of one of the exercises at the end of this chapter.

ROMBERG INTEGRATION

The Romberg scheme utilizes an extremely powerful technique known as Richardson's extrapolation. Suppose, for example, that we were able to estimate the error associated with one of our numerical quadrature procedures. Furthermore, suppose we then decreased the step size (increasing the number of intervals) and found the error associated with *that* estimate. If we extrapolated the diminishing error to zero, we might be able to obtain a high-quality estimate with far less numerical work than would need to be carried out otherwise. We will illustrate this concept with a simple example, using only the trapezoid rule to prepare our successive estimates. Let us consider the integral,

$$\int_1^5 x \ln x \, dx = 14.118 \text{ (from the analytic solution)}. \quad (4.29)$$

We begin by using the trapezoid rule and we will start with just one interval (such that $\Delta x = 4$. We then cut Δx in half and evaluate the integral again. We repeat this process several times, resulting in a series of estimates that appear in the following table:

N	Estimate for Integral, $I(x)$
1	16.094379
2	14.638863
4	14.250903
8	14.151423
16	14.126351
32	14.120069

In this procedure, we are cutting the interval in half, successively. An improved estimate for the definite integral is obtained using pairs of values from the table according to

$$I_{\text{improved}} = \frac{2^2 I_{\Delta x/2} - I_{\Delta x}}{2^2 - 1} = \frac{1}{3}(4I_{\Delta x/2} - I_{\Delta x}). \quad (4.30)$$

We will apply this procedure to the definite integral (4.29), using the subsequent pairs of values from our table.

Pair ID	I_{improved}
2–1	14.1537
4–2	13.4549
8–4	14.1183
16–8	14.1180
32–16	14.1180

Notice that by the time we get to the 8–4 pair, we have reduced the error of our estimate down to 0.002%. The attractive nature of the Romberg scheme is now apparent: We have dramatically improved the results of our numerical integration without much additional computational effort! You should be aware of the fact that if we simply apply the trapezoid rule with 32 intervals, our error is still about 0.015%. This example demonstrates the superiority of Romberg integration—at least for this case—very conclusively.

We would like to know if the outstanding results obtained in the preceding example can be expected more broadly (universally would be nice, but we will settle for less than that). Recall our exploration of the integral,

$$\int_0^{2.99} \frac{dx}{(3-x)^2} = 99.6667.$$

Let us apply the Romberg scheme here to find out if it is comparably effective for this more difficult case. The integrations will be carried out with Simpson's rule this time.

n	$I(x)$	Extrapolated Value
1	39,866.9658	–
2	4,984.2782	−6,643.28
4	2,493.8525	1,663.71
8	1,250.4297	835.95
16	632.0535	425.93
32	328.6733	227.55
64	185.9160	138.33
128	125.6151	105.52
256	105.2509	98.46
512	100.4610	98.86
1024	99.7431	99.50

By using the 1024–512 interval pair, we were able to use extrapolation to reduce the error of our estimate to just 0.17%. It should be apparent to you that for certain types of problems where the computational burden is really significant, Romberg integration can be used to great advantage.

ADAPTIVE INTEGRATION SCHEMES

The "difficult" integral we have examined previously,

$$\int_0^{2.99} \frac{dx}{(3-x)^2} = 99.6667,$$

is one for which the integrand varies by four orders of magnitude (1/3 to 10,000) as x varies from its lower limit (0) to the upper one (2.99). This kind of change in magnitude indicates that a different approach may be needed, namely, an *adaptive algorithm*. The underlying logic is straightforward: When we encounter a region in which the integrand begins to change dramatically, we decrease the interval size. There are many adaptive implementations available in commercial software packages. Mathcad's adaptive algorithm, for example, produces 99.667 for the integral in question.

Simpson's Rule

It is reasonable for us to wonder if we could make a few simple changes to one of the elementary quadrature schemes and produce an adaptive scheme that is effective. We will try this using Simpson's rule where we are using the variable, *sum*, to accumulate increments of area. We will look at the per-step change in *sum*; if it exceeds a set value, we will reduce Δx. In particular, we will set the threshold for change in sum at 10% initially, and if that size change occurs, we reduce the size of the interval by cutting it in half.

If Change Exceeds (%)	Divide dx by	Result, $I(x)$, Using Double Precision
10	2	109.9738
5	2	104.5676
2.5	2	99.66676
1.25	2	99.66676
0.625	2	99.66676

We quickly discover that our very crude modification to Simpson's rule will work for this case, as long as we select the *proper values*. The reader is invited to explore other combinations. If we do not know a lot about our integral in advance, it may be difficult to get this to work as we anticipate. After all, we have used two arbitrary choices: the threshold change for the accumulating sum and the size of the reduction of Δx. We must have something more reliable and much more broadly applicable.

Fortunately, a number of options are available to us designed specifically to deal with this type of problem.

Kuncir (1962) suggested a method using Simpson's rule in which an interval would be subdivided if the estimated error exceeds some threshold value. Consider the integral

$$\int_a^b f(x)dx = I(a,b). \tag{4.31}$$

The interval (a, b) has a midpoint, $(a + b)/2$; we rewrite the integral (eq. 4.31), splitting it into two pieces:

$$\int_a^b f(x)dx = \int_a^{\frac{a+b}{2}} f(x)dx + \int_{\frac{a+b}{2}}^b f(x)dx. \tag{4.32}$$

A strategy is now apparent: We compute the left-hand side of eq. (4.32) and compare it to the right-hand side, which was obtained by subdividing (a, b). If the estimates do not agree within a specified tolerance, further subdivision is required. We stop the successive subdividing process when the criterion suggested by Lyness (1969) is met:

$$\frac{\left|I\left(a, \frac{a+b}{2}\right) + I\left(\frac{a+b}{2}, b\right) - I(a,b)\right|}{15} < \varepsilon. \tag{4.33}$$

We can easily anticipate how this will work for elementary functions, but we should apply it to an integral that we know to be problematic. Once again, we consider

$$\int_0^{2.99} \frac{dx}{(3-x)^2} = 99.667; \tag{4.34}$$

we apply Simpson's rule with just three function evaluations (the minimum). Then we will bisect the interval (midpoint 1.495) and apply Simpson's rule to each half. We will put the results in a "tree" structure to illustrate the progress of the calculations:

$$I_0^{2.99} = 4984.278, \tag{4.35}$$

$$I_0^{1.495} = 0.33413, \quad I_{1.495}^{2.99} = 2493.518. \tag{4.36}$$

Of course, the sum in eq. (4.36) is very different from level 1 (2493.85 as opposed to 4984.28); subdivision is necessary:

$$I_0^{0.7475} = 0.11065, \quad I_{0.7475}^{1.495} = 0.22074, \quad I_{1.495}^{2.2425} 0.66153,$$

$$I_{2.2425}^{2.99} = 1249.437. \tag{4.37}$$

Note that the sum of the first two is 0.33139—this is a change of 0.8% from level 2, that is, the first half of eq. (4.36), and we will assume for this illustration that this is

adequate. We now subdivide the two quarters in the right half:

$$I_{1.495}^{1.86875} = 0.21959, \quad I_{1.86875}^{2.2425} = 0.43662, \quad I_{2.2425}^{2.61625} = 1.29678,$$

$$I_{2.61625}^{2.99} = 629.769.$$

$$\tag{4.38}$$

The sum of the first pair is 0.6562, which is a bit of a change from 0.66153 (once more about 0.8%). To fit the space available, we will again assume this is sufficient and subdivide the two intervals of the right half:

$$I_{2.2425}^{2.42938} = 0.43246, \quad I_{2.42938}^{2.61625} = 0.85425, \quad I_{2.61625}^{2.80313} = 2.49337,$$

$$I_{2.80313}^{2.99} = 323.899.$$

$$\tag{4.39}$$

The sum of the first pair is 1.28671 as opposed to 1.29678 from eq. (4.38). Once again, we assume this meets our (very relaxed) criterion and subdivide the right-half intervals:

$$I_{2.61625}^{2.70969} = 0.83894, \quad I_{2.70969}^{2.80313} = 1.63643, \quad I_{2.80313}^{2.89657} = 4.62063,$$

$$I_{2.89657}^{2.99} = 176.537.$$

$$\tag{4.40}$$

Compare the sum of the first pair (2.47537) with 2.49337 from the above-mentioned equation (the difference is once again about 0.8%); we will subdivide the right-half intervals again:

$$I_{2.80313}^{2.84985} = 1.58086, \quad I_{2.84985}^{2.89657} = 3.01076, \quad I_{2.89657}^{2.94329} = 8.00725,$$

$$I_{2.94329}^{2.99} = 108.261.$$

$$\tag{4.41}$$

We will repeat the process one more time, assuming that 4.59162 is sufficiently "close" to 4.62063:

$$I_{2.89657}^{2.91993} = 2.82120, \quad I_{2.91993}^{2.94329} = 5.14751, \quad I_{2.94329}^{2.96665} = 12.3914,$$

$$I_{2.96665}^{2.99} = 75.550.$$

$$\tag{4.42}$$

The sum of these four integrals is 95.9101 and we are rapidly closing in on a much better approximation for the definite integral (eq. 4.34). This detailed example reveals the power of the adaptive version of Simpson's rule. Since we stop making calculations in intervals where our accuracy is sufficient, we can significantly reduce the total number of operations relative to cases where we merely continually reduce Δx over the entire interval (a, b).

Gaussian Quadrature and the Gauss–Kronrod Procedure

It is important that the reader be aware of other recent developments in adaptive numerical quadrature. The Gauss–Kronrod (GK) procedure is a powerful method that can be used in this context; in fact, Kahaner et al. (1989) state that the GK algorithm "is currently one of the most effective methods for calculating general integrals." To set the framework for a useful description, we must begin with an n-point Gaussian quadrature, where a definite integral is approximated by the formula

$$\int_a^b f(x)dx \cong \sum_{j=1}^n \omega_j f(x_j). \qquad (4.43)$$

The ω_js are weight factors that are applied to specific x-positions. We can make Gaussian quadrature more transparent with an illustrative example. Consider the definite integral

$$\int_a^b f(x)dx = \int_0^5 (5 + 4x - x^2)dx = 33.3333. \qquad (4.44)$$

Our first task is to transform this integral to $\int_{-1}^{+1} F(t)dt$; we do this by setting $x = mt + c$. Therefore, we have the two equations, $0 = -m + c$ and $5 = m + c$. By adding the equations together, we find $c = 5/2$ and $m = 5/2$; thus,

$$x = \frac{5}{2}t + \frac{5}{2} \quad \text{and} \quad dx = \frac{5}{2}dt. \qquad (4.45)$$

Our definite integral is now written as

$$\frac{5}{2}\int_{-1}^{+1}\left(\frac{35}{4} - \frac{5}{2}t - \frac{25}{4}t^2\right)dt, \qquad (4.46)$$

and the reader can easily verify that this is also equal to 33.3333. Now we return to eq. (4.43) and fix $n = 2$. The specific t-positions are $-1/\sqrt{3}$ and $+1/\sqrt{3}$, and these locations correspond to zeros of the Legendre polynomial, P_2 (a limited table of Legendre polynomials is provided here for the reader's reference). Remember, Legendre polynomials are orthogonal on the interval $(-1, +1)$.

$n = 0$	Legendre Polynomials, P_n
0	1
1	x
2	$\frac{1}{2}(3x^2 - 1)$
3	$\frac{1}{2}(5x^3 - 3x)$
4	$\frac{1}{8}(35x^4 - 30x^2 + 3)$
5	$\frac{1}{8}(63x^5 - 70x^3 + 15x)$
6	$\frac{1}{16}(231x^6 - 315x^4 + 105x^2 - 5)$
7	$\frac{1}{16}(429x^7 - 693x^5 + 315x^3 - 35x)$

The weight factors we are to employ are both equal to 1, so our approximation for the definite integral is now written as

$$\int_a^b f(x)dx = \frac{5}{2}\sum_{j=1}^2 \omega_j F(t_j) = \frac{5}{2}\left[(1)F\left(-\frac{1}{\sqrt{3}}\right) + (1)F\left(+\frac{1}{\sqrt{3}}\right)\right]. \qquad (4.47)$$

The right-hand side is equal to 33.3333—one might ask, why does the method yield the exactly correct result? The answer is that an nth order Newton–Cotes formula will integrate a polynomial of order n, or less, *exactly*. With Gaussian quadrature, the exactness of the result is extended to polynomials of degree $2n + 1$.

For more general cases, we can improve the quality of the approximation by simply increasing the order, and the reader interested in using Gaussian quadrature should consult the extensive table (table 25.4) in Abramowitz and Stegun (1964, pp. 916–919). For example, if we choose $n = 6$, then t-positions are found from the roots of the appropriate entry in the previous table; these positions (nodes) and the corresponding weight factors are as follows:

t-Positions	Weight Factors
± 0.238619186083	0.467913934573
± 0.661209386466	0.360761573048
± 0.932469514203	0.171324492379

To apply this to

$$\int_0^{2.99} \frac{dx}{(3 - x)^2},$$

we would rewrite the integral as

$$\int_{-1}^{+1} \frac{1.495dt}{2.26503 - 4.49995t + 2.23503t^2}$$

and use the values in the previous table. This results in an estimate for the definite integral of 23.6594, which we know to be much too low (not at all surprising since n is only 6). Somewhat better results are obtained for $\int_0^1 x\ln(1 - x)dx = -3/4$, which we transform to

$$\frac{1}{2}\int_{-1}^{+1}\left(\frac{1}{2}t + \frac{1}{2}\right)\ln\left(-\frac{1}{2}t + \frac{1}{2}\right)dt.$$

In this case, Gaussian quadrature with $n = 6$ yields -0.734846, which is only about 2% smaller (in magnitude) than the correct value. And remember, this is achieved with only six function evaluations!

Now that we have a little familiarity with Gaussian quadrature, we can proceed to our real objective, which is to modify the process in such a way that it can be effective for "difficult" integrals. One clever way to accomplish this goal is with an *embedded algorithm* for which error estimation is automatically available (embedded schemes are also sometimes referred to as *nested* or *progressive*). Such an approach will be more efficient if we do not have to start over with respect to node placement; that is, we will make far fewer calculations if we can reuse some of the existing function evaluations. This is exactly what the GK procedure was designed to do. Each step requires $(2n + 1)$ function values, so for this example, we will consider the (7-15) GK nodal placements. We begin with Gaussian quadrature for $n = 7$ (see the table of Legendre polynomials). The Gaussian placements and weights are

±0.9491079123	0.1294849662
±0.7415311856	0.2797053915
±0.4058451514	0.3818300505
0.0000000000	0.4179591837

Since we have seven nodes here, the Kronrod modification will have $(2n + 1) = 15$, but seven of them are identically the Gaussian values. Note that the weight factors are different (they are about *half* of the Gaussian weights), however:

±0.9914553711	0.0229353220
±0.9491079123	0.0630920926
±0.8648644234	0.1047900103
±0.7415311856	0.1406532597
±0.5860872354	0.1690047266
±0.4058451514	0.1903505781
±0.2077849550	0.2044329401
0.0000000000	0.2094821411

We will explore the now familiar integral,

$$\int_0^{2.99} \frac{dx}{(3 - x)^2},$$

one more time. First, we apply the (7-15) GK to the entire integral remembering that because of the change in limits, we must make a variable change as $x = 1.495t + 1.495$. The results are

G	K
30.15765	84.13157

The discrepancy is enormous, so we cut the interval in half, applying the (7-15) GK to each (0-1.495) and (1.495-2.99). The results are as follows:

Left Half		Right Half	
G	K	G	K
0.331620	0.331371	39.730679	27.053353

As you can see, the significant error is in the right half, so we repeat the process for (1.495-2.2425) and (2.2425-2.99):

Left Half		Right Half	
G	K	G	K
0.656664	0.656175	71.875407	98.943049

Of course, we already know that the difficulty with this integral is a consequence of the upper limit, so the right half is divided into two once again (2.2425-2.61625) and (2.61625-2.99):

Left Half		Right Half	
G	K	G	K
1.2876488	1.2867079	88.4937846	98.0094297

You will observe that we are very rapidly acquiring a much-improved estimate for this definite integral; in fact, if we take these last two Kronrod values and add to them the left halves that have been dropped; we will see that we are not that far from the correct value (99.6667). The GK embedded procedure is powerful indeed, and it automatically provides us with a means to estimate our error. Consequently, it lends itself very nicely to adaptive quadrature for "difficult" integrals.

INTEGRATING DISCRETE DATA

Suppose we have the following set of discrete data:

x	$y(x)$
11	0.338
19	0.340
27	0.335
35	0.333
43	0.326

Our task is to produce an accurate integration of these data for $11 \leq x \leq 43$. We note immediately that the data are highly nonlinear as is evident in Figure 4.2.

We have several options for performing the necessary integration, including some crude methods discussed at the very beginning of the chapter. But because we have stipulated that we need a very accurate value, we should look for a better technique.

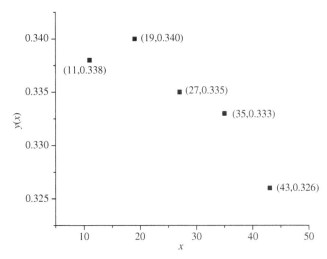

FIGURE 4.2. Nonlinear discrete data for numerical quadrature. Note that the ordinate, $y(x)$, has been greatly expanded and only a small segment of the axis has been plotted.

One powerful method for dealing with such situations is through the integration of a *cubic spline interpolation*. *Spline* refers to a draftsman's tool used to generate smooth curves connecting multiple points; for a picture of such a device, see Grandine (2005), who describes how extensively splines are being used at Boeing (to represent complicated geometries or geometric boundary conditions). The idea with spline interpolation is a very simple one: Polynomial *pieces* are fit together in such a way that the function is continuous and is continuously differentiable. Moreover, these pieces connect smoothly and the derivatives are continuous at the interval boundaries. The process employed is distinctly different from merely fitting a polynomial of higher degree to all of the available data—and if you have ever tried this, you already know that such curve fits are often unphysical. They usually show too much oscillation (curvature) for intermediate values and unbounded (nonasymptotic) behavior at the extremes.

The data set of interest to us has five points (so four interior intervals) with each separated by a Δx, or h, of 8. We will use a cubic polynomial to represent the data in each interval, so for $11 \leq x \leq 19$, we write

$$y = a_1(x - 11)^3 + b_1(x - 11)^2 + c_1(x - 11) + d_1. \quad (4.48)$$

Naturally, when $x = 11$, we see $d_1 = 0.338$. You will notice immediately that we have four sets of constants, one set for each interior interval. In every case, d will be given by the left-hand endpoint value of y, so for the second interval,

$$y = a_2(x - 19)^3 + b_2(x - 19)^2 + c_2(x - 19) + 0.340, \quad (4.49)$$

and so on. A key feature of spline interpolation is that the first and second derivatives will be continuous from one interval to the next. Accordingly, we will write the cubic polynomial in generic form:

$$y = a_i(x - x_i)^3 + b_i(x - x_i)^2 + c_i(x - x_i) + d_i. \quad (4.50)$$

Therefore,

$$y' = 3a_i(x - x_i)^2 + 2b_i(x - x_i) + c_i \quad (4.51)$$

and

$$y'' = 6a_i(x - x_i) + 2b_i. \quad (4.52)$$

It is apparent for our problem that we will have $(3) \times (4) = 12$ unknown coefficients. Of course, the relation (eq. 4.50) provides four equations. We will also require that the first and second derivatives match at the interval boundaries; that provides six more equations through application of eq. (4.51) and eq. (4.52), but we are still *two short*! One way we can obtain the final two relations is to specify the second derivative at the ends of the interval (i.e., for $x = 11$ and 43, in our case). If we just choose to set $y'' = 0$, we get what is referred to as a *linear spline*. Though this seems incredibly arbitrary, Hanna and Sandall (1995) note that it does not seem to lead to significant errors in actual application. We will now look at the complete set of equations for our cubic spline curve fit:

$$\begin{aligned} 512a_1 + 8c_1 &= 0.002 \\ 512a_2 + 64b_2 + 8c_2 &= -0.005 \\ 512a_3 + 64b_3 + 8c_3 &= -0.002 \\ 512a_4 + 64b_4 + 8c_4 &= -0.007 \end{aligned} \quad (4.53a,b,c,d)$$

$$\begin{aligned} 192a_1 + c_1 - c_2 &= 0, \\ 192a_2 + 16b_2 + c_2 - c_3 &= 0, \quad (4.54a,b,c) \\ 192a_3 + 16b_3 + c_3 - c_4 &= 0, \end{aligned}$$

$$\begin{aligned} 48a_1 - 2b_2 &= 0 \\ 48a_2 + 2b_2 - 2b_3 &= 0 \quad (4.55a,b,c) \\ 48a_3 + 2b_3 - 2b_4 &= 0. \end{aligned}$$

And finally, by setting the second derivative equal to zero at the endpoints,

$$b_1 = 0, \quad 48a_4 + 2b_4 = 0. \quad (4.56a,b)$$

Since b_1 was eliminated, we have 11 equations and the same number of unknowns. The solution for this set of equations is

		Each Multiplied by 10^6
1	a_1	-4.255
2	c_1	522.32
3	a_2	7.60323
4	b_2	-102.12
5	c_2	-294.643
6	a_3	-6.62667
7	b_3	80.3571
8	c_3	-468.75
9	a_4	3.27846
10	b_4	-78.683
11	c_4	-455.357

We can assess the quality of the representation in Figure 4.3.

Now let us see exactly how well this cubic spline curve fit performs. We can use eq. (4.48) as our example, integrating over interval 1:

$$\int_{11}^{19} y_1(x)dx = \frac{a_1}{4}(8)^4 + \frac{c_1}{2}(8)^2 + 0.338(8) = 2.71636. \quad (4.57)$$

Remember that b_1 is zero! The values for the remaining three integrals are 2.70093, 2.66514, and 2.63936, respectively, and the sum of all four is 10.7218. We have, through the use of cubic spline interpolation, obtained a very accurate value for the integral of the original discrete data set.

MULTIPLE INTEGRALS (CUBATURE)

Recall that the definite integral, $\int_a^b f(x)dx$, can be approximated by a Riemann sum, that is, the sum of areas of a

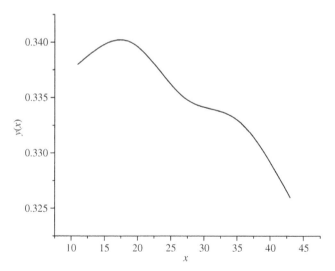

FIGURE 4.3. Cubic spline curve fit for the nonlinear data set. We now have a very nice spline-function representation of the original data, facilitating integration.

number of rectangles. For the case in which we use n rectangles (but allow n to become very large) and $x = a \le \phi \le x = b$,

$$\int_a^b f(x)dx = \lim_{n \to \infty}\left[\sum_{i=1}^n f(\phi_i)\Delta x\right]. \quad (4.58)$$

This process can also be extended to multiple integrals:

$$\iint_R f(x,y)dxdy = \lim_{\delta \to 0}\left[\sum_{i=1}^n f(x_i, y_i)\Delta A_i\right]. \quad (4.59)$$

Note that δ is the norm (diagonal) of the partition or rectangle; by allowing it to approach zero, we are employing an infinite number of rectangles. If $a \le x \le b$ and if the limits for the variable y are set by two functions of x, say, $g_1(x)$ and $g_2(x)$, then for this *type I* region, we have the *iterated integral*:

$$\int_a^b \int_{g_1(x)}^{g_2(x)} f(x,y)dydx. \quad (4.60)$$

For a *type II* region, the terminal values of y are constants, $c \le y \le d$, and the limits for the variable x are two functions of y. We can review the evaluation process for an iterated integral with a straightforward example:

$$\begin{aligned}
\int_3^5 \int_{-x}^{x^2} (4x+10y)dydx &= \int_3^5 (4xy+5y^2)\Big|_{-x}^{x^2} dx \\
&= \int_3^5 (4x^3 + 5x^4 - x^2)dx \quad (4.61) \\
&= 5^4 + 5^5 - \tfrac{1}{3}5^3 - 3^4 - 3^5 + \tfrac{1}{3}3^3 \\
&= 3393.33.
\end{aligned}$$

This was an extremely simple double integral that could be easily evaluated analytically. But our real interest here is to acquire some tools that might allow us to evaluate multiple integrals that defy elementary solution.

Multiple integrals can be handled, although somewhat inefficiently, using the one-dimensional numerical quadrature techniques that we have discussed previously in this chapter. Let us illustrate with the generic double integral:

$$\int_a^b \int_c^d f(x,y)dxdy. \quad (4.62)$$

We will proceed in the following way:

1. Set increment value for y, for example, $(b - a)/100$.
2. Set the increment value for x, for example, $(d - c)/100$.
3. Let y assume its first value, y_1.
4. Accumulate the incremental areas for $f(x, y_1)$ as x varies from c to d.
5. Set this sum equal to $F(y_1)$.
6. Increment y and repeat quadrature on x from c to d.
7. Continue the process until $y = b$, that is, until we find $F(y_n)$.
8. Perform quadrature on the complete set of $F(y_i)$s.

The process sketched here is referred to as a *product rule*. To illustrate its application, we will consider the elementary example,

$$\int_0^2 \int_1^4 (x + 2y)dxdy = 27. \tag{4.63}$$

We will start with 50 intervals for both x and y and use the trapezoid rule for the quadratures. Using single precision, this direct approach yields 27.72359 (an error of about 2.7%). If we merely increase the number of intervals in *both* the x- and y-directions to 100, the very same process yields 27.36090; with 200 intervals, the result is 27.00002. Although the product procedure we have described here works (very well for this particular example), it requires many function evaluations. In fact, Smyth (1998) notes that the number of evaluations required for product rules grows with the number of dimensions exponentially. And you will also observe that in our example we used the same number of intervals in both directions, which is generally a mistake; Kahaner et al. (1989) suggest that a rule of thumb for the application of one-dimensional algorithms to double integrals is that the inner integral be computed such that its accuracy is about 10 times better than the outer. We can demonstrate this using the trapezoid rule (with double precision) for another elementary double integral,

$$\int_0^2 \int_1^4 \frac{x^2}{4y} dydx = 0.924196.$$

We will increase the number of intervals for the inner integral while maintaining a constant number of evaluations for the outer integral. The results are summarized in the following table:

Number of Intervals, n		$\int_0^2 \int_1^4 (x^2/4y)dydx = 0.924196$
Inner	Outer	
10	10	1.3058895
20	10	1.2695594
40	10	1.2355796
80	10	1.2352861
160	10	1.2393796
100	100	0.92927096
200	100	0.92425417
400	100	0.92549427
800	100	0.92424318
1,600	100	0.92424263
1,000	1000	0.92419717
2,000	1000	0.92419682
4,000	1000	0.92419673
8,000	1000	0.92419671
16,000	1000	0.92419670

These data show that the accuracy of the estimate of the double integral improves (generally) as the number of intervals for the inner integral is increased. But these calculations also reveal that the application of a one-dimensional scheme using the trapezoid rule is not very efficient. An extremely large number of calculations is required to obtain the correct sixth decimal place.

In recent years, the search for modern *nonproduct methods* for multidimensional integrals has been intensive. A nonproduct method is a quadrature (or more appropriately, a cubature) rule that *does not* require us to apply a one-dimensional method successively to the different directions. As we stated earlier, it is computationally inefficient to apply a product rule to the evaluation of multiple integrals. You may recall our discussion of Gaussian quadrature; with proper selection of the nodes (the t-positions) and weight factors, we can exactly integrate a polynomial of degree $2n + 1$. An obvious question is whether such a process could be extended to polynomials of two (or three or more) dimensions. If this were possible, multiple integration could be made much more efficient. Stroud (1971) notes that the principal problem is identification of formulae of the form

$$\int \ldots \int w(x_1, x_2, \ldots x_n)f(x_1, x_2, \ldots x_n)dx_1 \ldots dx_n$$
$$\cong \sum_{i=1}^N B_i f(\phi_1, \phi_2, \ldots \phi_n). \tag{4.64}$$

The fifth-degree formula of Radon (1948) was one of the first results of this type, and Stroud (1971) developed Gauss–Legendre formulae for iterated integrals of two and three dimensions. His book includes FORTRAN programs for both in chapter 10. Also, in one important success in efforts to find efficient nonproduct methods, Laurie (1977) was able to extend the GK method to double integrals.

Monte Carlo Methods

Monte Carlo methods were first used extensively by John von Neumann and Stanislaw Ulam to estimate neutron diffusion in connection with the Manhattan Project of World War II. This technique has been used by many investigators since to evaluate multiple integrals. It is attractive because of its stark simplicity; all one really needs is a mechanism for the generation of random numbers. Originally, random numbers were obtained from tables, and for an example, one may consult table 26.11 in Abramowitz and Stegun (1964) (which was reprinted from numbers compiled by the RAND Corporation). These days high-level language compilers have built-in capability for generating "random" numbers, but the reader should recognize that these numbers are being generated typically by a pseudorandom algorithm. This means that the same seed value will produce the *exact same sequence* of "random" numbers. Commercial software packages like Mathcad function similarly—the seed value must be changed if a different sequence of random numbers is desired.

We begin this discussion with a familiar example that will illustrate just how straightforward this technique really is; consider the one-dimensional integral,

$$I = \int_a^b f(x)dx. \qquad (4.65)$$

Now envision a rectangle that extends from a to b on the x-axis, and from 0 to some value c on the y-axis; that is, we think of a box (an area) that corresponds to

$$a \le x \le b \quad \text{and} \quad 0 \le f(x) \le c.$$

The area of this box is of course merely $(b-a)c$. Suppose we begin selecting random locations inside the box—the probability that we pick a point *under* the curve, $f(x)$, will be just

$$p = \frac{I}{(b-a)c}. \qquad (4.66)$$

We obtain an estimate for p from the ratio of the number of successes divided by the number of trials, n_{hits}/n. For obvious reasons, this approach is referred to as the "hit-or-miss" method:

$$I \cong (b-a)c \frac{n_{\text{hits}}}{n}. \qquad (4.67)$$

Suppose we were interested in obtaining a value for

$$\int_1^4 \left(2 + x + \frac{1}{3}x^2\right)dx$$

(we know this integral has the value, 20.5). We will conduct our trials on the rectangular area that is bounded by $1 \le x \le 4$ and $0 \le y = f(x) \le 20$. We use random numbers to "shoot" at the box and just keep track of the number of "hits" that fall under $f(x)$; our estimate then comes from eq. (4.62). Here is a typical sequence of results from such an experiment:

Number of Trials	Estimate for I
100	16.2
1,000	20.7
10,000	20.676
100,000	20.4888
1,000,000	20.53326
10,000,000	20.51026

It is extremely easy to obtain a *rough* estimate in this fashion, but it is apparent that an accurate result will require *many* trials. The reader can obtain greater detail regarding the use of the Monte Carlo method in this context from Rubinstein (1981), for example.

Now let us focus on the use of the Monte Carlo technique for the evaluation of multiple integrals; we should emphasize the important difference between the "hit-or-miss" method used earlier and the approach employed here (we will be using the Monte Carlo method to *place* the function evaluations). To illustrate, we use the simple double integral,

$$\int_0^2 \int_0^3 (x^2 + 2xy^2)dxdy = 42. \qquad (4.68)$$

We proceed in the following way: We select the number of trials (n), then use random numbers to select the (x, y) location for integrand evaluation. This function evaluation, $f(x, y)$, is multiplied by the area determined from the limits of the integration and the number of trials chosen, and the results are accumulated. For the double integral that we are using as our example, this accumulation will be performed n^2 times. A simple code written for eq. (4.68) is provided as follows:

```
#COMPILE EXE
#DIM ALL
   REM *** Monte Carlo integration of double integral
   GLOBAL n,ntrial,dx,dy,rnx,rny,x,y,sum,fxy,xlimit,
      ylimit,ZZ AS SINGLE
```

```
FUNCTION PBMAIN
        n=0:xlimit=3:ylimit=2
        ntrial=100
        dx=xlimit/ntrial:dy=ylimit/ntrial
        RANDOMIZE TIMER
  50 REM *** continue
        rnx=RND
        rny=RND
          x=rnx*xlimit
          y=rny*ylimit
            GOSUB 200
          sum=sum+fxy*dx*dy
            n=n+1
            PRINT n,sum
            IF n>ntrial^2 THEN 100 ELSE 50
 100 REM *** continue
        INPUT "Shall we continue?";ZZ
            IF ZZ>0 THEN END
 200 REM *** subroutine for function evaluation
        fxy=x^2+2*x*y^2
        RETURN

END FUNCTION
```

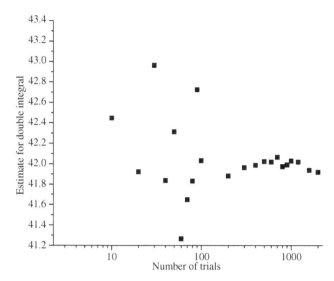

FIGURE 4.4. Progress of the Monte Carlo integration of the double integral, eq. (4.68), as a function of the number of trials. The result should be 42, of course.

Please make note of the ninth line down from the top, *RANDOMIZE TIMER*. This ensures that the random number generator obtains a different seed each time it is invoked. You will also observe that the number of trials, *ntrial*, is set to 100; this is a ridiculously small number for a Monte Carlo integration (just 10^4 function evaluations). To illustrate, we ran this program five times, just as it appears here, and we obtained the set of estimates: 41.95124, 41.96887, 42.31614, 41.66427, and 42.33305. The average of the set is 42.0467, which again reveals the unpleasant truth that a Monte Carlo integration—if accuracy is required—is computationally expensive. In fact, we can monitor the results from a sequence of realizations in which we increase the number of trials each time. Such an experiment results in a graph similar to Figure 4.4.

It is known that the error for this method diminishes as $\approx 1/\sqrt{ntrial}$; if we need a high level of decimal precision, then *many* trials will be necessary. To illustrate, we can see from Figure 4.4 that $2000^2 = 4 \times 10^6$ function evaluations produce just 41.91697 for the double integral—an error of about 0.2%. Just as we observed previously, a rough result is often very easy to achieve with the Monte Carlo technique, but the computational burden may be prohibitive if we need an extremely accurate answer.

Lattice methods (sometimes referred to as number theoretic, or quasi-random) have been applied to multiple integration and a useful reference is Sloan and Joe (1994).

With a lattice method, the integration region is transformed to a unit cube and a multiple sum yields an unweighted mean of the integrand evaluated over a regular lattice. Smyth (1998) notes that lattice methods function best when the integrand is transformed so that it is periodic over the cube. There is some evidence (including Sloan and Joe) indicating that lattice techniques may outperform other methods when the number of dimensions is large.

Should one have a critical need to evaluate multidimensional integrals accurately (and particularly if this must be done often), the best starting point is the *Cuba Library*, which has been developed at the Max Planck Institut fur Physik (München, Germany). *Cuba* has four routines, Vegas, Suave, Divonne, and Cuhre, which provide the analyst with the choice between quasi- and pseudo-Monte Carlo methods as well as lattice and deterministic approaches too. Several interfaces are available (in FORTRAN, C/C++, and Mathematica), and the open-source package is available at http://www.feynarts.de/cuba. Because the algorithms are invoked similarly, the package can be used to very quickly compare methods and thereby assess likely error. The *Cuba* website is maintained by Thomas Hahn and it is updated frequently.

CONCLUSION

Our focus in this chapter was the numerical evaluation of definite integrals. We have not discussed cases in which the integrand is undefined at either one of the limits (a singular endpoint), nor have we given very much consideration to cases in which one of the limits is infinite.

A singular endpoint may sometimes be handled by using an *open* Newton–Cotes formula such as Milne's rule. Recall that most of the simple quadrature techniques we discussed previously made use of the interval endpoints for function evaluation; such quadrature procedures are said to be *closed*. An *open* algorithm such as the fourth-degree Milne rule does not require function evaluations at the interval endpoints, thereby avoiding the problem of an endpoint singularity. With Milne's rule the function evaluations required for the integration of $\int_a^b f(x)dx$ are placed at locations denoted by

$$x_i = a + \frac{i(b-a)}{n}, \qquad (4.69)$$

where $n = 4$ and the index i assumes the values 1, 2, and 3. Therefore, if we wished to evaluate $\int_1^5 (x^2 - 2x)dx$ using Milne's rule just once, we would place the integrand evaluations at positions $x = 2, 3,$ and 4, and "estimate" the definite integral with

$$\frac{b-a}{3}(2f_1 - f_2 + 2f_3) = \frac{5-1}{3}[2(0) - 3 + 2(8)] = \frac{52}{3}$$
$$= 17.3333.... \qquad (4.70)$$

Of course, this is precisely the value of the definite integral. Let us apply the technique to an integral we examined much earlier that does *not* have a singular endpoint, $\int_1^5 x\ln(x)dx = 14.118$. This will provide us with a performance comparison relative to conventional, closed methods. Since the interval is the same as that mentioned earlier (1, 5), the function evaluations are again placed at 2, 3, and 4, resulting in

$$\frac{5-1}{3}[2(1.38629) - (3.29584) + 2(5.54518)] = 14.0895,$$
$$(4.71)$$

a surprisingly good result (0.2% low) obtained with only a single application of the rule. Obviously, we could subdivide the interval, applying Milne's rule to each piece, then add the results together (this is called a *composite* rule) to improve our estimate. We will try this using eight pieces:

```
#COMPILE EXE
#DIM ALL
   REM *** Open Newton-Cotes for quadrature
   REM *** This is the 4th degree Milne's rule for the
interval (a,b)
         GLOBAL a,b,i,fx,del,n,area,zz,bf,sum AS
         DOUBLE
FUNCTION PBMAIN
         DIM x(3) AS DOUBLE
         DIM f(3) AS DOUBLE
         a=1:bf=5
         b=bf:sum=0
```

```
   REM *** refine estimate by splitting interval into
      n-pieces
            n=8
            del=(b-a)/n
            b=a+del
50 REM *** continue
         FOR i=1 TO 3
            x(i)=a+i*(b-a)/4
         NEXT i
         FOR i=1 TO 3
         GOSUB 200
         f(i)=fx
         NEXT i
         area=(b-a)/3*(2*f(1)-f(2)+2*f(3))
         sum=sum+area
         PRINT area,sum
         a=b
         b=a+del
         IF b>bf THEN 100 ELSE 50
100 REM *** continue
         INPUT "Shall we continue?";zz
         IF zz>0 THEN END
200 REM *** subroutine for function evaluation
         fx=x(i)*LOG(x(i))
         RETURN

END FUNCTION
```

This code produces a value for our example integral of 14.117957. Now we can try Milne's algorithm on an integral with a singular endpoint; consider

$$\int_0^1 \frac{\ln(x)}{1+x}dx = -\frac{\pi^2}{12} = -0.822467. \qquad (4.72)$$

We will apply Milne's method in composite form to this integral by subdividing the interval (0, 1). We will begin with eight intervals and successively double n to gauge the approach to the correct value.

8	−0.8078678
16	−0.8152081
32	−0.8188470
64	−0.8206596
128	−0.8215639
256	−0.8220156
512	−0.8222414
1,024	−0.8223542
2,048	−0.8224106
4,096	−0.8224388
8,192	−0.8224529
16,384	−0.8224600
32,768	−0.8224635

Note that we obtain the first four correct decimal places with $n = 2048$, and we get five decimal places with $n = 16,384$; the open Milne's rule is able to cope with this much more difficult case. We should make one final observation in the context of this example: This integration would appear to be an ideal application for Richardson's extrapolation, where we would obtain improved accuracy with far fewer computations.

In the case of a definite integral with an infinite limit, you may recall that we used truncation (of the upper limit) for the very first example in this chapter:

$$\int_0^\infty x^2 \exp(-x^2)dx = \frac{\sqrt{\pi}}{4}.$$

We will assess the error involved in a similar situation in one of the exercises at the end of the chapter. Occasionally, difficulties like this can be handled through transformation. Given $\int_0^\infty f(x)dx$, one might let $x = t/(1 - t)$, or $x = -\ln(t)$. For the former, we would integrate from 0 to 1, and for the latter, from 1 to 0. Kahaner et al. (1989) point out that this kind of transformation may succeed in producing finite limits but may result in a significantly more difficult integrand.

PROBLEMS

4.1. The volume of a solid of revolution (V) generated by rotation of a curve about the y-axis is given by

$$V = 2\pi \int_a^b xf(x)dx.$$

Let $f(x)$ be the continuous function of x, $f(x) = 10 - (1/4)x^2$, defined for $1 \le x \le 5$. Find V by numerical quadrature.

4.2. Consider a horizontal cylindrical tank with a capacity of 600 gal. The tank is 6 ft long with a radius of 2.065 ft (ID = 4.13 ft). Let h be the depth of liquid in the tank and R be the inside radius. We know that the (area of a) segment of a circle is

$$A_S = R^2 \cos^{-1}\left(\frac{R-h}{R}\right) - (R-h)\sqrt{2Rh - h^2}.$$

Therefore, when $h = R$, the occupied portion of the circle is $\pi R^2/2$. If fluid enters the initially empty tank at a rate of $\dot{V} = (30 - 9\sqrt{h})$ gpm through a small orifice at the bottom of one end, how long will it take for the tank to fill (within 1 in. of the top)?

4.3. The moment of inertia about the y-axis for a continuous function, $f(x)$, defined for $a \le x \le b$, is

$$MI_y = \int_a^b x^2 f(x)dx.$$

Let $f(x) = 1 + x^4 \exp(-2x)$; find MI_y for $0 \le x \le 5$ by numerical quadrature.

4.4. We want to find the work required to compress nitrogen (N_2) adiabatically from atmospheric conditions to a pressure of 8 atm, at a rate of 50 gmol/min. Assume that nitrogen behaves ideally such that $PV = nRT$. We know that $W = \int PdV$, and for an adiabatic process, $PV^\gamma = C_1$. Since the initial and terminal pressures are specified, we will rearrange the latter equation to yield

$$dV = -\frac{1}{\gamma}\left(\frac{C_1}{P}\right)^{\frac{1-\gamma}{\gamma}} \frac{dP}{P^2}.$$

Find the work of compression by numerical quadrature using a normal ambient temperature.

4.5. Evaluate the definite integral,

$$\int_0^{5.95} \frac{1+x}{(6-x)^3}dx.$$

Mathcad's adaptive algorithm yields 1380 and the TI-89™ produces 1380.07. These results are certainly very close, but which is more nearly correct?

4.6. An initially empty 2000-gal tank is being filled by gravity; as the available head in the supply diminishes, the flow rate into the tank decreases. The flow rates were observed periodically resulting in the following table of data:

Time (min)	Flow Rate (gpm)
1	348
3	195
5	117
7	72
9	49
11	36
13	26

If the filling process started at $t = 0$, when will the tank contain precisely 1515 gal?

4.7. The definition for the gamma function is

$$\Gamma(n) = \int_0^\infty x^{n-1} \exp(-x)dx.$$

Find $\Gamma(1.525)$ and $\Gamma(0.525)$ by numerical quadrature. The values you obtain should be related by the recurrence formula:

$$\Gamma(n+1) = n\Gamma(n).$$

Are they?

4.8. Use Gaussian quadrature with $n = 6$ to find the approximate value for the definite integral,

$$\int_0^4 x^{2.25} \exp(-2x)dx.$$

4.9. Use the Gauss–Kronrod (GK) (7-15) procedure to evaluate the definite integral,

$$\int_0^{10} \frac{100dx}{(x-2.5)^2 + 1}.$$

Will it be necessary in this case to use an adaptive modification of GK (7-15)? You can see by inspection that the integrand will have a fairly strong peak centered about $x = 2.5$.

4.10. Consider the definite integral,

$$\int_0^\infty \exp\left(-\frac{1}{4}x\right)\cos(3x)dx.$$

This integral has the value 0.027586 and the integrand behaves as illustrated in Figure 4.5.

We want to determine the error associated with truncating the upper limit of the integral; use the quadrature routine of your choice and prepare a plot that shows the absolute value of the error as a function of the finite upper limit.

4.11. The definite integral

$$\int_0^\infty \frac{dx}{(1+x)\sqrt{x}} = \pi\csc(\pi) = 3.1415926\ldots$$

illustrates both of the difficulties we described at the very end of the chapter: We have a singular endpoint at $x = 0$ and

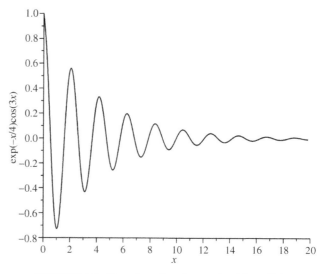

FIGURE 4.5. Plot of the function $\exp(-x/4)\cos(3x)$.

an infinite upper limit. Use the open fourth-degree Newton–Cotes formula (composite Milne's rule) to evaluate the integral with an appropriately truncated upper limit. Can you actually obtain as answer close to π? If you cannot, what quadrature method would you recommend?

4.12. Consider the definite integral

$$\int_0^1 \sqrt{\ln\left(\frac{1}{x}\right)}dx = \frac{\sqrt{\pi}}{2} = 0.8862269\ldots.$$

Use an open composite Newton–Cotes formula to evaluate this integral. Employing the example at the end of the chapter as your guide, determine how many subdivisions will be necessary (using the fourth-degree Milne's rule) to obtain the correct value for the seventh decimal place (9).

4.13. We want to evaluate the double integral,

$$\int_0^\pi \int_{-1}^{+1} x^2 y^2 dxdy = 6.89028,$$

using Simpson's rule (one-dimensional quadrature applied sequentially in the two directions). Explore the accuracy of your technique by adjusting the number of intervals employed for both the inner and outer evaluations. Does Kahaner et al.'s (1989) observation that the accuracy of the inner integral is critical to the success of this approach seem to be upheld?

4.14. Use the Monte Carlo "hit-or-miss" method to evaluate the integrals,

$$\int_0^{2\pi} \frac{dx}{1 + \frac{1}{4}\cos x} = 6.48925 \quad \text{and} \quad \int_0^{\pi} \cos^2\left(\frac{x}{3}\right) dx = 1.5708.$$

4.15. Use the Monte Carlo method to confirm the given value for the double integral,

$$\int_1^2 \int_0^1 \left(yx^{1/2} + \frac{x^2}{y} \right) dx\, dy = 1.23105.$$

How many function evaluations are necessary to verify the second and third decimal places (3 and 1)? If you model your procedure on the sample code provided earlier, you must set *ylimit* = 2, *dy* = (*ylimit* − 1)/*ntrial*, and *y* = *rny**(*ylimit* − 1) + 1. A typical result obtained with these changes for *ntrial* = 400 is 1.230775.

4.16. Consider the definite integral

$$\int_{0.01}^9 \frac{dx}{x\sqrt{1 + 2x}} = 4.8411.$$

Our objective is to learn about the error produced with the estimation of this integral using different quadrature techniques. First, solve the definite integral analytically, verifying that 4.8411 is the correct value. Then, run through the list of Newton–Cotes formulae given by eq. (4.10), eq. (4.11), eq. (4.12), eq. (4.13), and eq. (4.14) generating five different approximations for the integral. Use the minimum number of function evaluations for each, that is, two for the trapezoid rule and three for Simpson's rule. Assess the error for each approximation using eq. (4.28). Are these estimates in accord with your numerical experiments?

4.17. We have two points in the *x*-*y* plane located at (1, 1) and (6, 3), and these two points are to be connected by a function, *y*(*x*). *If* we take this function to be a straight line, then *y* = (2/5)*x* + (3/5). From elementary calculus, the distance between these two points is given by

$$L = \int_1^6 \left[1 + \left(\frac{dy}{dx} \right)^2 \right]^{1/2} dx,$$

and we can easily verify that for this straight line, *L* = 5.38516. But now suppose we want to create a volume by rotating *y*(*x*) about the *x*-axis:

$$V = \pi \int_{x_1}^{x_2} [y(x)]^2 \, dx,$$

and furthermore, suppose we require that the function—while still connecting the points (1, 1) and (6, 3)—be of the form $y(x) = ax^2 + bx + c$. We would like to determine what function of this class yields the smallest volume, *V*, with the constraint that it passes through (x_1, y_1) and (x_2, y_2) while touching the *x*-axis *at only one point* (such that $y(x) \geq 0$ of course). First, determine if there is more than one possibility for *f*(*x*); for example, if *y*(*x*) is zero at *x* = 3, then $y(x) = 0.3x^2 - 1.7x + 2.4$, but we see immediately that $(dy/dx) = 0.6x - 1.7 \neq 0$ at *x* = 3 (as it should). Then find the swept volume, *V*, and the length, *L*, by numerical quadrature. Repeat, but using the function type, $y(x) = ax^3 + bx^2 + cx + d$. With this problem type, we are anticipating one of the uses for the calculus of variations (COV) which will be introduced in Chapter 11.

REFERENCES

Abramowitz, M. and I. Stegun. *Handbook of Mathematical Functions*, Dover Publications, New York (1964).

Grandine, T. A. The Extensive Use of Splines at Boeing. *SIAM News*, 38, no. 4 (2005).

Hanna, O. T. and O. C. Sandall. *Computational Methods in Chemical Engineering*, Prentice Hall, Upper Saddle River, NJ (1995).

James, M. L., Smith, G. M., and J. C. Wolford. *Applied Numerical Methods for Digital Computation*, 2nd edition, Harper & Row, New York (1977).

Kahaner, D., Moler, C., and S. Nash. *Numerical Methods and Software*, Prentice Hall, Englewood Cliffs, NJ (1989).

Kuncir, G. F. Algorithm 102: Simpson's Rule Integrator. *Communications of the ACM*, 5:347 (1962).

Laurie, D. P. Automatic Numerical Integration over a Triangle. *Special Report WISK 273*, NRC Mathematical Sciences, CSIR (1977).

Lyness, J. N. Notes on the Adaptive Simpson Quadrature Routine. *Journal of the ACM*, 16:483 (1969).

Radon, J. Zur mechanische Kubatur. *Monatshefte fur Mathematik*, 52:286 (1948).

Rubinstein, R. Y. *Simulation and the Monte Carlo Method*, John Wiley & Sons, New York (1981).

Sloan, I. H. and S. Joe. *Lattice Methods for Multiple Integration*, Oxford University Press, Oxford (1994).

Smyth, G. K. Numerical Integration. In: *Encyclopedia of Biostatistics* (P. Armitage and T. Colton, editors), Wiley & Sons, Chichester (1998).

Stroud, A. *Approximate Calculation of Multiple Integrals*, Prentice-Hall, Englewood Cliffs, NJ (1971).

5

ANALYTIC SOLUTION OF ORDINARY DIFFERENTIAL EQUATIONS

AN INTRODUCTORY EXAMPLE

Suppose we place about 500 mL of water in a beaker and heat the contents with a candle flame until the water begins to boil. We then remove the heat source and allow the contents of the beaker to cool simply by exposure to the ambient air. An approximate energy balance for this situation (the cooling process) can be written as

$$\{\text{Accumulation}\} = -\{\text{Rate of loss to surroundings}\},$$

or symbolically as

$$mC_p \frac{dT}{dt} = -hA(T - T_\infty). \tag{5.1}$$

In this first-order ordinary differential equation (ODE), m is the mass of heated water, C_p is the heat capacity of the water, h is the heat transfer coefficient, and A is the surface area for heat transfer. T_∞, of course, is the temperature of the air surrounding the beaker. Actual experimental data for this process are shown in Figure 5.1.

We want to solve eq. (5.1) and then see if the result is capable of describing the cooling process shown in Figure 5.1. It will be convenient to let $\theta = T - T_\infty$, such that

$$\frac{d\theta}{dt} = -\frac{hA}{mC_p} \theta. \tag{5.2}$$

This equation is separable and easily integrated to yield

$$\ln\theta = -\frac{hA}{mC_p} t + C_1 \tag{5.3}$$

or, more conveniently,

$$\theta = C_1 \exp\left(-\frac{hA}{mC_p} t\right). \tag{5.4}$$

We should make note of the fact that hA/mC_p is the reciprocal time constant $(1/\tau)$ for this first-order system. Moreover, after an elapsed time of 1τ, about 63% of the change has been accomplished; after 2τ, about 86%; and after 3τ, about 95%. In our example, the cooling process was initiated at $t = 865$ s when the temperature of the water was 99.8+°C. Since the ambient temperature was 22°C, we have

$$T = 77.8 \exp\left(-\frac{hA}{mC_p} t\right) + 22. \tag{5.5}$$

We can use the experimental data shown in Figure 5.1 to estimate the unknown parameters in the quotient (hA/mC_p); at $t = 1200$ s, the water temperature had fallen to 83.75°C. So when $t = 335$ s $(1200 - 865)$, we find

$$\ln\left(\frac{83.75 - 22}{77.8}\right) = -\frac{hA}{mC_p}(335), \tag{5.6}$$

Applied Mathematics for Science and Engineering, First Edition. Larry A. Glasgow.
© 2014 John Wiley & Sons, Inc. Published 2014 by John Wiley & Sons, Inc.

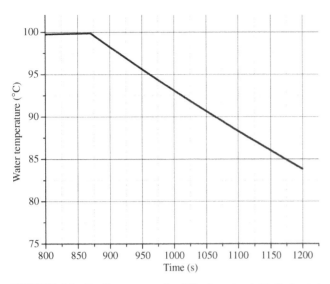

FIGURE 5.1. Cooling process for 500 mL water, initially heated to incipient boiling. The heat source (flame) was removed at $t = 865$ s.

which produces a value for hA/mC_p (0.00069 1/s). How good is our crude model? At $t = 1050$ s, the data in Figure 5.1 indicate that $T = 90.7°$C; the model shows that at $t = 185$ s ($1050 - 865$), $T = 90.5°$C.

Let us make some concluding remarks about this example. The differential equation we formulated by energy balance, eq. (5.1), was an extremely simple first-order linear ODE. However, it was *not homogeneous* due to the presence of the constant, T_∞. Notice that we eliminated the inhomogeneity by redefining the dependent variable: $\theta = T - T_\infty$. Of course, this allowed us to follow the sequence: Separate the variables, integrate the equation, and evaluate the constant of integration, C_1, through the use of the initial condition. The procedure worked well for us in this case, but we need a more general approach that will enable us to solve a broader array of problems.

FIRST-ORDER ORDINARY DIFFERENTIAL EQUATIONS

We begin by considering equations of the type

$$\frac{dy}{dx} = f(x, y). \qquad (5.7)$$

Let us assume that this equation can be written as $Mdx + Ndy = 0$. Such an equation is said to be *exact* if

$$\frac{\partial M}{\partial y} = \frac{\partial N}{\partial x}. \qquad (5.8)$$

Now suppose

$$\frac{dy}{dx} = f(x, y) = \frac{y + 2}{x^2}, \qquad (5.9)$$

which we rewrite as $(y + 2)dx - x^2 dy = 0$. Therefore,

$$\frac{\partial M}{\partial y} = 1 \quad \text{and} \quad \frac{\partial N}{\partial x} = -2x; \qquad (5.10)$$

this equation is not exact. However, it is separable and we can easily show that

$$y = C_1 \exp\left(-\frac{1}{x}\right) - 2. \qquad (5.11)$$

We contrast this case with the equation

$$\frac{dy}{dx} = -\frac{3x^2 y^2 + 2xy}{2x^3 y + x^2}; \qquad (5.12)$$

this time, separation is not possible. We rewrite eq. (5.12) as $(2x^3 y + x^2)dy + (3x^2 y^2 + 2xy)dx = 0$, and observe

$$\frac{\partial M}{\partial y} = 6x^2 y + 2x \quad \text{and} \quad \frac{\partial N}{\partial y} = 6x^2 y + 2x. \qquad (5.13)$$

We see that eq. (5.12) is exact and because we have an exact differential, $Mdx + Ndy = dU$, where

$$\frac{\partial U}{\partial x} = M \quad \text{and} \quad \frac{\partial U}{\partial y} = N. \qquad (5.14)$$

Thus, $\partial U/\partial x = 3x^2 y^2 + 2xy$, and integrating with respect to x,

$$U = x^3 y^2 + x^2 y + F(y), \qquad (5.15)$$

since F cannot be a function of x. From the second part of eq. (5.14), we differentiate U with respect to y and set the result equal to N:

$$2x^3 y + x^2 + F'(y) = 2x^3 y + x^2. \qquad (5.16)$$

Clearly, $F'(y) = 0$, so $F = C$, and we note

$$x^3 y^2 + x^2 y = C. \qquad (5.17)$$

When a first-order ODE is *not exact*, we look for an integrating factor. For example, if

$$\frac{\dfrac{\partial M}{\partial y} - \dfrac{\partial N}{\partial x}}{N} = f(x), \qquad (5.18)$$

where $f(x)$ is a function only of x, then an integrating factor (IF) is $IF = e^{\int f(x)dx}$. Let us illustrate with an example; suppose we have

$$\frac{dy}{dx} + \frac{1}{x}y = \frac{2}{x} + x. \qquad (5.19)$$

Therefore,

$$dy = \left(\frac{2}{x} + x - \frac{y}{x}\right)dx,$$

so that $M = (2/x) + x - (y/x)$ and $N = -1$; the equation is not exact. By eq. (5.18), $f(x) = 1/x$, and $e^{\int (1/x)dx} = e^{\ln x} = x$. We multiply eq. (5.19) by the IF:

$$x\left(\frac{dy}{dx} + \frac{y}{x}\right) = (xy)' = \frac{2}{x} + x. \qquad (5.20)$$

We integrate and divide by x:

$$y = \frac{2}{x}\ln x + \frac{x}{2} + \frac{C_1}{x}. \qquad (5.21)$$

We need to see this process employed for something that is more characteristic of a real problem encountered in applied science. Suppose we have two identical stirred-tank reactors in series, each with a capacity of 100 gal. Both tanks are perfectly mixed, and tank 1 contains a solute at a concentration of 0.8 lb/gal, initially. Pure solvent flows into the first tank at a rate of 4 gal/min (gpm), then the discharge (overflow) from tank 1 flows into tank 2. Our interest is the concentration of solute in tank 2 as a function of time. The situation we are describing corresponds to two, non-interacting first-order systems in series. By mass balance, we formulate two first-order ODEs: For tank 1,

$$\dot{V}C_{in} - \dot{V}C_1 = V_T\frac{dC_1}{dt}, \qquad (5.22)$$

and for tank 2,

$$\dot{V}C_1 - \dot{V}C_2 = V_T\frac{dC_2}{dt}. \qquad (5.23)$$

Of course, $C_{in} = 0$ since pure solvent is being fed to the first tank. Also, if we divide both equations by the volumetric flow rate, $\dot{V}$, and note that the volume of each tank, V_T, divided by the volumetric flow rate is the time constant, $\tau\left(\tau = V_T/\dot{V}\right)$, then we can immediately find the solution for eq. (5.22):

$$C_1 = C_0\exp\left(-\frac{t}{\tau}\right), \qquad (5.24)$$

where C_0 is the initial concentration in tank 1, 0.8 lb/gal. We take this result back to the ODE (eq. 5.23) and rearrange the equation so that

$$\frac{dC_2}{dt} + \frac{1}{\tau}C_2 = \frac{C_0}{\tau}\exp\left(-\frac{t}{\tau}\right). \qquad (5.25)$$

Now we recognize that eq. (5.25) is of the form $y' + a(x)y = b(x)$ so that the integrating factor (IF) can be determined from $e^{\int a(x)dx}$, which in our case is $e^{\int (1/\tau)dt} = e^{t/\tau}$. Therefore,

$$\frac{d}{dt}\left(e^{t/\tau}C_2\right) = \frac{C_0}{\tau}\exp\left(-\frac{t}{\tau}\right)\exp\left(\frac{t}{\tau}\right). \qquad (5.26)$$

The concentration in tank 2 then is simply $C_2 = (C_0/\tau)te^{-t/\tau}$ (remember, the concentration in tank 2 was zero at $t = 0$). So for this example, after 1 hour ($t = 60$ min), the concentration of solute in tank 2 will be 0.1742 lb/gal.

NONLINEAR FIRST-ORDER ORDINARY DIFFERENTIAL EQUATIONS

A *nonlinear* first-order ODE is one that has the form

$$\frac{dy}{dx} = f(x, y), \qquad (5.27)$$

where $f(x, y)$ includes a term in which the dependent variable, y, occurs to some power other than one. Examples could include

$$\frac{dy}{dx} = x^2 + y^2 \quad \text{or} \quad \frac{dy}{dx} = xy\sin(y). \qquad (5.28)$$

The second equation of this pair is particularly interesting and the reader may want to compare what happens when $y(x = 0) = \frac{1}{4}$ with the case where $y(x = 0) = 4$. Certain nonlinear first-order ODEs have seen a great deal of study. For example, Bernoulli's equation, which is usually written as

$$\frac{dy}{dx} + p(x)y = q(x)y^n, \qquad (5.29)$$

where $n \neq 0, 1$, has attracted the attention of mathematicians for more than 300 years. One of the reasons this equation is interesting is because of the effect of the transformation, $z = y^{1-n}$; we will illustrate with an example. Let us consider

$$\frac{dy}{dx} + \frac{y}{x} = y^3. \qquad (5.30)$$

Applying the transformation ($z = y^{-2}$), we find

$$-\tfrac{1}{2}z^{-3/2}\frac{dz}{dx} + \frac{z^{-1/2}}{x} = z^{-3/2},$$

which leads directly to

$$\frac{dz}{dx} - 2\frac{z}{x} + 2 = 0. \qquad (5.31)$$

This is a familiar form and we know that the integrating factor (*IF*) is just $e^{\int -(2/x)dx} = e^{-2\ln x} = 1/x^2$. It is straightforward to show that

$$z = 2x + C_1 x^2, \qquad (5.32)$$

and therefore,

$$y = \sqrt{\frac{1}{2x + C_1 x^2}}. \qquad (5.33)$$

One of the best-known examples of a nonlinear first-order ODE is the *Riccati equation*, which is generally written as

$$\frac{dy}{dx} + a(x)y + b(x)y^2 + c(x) = 0. \qquad (5.34)$$

This equation is reducible to the form

$$\frac{dw}{dx} + c(x)w^2 + d(x) \qquad (5.35)$$

by use of the transformation $w = e^{\int_a^x a(x)dx} y$. Riccati equations arise regularly in applied mathematics and we will look at two examples for illustration.

Consider a constant volume reactor in which the reactant species A is consumed by a second-order (bimolecular) process. The reactant is added to the process vessel continuously but in a concentrated form such that the volume (V) is nearly unaffected. A mass balance on the reactant A yields

$$V\frac{dC_A}{dt} = \dot{n}_A - Vk_2 C_A^2. \qquad (5.36)$$

Therefore,

$$\frac{dC_A}{dt} = \frac{\dot{n}_A}{V} - k_2 C_A^2 \quad \text{or, equivalently,} \quad \frac{dy}{dx} = a - by^2.$$
$$\qquad (5.37)$$

In this case, the solution can be found very easily using a table of integrals (see Selby, 1975, p. 546):

$$y = \frac{a}{\sqrt{ab}}\tanh\left[\sqrt{ab}(x + C_1)\right]. \qquad (5.38)$$

The initial concentration of the reactant is zero, so $C_1 = 0$. We assign the constants a and b the values 3 and 0.475, respectively, resulting in

$$C_A = 2.5131\tanh(1.19373t). \qquad (5.39)$$

Thus, the concentration of A in the reactor, which will approach 2.52312 as $t \to \infty$, behaves as shown in the following table:

Time	0	0.1	0.2	0.6	1.0	2.0
Concentration	0.00	0.2986	0.5888	1.5445	2.0902	2.4710

For our second example, suppose a pilot must leave an airplane at very high altitude. The pilot will begin to fall vertically under the influence of gravity, and that motion will be opposed by drag. We are interested in the pilot's approach to terminal velocity, what that velocity will be, and how long it will take the pilot to attain that velocity. An approximate model for this free fall will be

$$m\frac{dV}{dt} = mg - KV^2. \qquad (5.40)$$

m is the mass of the pilot (plus gear) and K is the product of the frontal area of the falling pilot, the fluid density, and the drag coefficient. Generally, the drag coefficient is a function of velocity, but we are taking it to be approximately constant. Dividing by the mass, we get

$$\frac{dV}{dt} = g - \frac{K}{m}V^2 \quad \text{or, alternatively,} \quad \frac{dy}{dx} = a - by^2, \quad (5.41)$$

which, conveniently, is the same form as the constant volume chemical reactor. Experiments conducted in the World War II era indicated that the terminal velocity of the falling pilot from high altitude would be roughly 250 ft/s (about 170 mph); therefore, $a = 32.2$ and $b = 0.000515$ (1/ft). The solution is exactly the same as given earlier by eq. (5.38), and the initial (vertical) velocity is zero.

Time (s)	0.5	1	5	10	20	50
Velocity (ft/s)	16.07	32.02	141.88	214.63	247.12	249.99+

Thus, the falling pilot attains terminal velocity in about 50 s.

In addition to Bernoulli and Riccati equations, many other nonlinear ODEs have attracted attention. For example, in the first third of the twentieth century, there was great interest in predator–prey problems (populations in conflict). Generally, such problems were formulated in terms of two simultaneous ODEs, one for population "1" and one for population "2." Nonlinear terms come about through interaction between the animals (or species). In some cases, the two equations could be combined to produce one nonlinear second-order ODE. In a typical formulation for populations in conflict, we might see

$$\frac{dN_1}{dt} = aN_1 - bN_1N_2 \quad \text{and} \quad \frac{dN_2}{dt} = -cN_2 + dN_1N_2. \quad (5.42)$$

Of course, it is possible to rewrite the former as

$$N_2 = \frac{\dfrac{dN_1}{dt} - aN_1}{-bN_1}$$

and to use that to eliminate N_2 from the second of the pair of ODEs. We note in passing that the model (eq. 5.42) for the predator–prey problem is not very realistic for reasons that will be made clear in Chapter 9.

Davis (1962) points out that it can often be exceedingly difficult to find an analytic solution for a *nonlinear* ODE. But the analyst may learn a great deal about the behavior of such an equation through simple graphical interpretation as follows. First, we recognize that given a nonlinear first-order ODE,

$$\frac{dy}{dx} = f(x, y), \quad (5.43)$$

we have a means for finding the slope of tangents at any point(s) we wish. Let us examine the first quadrant (on the diagonal) for the equation,

$$\frac{dy}{dx} = xy(y - 2). \quad (5.44)$$

x	0.125	0.250	0.500	1	1.95	2	3	4	5
y	0.125	0.250	0.500	1	1.95	2	3	4	5
Slope	−0.0293	−0.1094	−0.375	−1	−0.190	0	9	32	75

It is a simple matter for us to construct line segments with the indicated slope at the points listed in the table. We extend those segments a short distance, estimate the new position (x, y), calculate a new slope, and repeat. To illustrate, we will take the point (1, 1) and try this.

x position	y position	Slope
1	1	−1
1.1	0.9	−1.089
1.2	0.791	−1.148
1.3	0.676	−1.164
1.4	0.560	−1.129
1.5	0.447	−1.041
1.6	0.343	−0.909
1.7	0.252	−0.749
1.8	0.177	−0.581
1.9	0.119	−0.425
2.0	0.076	−0.294

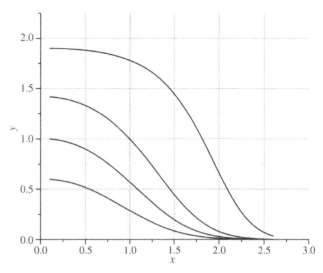

FIGURE 5.2. Local behavior of $y(x)$ for eq. (5.44) as estimated through the construction of tangents. Please note that the slope is always zero for $y = 2$; one of the loci is a horizontal line passing through $y = 2$. What would we see if we started at $(x = 0, y = 1.99)$?

We repeat this process for various starting positions and plot the results in Figure 5.2.

The reader is encouraged to apply this method in the first quadrant, but *above* $y = 2$; in fact, we will make it one of the exercises at the end of this chapter. We note from our original table of slopes that for large (x, y) pairs, the slopes are very large—the tangents are essentially vertical lines. Furthermore, for very small values of x (but $y > 2$), the slopes are very small (the tangents are nearly horizontal). Thus, the behavior of $y(x)$ *above* $y = 2$ is very different from what we observe in Figure 5.2 for values of $y < 2$.

We should also point out that we can always differentiate our first-order nonlinear ODE and identify the locus of points of inflection by setting the second derivative equal to zero. Thus, a great deal of information regarding the behavior of the nonlinear first-order ODE may be obtained rather easily even though we may not be able to find a solution for the equation itself.

Solutions with Elliptic Integrals and Elliptic Functions

We want to consider a class of nonlinear ODEs that can be initially represented by the equation

$$\left(\frac{dy}{dx}\right)^2 = a_0 + a_1 y + a_2 y^2 + a_3 y^3 + a_4 y^4. \quad (5.45)$$

We now restrict our attention (for the right-hand side) to quartic polynomials that can be written in the form

$$1 - (1 + k^2)y^2 + k^2 y^4 \quad (5.46)$$

such that $a_0 = 1$, $a_2 = -(1 + k^2)$, and $a_4 = k^2$ by comparison with eq. (5.45). We also limit consideration to values for k less than 1. By taking the square root of the modified ODE,

$$\frac{dy}{dx} = \sqrt{(1 - y^2)(1 - k^2 y^2)}. \qquad (5.47)$$

It is a simple matter for us to rearrange this equation so that

$$x = \int_0^\beta \frac{dy}{\sqrt{(1 - y^2)(1 - k^2 y^2)}}. \qquad (5.48)$$

This is an *elliptic integral of the first kind* (see Dwight, 1957, p. 170), which can be written equivalently as an inverse Jacobi elliptic sine, sn^{-1}. If we now substitute $y = \sin \phi$, we obtain a more compact version of the right-hand side:

$$x = \int_0^\theta \frac{d\phi}{\sqrt{1 - k^2 \sin^2 \phi}}. \qquad (5.49)$$

k is the modulus of the elliptic integral and θ is the amplitude. Extensive tables of values for elliptic integrals are available; for example, if we set $k = 1/4$ and $\theta = \pi/4$, we can find the value for the elliptic integral, 0.8044 (Davis, 1962, has a small table and *CRC Handbook of Tables for Mathematics* a much larger one). The reader may want to confirm his or her facility with such tables by checking the following value:

$$F(k, \theta) = F(\tfrac{1}{4}, \tfrac{\pi}{6}) = \int_0^{\pi/6} \frac{d\phi}{\sqrt{1 - \tfrac{1}{16} \sin^2 \phi}} = 0.525. \quad (5.50)$$

At this point, you might be thinking that the scope of application is strictly limited to equations where the right-hand side is similar to eq. (5.46). However, let us go back to eq. (5.45) and differentiate, obtaining

$$2\frac{dy}{dx}\frac{d^2 y}{dx^2} = a_1 \frac{dy}{dx} + 2a_2 y \frac{dy}{dx} + 3a_3 y^2 \frac{dy}{dx} + 4a_4 y^3 \frac{dy}{dx}. \quad (5.51)$$

If we divide by $2dy/dx$, we see a nonlinear second-order ODE of the type that may be solved using elliptic functions; Davis provides a number of examples and Milne-Thomson (1950) points out that application of elliptic functions for solution of nonlinear ODEs includes analysis of pendulum oscillations, capillary phenomena, bending of an elastic rod, viscous flow in a convergent channel, and the potential of an electrified ellipsoid, among others. We will look at an elementary example to illustrate this. Consider a simple, *frictionless* pendulum with oscillatory motion governed by

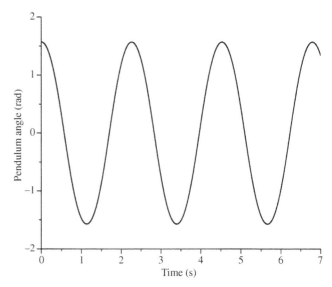

FIGURE 5.3. Behavior of frictionless pendulum with a starting position of 90° (1.5708 rad) and $L = 3$ ft.

$$\frac{d^2\theta}{dt^2} + \frac{g}{L}\sin\theta = 0. \qquad (5.52)$$

L is the length of the pendulum and g is the acceleration of gravity. Let us suppose that $\theta(t = 0) = 90°$ and the $d\theta/dt = 0$. We set $g = 32.17$ ft/s and take $L = 3$ ft. First, we will find the solution for this problem numerically, and the resulting behavior is shown in Figure 5.3.

Now we turn our attention back to eq. (5.52), noting that it can be integrated once to yield

$$\frac{1}{2}\left(\frac{d\theta}{dt}\right)^2 - \frac{g}{L}\cos\theta = C_1. \qquad (5.53)$$

Since $d\theta/dt|_{t=0} = 0$, $C_1 = -(g/L)\cos\theta_0$ and therefore

$$\frac{d\theta}{dt} = \sqrt{\frac{2g}{L}(\cos\theta - \cos\theta_0)}. \qquad (5.54)$$

If we now let $\cos\theta = 1 - 2k^2 \sin^2 \phi$ and take $k = \sin(\theta_0/2)$, we find

$$dt = \sqrt{\frac{L}{g}} \frac{d\phi}{\sqrt{1 - k^2 \sin^2 \phi}}. \qquad (5.55)$$

Of course, when integrated, this is exactly the form of eq. (5.49). We can use this elliptic integral to find the period of the pendulum's motion; we will integrate from the zero position to our maximum angle of $\pi/2$ rad (which is one-quarter of a cycle) and then multiply by 4:

$$T_P = 4\sqrt{\frac{L}{g}} \int_0^{\pi/2} \frac{d\phi}{\sqrt{1 - k^2 \sin^2 \phi}}. \qquad (5.56)$$

Since the initial displacement (angle) is 90°, this means that our modulus is sin(45°) and our amplitude is $\pi/2$. Consulting a table for elliptic integrals, we find a value of 1.8541 and therefore,

$$T_P = (4)(3/32.17)^{1/2}(1.8541) = 2.265.$$

Please compare this value with the results presented in Figure 5.3; you will find that our analytic solution (2.265) corresponds exactly with the numerical calculation.

HIGHER-ORDER LINEAR ODEs WITH CONSTANT COEFFICIENTS

We want to examine some higher-order ODEs but initially with the stipulations that they be linear, homogeneous, and have constant coefficients. Let us begin with a set of three second-order equations:

$$\frac{d^2y}{dx^2} + 3\frac{dy}{dx} + 2y = 0 \qquad (5.57)$$

$$\frac{d^2y}{dx^2} + 4\frac{dy}{dx} + 4y = 0 \qquad (5.58)$$

$$\frac{d^2y}{dx^2} + 3\frac{dy}{dx} + 4y = 0. \qquad (5.59)$$

We rewrite eq. (5.57) using linear differential operator notation:

$$(D^2 + 3D + 2)y = 0, \qquad (5.60)$$

which we can readily factor as

$$(D+2)(D+1)y = 0. \qquad (5.61)$$

Anytime we can successfully factor the linear operator for an ODE in this manner, we can immediately write down the *primitive*:

$$y = C_1\exp(-2x) + C_2\exp(-x). \qquad (5.62)$$

Values for the two constants of integration can be obtained from the initial (or boundary) conditions. For an equation such as eq. (5.57), these might take the form

$$y(x=0) = a \quad \text{and} \quad \left.\frac{dy}{dx}\right|_{x=0} = b. \qquad (5.63)$$

In the case of ODE (eq. 5.58), we have

$$(D^2 + 4D + 4)y = (D+2)(D+2)y = 0, \qquad (5.64)$$

where we see *repeated* roots. In a situation like this, the primitive is written as

$$y = C_1\exp(-2x) + C_2 x\exp(-2x). \qquad (5.65)$$

For an equation of higher order, it is certainly possible to obtain

$$(D+\alpha)^n y = 0. \qquad (5.66)$$

The primitive in this case is written as

$$\begin{aligned}y = C_1\exp(-\alpha x) &+ C_2 x\exp(-\alpha x) + C_3 x^2\exp(-\alpha x)\\ &+ \cdots C_n x^{n-1}\exp(-\alpha x).\end{aligned} \qquad (5.67)$$

Equation (5.59) presents a different challenge; in linear differential operator notation, we find

$$(D^2 + 3D + 4)y = 0. \qquad (5.68)$$

When we look at $b^2 - 4ac$ for the quadratic formula, we see $9 - 16 = -7$. Consequently,

$$\left(D + \frac{3}{2} + \frac{\sqrt{7}}{2}i\right)\left(D + \frac{3}{2} - \frac{\sqrt{7}}{2}i\right)y = 0. \qquad (5.69)$$

We can, of course, write down the primitive just as we have done for both eq. (5.57) and eq. (5.58). However, this is a case where Euler's relation will be useful; you may recall

$$e^{(a+ib)x} = e^{ax}(\cos bx + i\sin bx). \qquad (5.70)$$

This will allow us to put our result into a more appropriate form:

$$\begin{aligned}y = C_1 e^{-\frac{3}{2}t}&\left(\cos\frac{\sqrt{7}}{2}x + i\sin\frac{\sqrt{7}}{2}x\right)\\ &+ C_2 e^{-\frac{3}{2}x}\left(\cos\left(-\frac{\sqrt{7}}{2}x\right) + i\sin\left(-\frac{\sqrt{7}}{2}x\right)\right).\end{aligned} \qquad (5.71)$$

Since *cosine* is an even function and *sine* is odd, we can write

$$\begin{aligned}y = C_1 e^{-\frac{3}{2}x}&\left(\cos\frac{\sqrt{7}}{2}x + i\sin\frac{\sqrt{7}}{2}x\right)\\ &+ C_2 e^{-\frac{3}{2}x}\left(\cos\frac{\sqrt{7}}{2}x - i\sin\frac{\sqrt{7}}{2}x\right).\end{aligned} \qquad (5.72)$$

For a real (physical) problem such as one arising through a force balance, the imaginary parts will cancel, leaving a real result. We can very conveniently confirm the result, eq. (5.72), by using *deSolve*(...) on the Texas Instruments TI-89™; we obtain

$$y = C_1 \exp\left(-\frac{3x}{2}\right)\cos\frac{\sqrt{7}x}{2} + C_2 \exp\left(-\frac{3x}{2}\right)\sin\frac{\sqrt{7}x}{2}.$$

The procedure we have sketched previously is valid for comparable equations of any order. For example, consider

$$\frac{d^4y}{dx^4} + 2\frac{d^3y}{dx^3} + 3\frac{d^2y}{dx^2} + 5\frac{dy}{dx} + 8y = 0. \qquad (5.73)$$

In this case, we have two pairs of complex conjugates (roots), which are $D + 0.457547 \pm 1.56529i$ and $D - 1.45755 \pm 0.940024i$. And again we can immediately write down the primitive by inspection.

Unfortunately, not every ODE of higher order will be homogeneous; often, if the differential equation represents some dynamic process, a forcing function will be applied on the right-hand side to drive the model. For example, we might have

$$\frac{d^2y}{dx^2} + 3\frac{dy}{dx} + 2y = 5. \qquad (5.74)$$

In such cases, the solution is the sum of the primitive for the *homogeneous* equation—which we refer to as the complementary function (*CF*)—plus a particular integral (*PI*). Thus, for eq. (5.74), we have

$$y = C_1 \exp(-2x) + C_2 \exp(-x) + PI. \qquad (5.75)$$

Many times, the form of the particular integral can be determined by inspection. For example, for eq. (5.74), once the *dynamic* behavior is completed, the terms involving derivatives will be zero (for large x, the exponentials disappear, leaving just the *PI*). Therefore, the *PI* must be 5/2, and the solution can be written as

$$y = C_1 \exp(-2x) + C_2 \exp(-x) + \frac{5}{2}. \qquad (5.76)$$

We should actually complete this problem by including boundary (or initial) conditions. We assume that the physical process described by this model was drifting along at an equilibrium state until the independent variable, x, or very commonly time, t, was zero, when the forcing function (the constant 5) was applied. Of course, this means that both y and its derivative can be set to zero for $x = 0$—we can always do this for models of this type by using *deviation* variables. For example, if y_{eq} is the equilibrium (or steady) value of the dependent variable, then we simply write

$$Y = y - y_{eq}. \qquad (5.77)$$

This guarantees that the dependent variable was zero at the instant the forcing function was applied. Consequently, we will now assume that the dependent variable was in fact written in this form, such that

$$y(x = 0) = 0 \quad \text{and} \quad \left.\frac{dy}{dx}\right|_{x=0} = 0. \qquad (5.78)$$

Therefore,

$$C_1 + C_2 + \frac{5}{2} = 0 \quad \text{and} \quad C_2 = -2C_1. \qquad (5.79)$$

For these conditions, the solution for our problem becomes

$$y = \frac{5}{2}\exp(-2x) - \frac{10}{2}\exp(-x) + \frac{5}{2}. \qquad (5.80)$$

Of course, the inhomogeneity does not have to be a simple constant as it was in the previous example. Let us examine a method that allows us to deal with cases in which the inhomogeneity is a function of x. For example,

$$\frac{d^2y}{dx^2} + 4\frac{dy}{dx} + 4y = x. \qquad (5.81)$$

Again, we find the complementary function from the *homogeneous version* of eq. (5.81) and add the particular integral:

$$y = C_1 \exp(-2x) + C_2 x \exp(-2x) + PI. \qquad (5.82)$$

The technique we will employ for the determination of the particular integral is called *variation of parameters*. We begin by rewriting the complementary function, but we insert unknown functions of x in place of the constants of integration that appear in eq. (5.82):

$$y = L_1(x)\exp(-2x) + L_2(x)x\exp(-2x). \qquad (5.83)$$

We now differentiate with respect to x, obtaining

$$\begin{aligned} Dy = {} & L_1'\exp(-2x) - 2L_1\exp(-2x) + L_2'x\exp(-2x) \\ & + L_2\exp(-2x) - 2L_2 x\exp(-2x). \end{aligned} \qquad (5.84)$$

The terms involving derivatives (of Ls) are set equal to zero:

$$L_1'\exp(-2x) + L_2'x\exp(-2x) = 0. \qquad (5.85)$$

Now we differentiate again, retaining only those terms involving derivatives of the Ls. But this time, the order of the derivative (2) matches the order of the ODE, so the sum is set equal to the inhomogeneity (which is x); thus,

$$-2L_1{}'\exp(-2x)+L_2{}'\exp(-2x)-2L_2{}'x\exp(-2x)=x. \tag{5.86}$$

By virtue of eq. (5.85), the sum of the first and third terms is zero, leaving us with two equations to solve:

$$L_2{}'=x\exp(+2x) \quad \text{and} \quad L_1{}'=-x^2\exp(+2x). \tag{5.87}$$

These equations are integrated and taken back to eq. (5.83). The reader should show that the particular integral turns out to be $PI = (x/4) - (1/4)$; therefore the solution we were seeking is written as

$$y=C_1\exp(-2x)+C_2 x\exp(-2x)+\frac{1}{4}(x-1). \tag{5.88}$$

Since $y(x=0)=0$, $C_1 = 1/4$. The first derivative is also zero at $x = 0$, so it is easy to show that $C_2 = 1/4$ as well. Problems of a type similar to the example immediately above are excellent candidates for solution through use of the Laplace transform, and it is worthwhile for us to spend a little effort reviewing that technique. But keep in mind that the procedure we are about to explore can only be used to solve a very limited class of ODEs—we are restricted to *linear* ODEs with *constant coefficients*.

Use of the Laplace Transform for Solution of ODEs

We normally think of the Laplace transform within the context of problems involving change or evolution in time. Accordingly, for decades, the Laplace transform was a staple of classical linear process control as it allowed the control engineer to explore the dynamic behavior of a system in the s-plane using only algebraic manipulations. With this perspective in mind, we will use t as our standard independent variable in this section. The formal definition of the Laplace transform of a function, $f(t)$, follows:

$$f(s)=\int_0^\infty f(t)e^{-st}dt. \tag{5.89}$$

This transform is often written more compactly as $f(s) = L\{f(t)\}$. Notice that the transformation takes us from the time domain to the s-plane. Let us actually apply the definition to something—a numerical constant, for example: We let $f(t) = 1$, then

$$f(s)=\int_o^\infty (1)e^{-st}dt=-\frac{1}{s}e^{-st}\Big|_0^\infty=-\frac{1}{s}(0-1)=\frac{1}{s}. \tag{5.90}$$

We envision needing to find the transform for many different types of terms, and we will use the formal definition one more time, applying it to the independent variable, t:

$$f(s)=\int_0^\infty te^{-st}dt=\frac{e^{-st}}{s^2}(-st-1)\Big|_0^\infty=\frac{1}{s^2}. \tag{5.91}$$

If our focus is the solution of ODEs, then we must also be able to transform derivatives of various orders. For example,

$$L\left\{\frac{df(t)}{dt}\right\}=sf(s)-f(t=0), \tag{5.92}$$

$$L\left\{\frac{d^2f(t)}{dt^2}\right\}=s^2f(s)-sf(t=0)-f'(t=0), \tag{5.93}$$

and more generally,

$$L\left\{\frac{d^nf(t)}{dt^n}\right\}=s^nf(s)-s^{n-1}f(t=0)-s^{n-2}f'(t=0)$$
$$-s^{n-3}f''(t=0)\dots.$$

The reader will observe that in the transformation of derivatives, differentiation is replaced by multiplication by s. However, we must also subtract off the initial conditions (in the time domain). This makes it clear why it is so important (and convenient) to put problems into deviation form whenever possible. Of course, we do not actually use the formal definition of the Laplace transform, eq. (5.89), in practice since nearly every conceivable form of function has already been transformed. We merely need to look them up in a suitable reference; one very useful source is Abramowitz and Stegun (1972). Here is an abbreviated listing of some very common transforms.

$f(t)$	$L\{f(t)\}$
1	$\dfrac{1}{s}$
t	$\dfrac{1}{s^2}$
$\exp(-at)$	$\dfrac{1}{s+a}$
$t^n\exp(-at)$	$\dfrac{n!}{(s+a)^{n+1}}$
$\exp(-at)\sin(bt)$	$\dfrac{b}{(s+a)^2+b^2}$
$\sin(at)$	$\dfrac{a}{s^2+a^2}$
$\cos(at)$	$\dfrac{s}{s^2+a^2}$

$f(t)$	$L\{f(t)\}$
$t\cos(at)$	$\dfrac{s^2-a^2}{(s^2+a^2)^2}$
$\exp(-at)\cos(bt)$	$\dfrac{s+a}{(s+a)^2+b^2}$
$\cosh(at)$	$\dfrac{s}{s^2-a^2}$
$\dfrac{1}{2a^2}\sin(at)\sinh(at)$	$\dfrac{s}{s^4+4a^4}$
$\dfrac{1}{\sqrt{\pi t}}\cosh\left(\sqrt{4kt}\right)$	$\dfrac{1}{\sqrt{s}}\exp\left(\dfrac{k}{s}\right)$
$J_0\left(\sqrt{4kt}\right)$	$\dfrac{1}{s}\exp\left(-\dfrac{k}{s}\right)$
$erfc\left(\dfrac{k}{\sqrt{4t}}\right)$	$\dfrac{1}{s}\exp(-k\sqrt{s}),\quad$ where $k\ge0$
$\sqrt{\dfrac{4t}{\pi}}\exp\left(-\dfrac{k^2}{4t}\right)-(k)erfc\left(\dfrac{k}{\sqrt{4t}}\right)$	$\dfrac{1}{s^{3/2}}\exp(-k\sqrt{s}),\quad$ where $k\ge0$
$\dfrac{1}{\sqrt{\pi t}}\exp\left(-\dfrac{k^2}{4t}\right)$	$\dfrac{1}{\sqrt{s}}\exp(-k\sqrt{s}),\quad$ where $k\ge0$

We will illustrate the application of the Laplace transform to the solution of ODEs (with constant coefficients) by reconsidering the previous example, eq. (5.81). However, we will change the independent variable to t (from x) to be consistent with the previous table of transforms. Accordingly,

$$\frac{d^2y}{dt^2}+4\frac{dy}{dt}+4y=t. \qquad (5.94)$$

Applying the transform to each term, we get

$$y(s)(s^2+4s+4)=\frac{1}{s^2}, \qquad (5.95)$$

and therefore,

$$y(s)=\frac{1}{s^2(s+2)(s+2)}. \qquad (5.96)$$

To solve such problems the old-fashioned way (manually), we use partial fraction expansion. There is a slight complication in this case since we have repeated roots that will require us to write

$$\frac{1}{s^2(s+2)^2}=\frac{A}{s^2}+\frac{B}{s}+\frac{C}{(s+2)^2}+\frac{D}{s+2}. \qquad (5.97)$$

We begin by multiplying by s^2 and setting $s=0$; we find immediately that $A=1/4$. To find B, we return to eq. (5.97) and multiply by s^2 again. But this time, we differentiate with respect to s, which isolates B, and we find $B=-1/4$. We proceed analogously for the remaining two terms and find

$C=1/4$ and $D=1/4$. Therefore, we now know the right-hand side of eq. (5.97):

$$y(s)=\frac{\frac{1}{4}}{s^2}-\frac{\frac{1}{4}}{s}+\frac{\frac{1}{4}}{(s+2)^2}+\frac{\frac{1}{4}}{s+2}. \qquad (5.98)$$

Our work is almost complete; we need only to consult the abbreviated list of transforms provided earlier so that we can write $y(t)$:

$$y(t)=\frac{1}{4}t-\frac{1}{4}+\frac{1}{4}t\exp(-2t)+\frac{1}{4}\exp(-2t). \qquad (5.99)$$

Compare this result with eq. (5.88); you will see that they are identical (except for the change in independent variable). Our use of the Laplace transform has significantly reduced the work necessary to find the solution for the ODE (eq. 5.81).

We also want to illustrate the procedure with a variation that arises regularly with this type of ODE. Consider the second-order ODE:

$$\frac{d^2y}{dt^2}+3\frac{dy}{dt}+5y=2 \qquad (5.100)$$

with $y(t=0)=0$ and $y'(t=0)=0$. Proceeding as before,

$$s^2y(s)+3sy(s)+5y(s)=\frac{2}{s} \qquad (5.101)$$

and

$$y(s)=\frac{2}{s(s^2+3s+5)}=\frac{2}{s\left(s+\frac{3}{2}+\frac{\sqrt{11}}{2}i\right)\left(s+\frac{3}{2}-\frac{\sqrt{11}}{2}i\right)}. \qquad (5.102)$$

Just as we did previously, we expand the right-hand side:

$$y(s)=\frac{A}{s}+\frac{B}{s+\frac{3}{2}+\frac{\sqrt{11}}{2}i}+\frac{C}{s+\frac{3}{2}-\frac{\sqrt{11}}{2}i}. \qquad (5.103)$$

The reader may wish to verify that $A=2/5$, $B=-0.150943-0.068267i$, and $C=-0.150943+0.068267i$. It remains for us to take the following three terms back to the time domain:

$$y(s)=\frac{\frac{2}{5}}{s}+\frac{-0.150943-0.068267i}{s+\frac{3}{2}+\frac{\sqrt{11}}{2}i}$$
$$+\frac{-0.150943+0.068267i}{s+\frac{3}{2}-\frac{\sqrt{11}}{2}i}. \qquad (5.104)$$

At this point, you should be able to identify the essentials of this system's behavior in the time domain: We will obtain

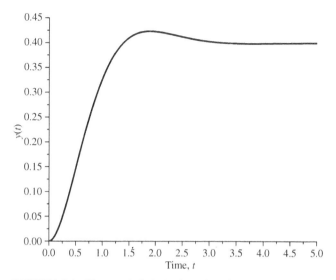

FIGURE 5.4. Characteristic behavior of a slightly underdamped second-order system as produced by the solution, eq. (5.105). Note the limited overshoot and the rapidly damped oscillation about the ultimate value for $y(t)$ of 0.4.

a constant (2/5) combined with an exponentially damped $(\exp(-\frac{3}{2}t))$ oscillatory response. Specifically, we find

$$y(t) = \frac{2}{5} - 0.4\exp\left(-\frac{3}{2}t\right)\cos\frac{\sqrt{11}}{2}t$$
$$- 0.361814\exp\left(-\frac{3}{2}t\right)\sin\frac{\sqrt{11}}{2}t. \qquad (5.105)$$

The behavior produced by eq. (5.105) is illustrated in Figure 5.4.

To conclude this section, it is essential that we reiterate that the Laplace transform is a linear operator and it can only be used to *solve linear ODEs with constant coefficients*. Historically, this method was of enormous importance in the development of automatic process control because *idealized* models for the dynamic behavior of processes could be dealt with algebraically (and the computational solution of differential equations could be avoided). But the restriction of the Laplace transform method to linear ODEs with constant coefficients is a significant one since so many of the equations of interest to us in applied mathematics have variable coefficients.

HIGHER-ORDER EQUATIONS WITH VARIABLE COEFFICIENTS

We want to focus on second-order ODEs of the form

$$\frac{d^2y}{dx^2} + P(x)\frac{dy}{dx} + Q(x)y = R(x). \qquad (5.106)$$

This differential equation is linear, but the coefficients are not constant. We must preface this part of our discussion with a warning: *There is no general procedure that will permit the analyst to solve every differential equation of this type.* Ayres (1952) laid out a plan of attack for such equations that will sometimes yield success and we will use the strategy he recommends on the following example. Let us suppose the equation of interest is

$$\frac{d^2y}{dx^2} - \frac{3}{x}\frac{dy}{dx} + \frac{3}{x^2}y = 2x - 1. \qquad (5.107)$$

Our plan is to render the equation homogeneous (removing the $2x - 1$ from the right-hand side) and then to see if we identify a particular integral, which we will call $u(x)$. Consider trying $PI = x$; the second derivative is zero, the first derivative is 1, and we find $(-3/x) + (3/x^2)x = 0$, so the function x is a particular integral for this equation. We now let y be rewritten as the product of the particular integral and a new dependent variable, v: $y = xv$. Therefore,

$$\frac{dy}{dx} = v + x\frac{dv}{dx} \quad \text{and} \quad \frac{d^2y}{dx^2} = 2\frac{dv}{dx} + x\frac{d^2v}{dx^2}. \qquad (5.108)$$

Of course, this leads to

$$2\frac{dv}{dx} + x\frac{d^2v}{dx^2} - \frac{3}{x}\left(v + x\frac{dv}{dx}\right) + \frac{3}{x^2}(xv) = 2x - 1, \qquad (5.109)$$

which means that the terms in which v is divided by x cancel, leaving us with just

$$x\frac{d^2v}{dx^2} - \frac{dv}{dx} = 2x - 1. \qquad (5.110)$$

Anytime we are confronted by an ODE that includes *only* derivatives, we should immediately think about reducing the order. In this case, we achieve this by letting $\beta = dv/dx$, resulting in

$$\frac{d\beta}{dx} - \frac{1}{x}\beta = 2 - \frac{1}{x}. \qquad (5.111)$$

This first-order ODE is familiar—see eq. (5.31), for example. We know that an integrating factor for this equation is $e^{\int -(1/x)dx} = e^{-\ln x} = 1/x$. Consequently,

$$\frac{d}{dx}\left(\frac{1}{x}\beta\right) = \frac{2}{x} - \frac{1}{x^2}, \qquad (5.112)$$

and therefore,

$$\beta = \frac{dv}{dx} = 2x\ln x + C_1 x + 1. \qquad (5.113)$$

We integrate with respect to x:

$$v = 2\left[\frac{x^2}{2}\ln x - \frac{x^2}{4}\right] + \frac{C_1}{2}x^2 + x + C_2, \qquad (5.114)$$

and since $y = xv$,

$$y = x^3[\ln x + C_1] + x^2 + C_2 x. \qquad (5.115)$$

The procedure we sketched immediately above will *only work* if we can identify a particular integral for the homogeneous equation. Obviously, we might fail to find a *PI*, and in that case, Ayres suggests the following: Return to eq. (5.106) and compute $Q - (1/4) P^2 - (1/2) (dP/dx)$; if this difference is a constant or alternatively a constant divided by x, then the transformation

$$y = v\exp\left(-\frac{1}{2}\int P(x)dx\right) \qquad (5.116)$$

will reduce the ODE to a linear equation with constant coefficients (or to a Cauchy equation of the form

$$A_n x^n \frac{d^n y}{dx^n} + A_{n-1}x^{n-1}\frac{d^{n-1}y}{dx^{n-1}} + \ldots = F(x).$$

Let us see how this might work in practice using the following equation for our example:

$$\frac{d^2 y}{dx^2} - 4x\frac{dy}{dx} + 4x^2 y = x\exp(x^2). \qquad (5.117)$$

We see $Q = 4x^2$ and $P = -4x$; therefore,

$$Q - \frac{1}{4}P^2 - \left(\frac{1}{2}\right)\frac{dP}{dx} = 4x^2 - \frac{1}{4}(16x^2) - \frac{1}{2}(-4) = 2.$$

Accordingly, we let

$$y = v\exp\left(-\frac{1}{2}\int Pdx\right) = v\exp(x^2).$$

Proceeding, we differentiate appropriately and substitute into eq. (5.117). All terms involving dv/dx cancel, as well as the terms that involve $x^2 v$, leaving us with

$$\frac{d^2 v}{dx^2} + 2v = x. \qquad (5.118)$$

By inspection, the *PI* is $x/2$, and from the homogeneous equation we observe

$$(D^2 + 2)v = (D + \sqrt{2}i)(D - \sqrt{2}i)v = 0. \qquad (5.119)$$

Thus, the solution for v is

$$v = C_1 \exp(-\sqrt{2}ix) + C_2 \exp(+\sqrt{2}ix) + \frac{x}{2}. \qquad (5.120)$$

Of course, we must remember to multiply eq. (5.120) by $\exp(x^2)$ to obtain our original dependent variable, y. We can also use the Euler relation to rewrite the solution in terms of *cosine* and *sine*, should we desire to do so; for example, since $e^{(a+ib)t} = e^{at}(\cos bt + i\sin bt)$, then $\exp(-\sqrt{2}ix) = \cos(\sqrt{2}x) - i\sin(\sqrt{2}x)$.

BESSEL'S EQUATION AND BESSEL FUNCTIONS

Among the more important second-order ODEs with variable coefficients appearing in applied mathematics is *Bessel's equation*. The solution for Bessel-type ODEs can be written in terms of Bessel functions, and *the definitive source* of information regarding Bessel functions is *A Treatise on Bessel Functions and Their Applications to Physics* by Gray et al. (1931, reprinted 1966); Carslaw and Jaeger (1959) can also be very useful. The reader should be aware of the fact that Bessel functions are often referred to—quite appropriately—as "cylinder functions" in the older German literature. The form for Bessel's ODE seen frequently throughout applied mathematics is

$$x^2 \frac{d^2 y}{dx^2} + x\frac{dy}{dx} + (x^2 - n^2)y = 0, \qquad (5.121)$$

where $n \geq 0$. The general solution for this particular form is

$$y = C_1 J_n(x) + C_2 Y_n(x), \qquad (5.122)$$

where J is the Bessel function of the first kind and Y is Neumann's Bessel function of the second kind (both of order n). However, the form of eq. (5.121) is quite restrictive and we should recognize that many ODEs with forms *similar* to eq. (5.121) have solutions that can be written in terms of Bessel functions. In fact, we need to make the following observation: *Whenever we encounter a radially directed flux in cylindrical coordinates*, the operator

$$\frac{1}{r}\frac{d}{dr}\left(r\frac{d\phi}{dr}\right)$$

will arise. Depending on the exact nature of the problem, this can result in some form of Bessel's differential equation, which, for the generalized case, can be written as shown by Mickley et al. (1957):

$$r^2\frac{d^2\phi}{dr^2} + r(a + 2br^v)\frac{d\phi}{dr} \qquad (5.123)$$
$$+ [c + dr^{2s} - b(1-a-v)r^v + b^2 r^{2v}]\phi = 0.$$

For many real, physical problems in applied mathematics, we find that $a = 1$, $b = 0$, and $c = 0$. The nature of the solution is then determined by the sign of d. If the $\sqrt{d}$ is real, then the solution is written in terms of J_n or J_n plus Y_n.

If $\sqrt{d}$ is imaginary, then the solution will be either I_n or I_n plus K_n. The order, n, is determined by

$$n = \frac{1}{s}\sqrt{\left(\frac{1-a}{2}\right)^2 - c}.$$

As an illustration, consider steady conduction of thermal energy in an infinitely long cylinder with a production term that is linear with respect to temperature. The governing differential equation has the form

$$r^2\frac{d^2T}{dr^2} + r\frac{dT}{dr} + r^2\frac{\gamma T}{k} = 0, \qquad (5.124)$$

where γ is a positive constant. Note that $a = 1$, $b = 0$, $c = 0$, $s = 1$, and $d = \gamma/k$. In this case, the solution is

$$T = AJ_0\left(\sqrt{\frac{\gamma}{k}}r\right) + BY_0\left(\sqrt{\frac{\gamma}{k}}r\right). \qquad (5.125)$$

For a solid cylindrical domain, $T(r = 0)$ would have to be finite and therefore $B = 0$ (since $Y_0(0) = -\infty$). But of course, for an *annular region*, no boundary condition could be written for $r = 0$ and both terms (A and B) would remain in the solution. Note that if the production term in eq. (5.124) were replaced by a sink (disappearance) term, then γ/k would have been negative and the solution would have been written in terms of the modified Bessel functions, I_0 and K_0. To illustrate this, consider a catalytic reaction in a long, cylindrical pellet; the reactant species, A, is being consumed by a first-order reaction. A homogeneous model (for heterogeneous catalysis) results in the differential equation

$$r^2\frac{d^2C_A}{dr^2} + r\frac{dC_A}{dr} - r^2\frac{k_1 a}{D_{\text{eff}}}C_A = 0, \qquad (5.126)$$

with the solution

$$C_A = AI_0\left(\sqrt{\frac{k_1 a}{D_{\text{eff}}}}r\right) + BK_0\left(\sqrt{\frac{k_1 a}{D_{\text{eff}}}}r\right). \qquad (5.127)$$

Note that k_1 is the rate constant and a is the specific surface area of the catalyst.

Let us apply eq. (5.127) to a situation where the concentration of the reactant at the surface ($r = R$) is 1 and the concentration at the center of the cylindrical catalyst pellet is finite. We will assume that $\sqrt{(k_1 a)/D_{\text{eff}}} = 6$ and that $R = 1$. Before we begin to work on the constants of integration, A and B, we ought to know a little about the behavior of these Bessel functions. Therefore, a short table of numerical values is provided as follows for zero-order Bessel function of the first and second kinds, as well as the modified Bessel functions, I_0 and K_0; more extensive tables are provided by Abramowitz and Stegun (1972). *Note that neither*

Y_0 *nor* K_0 *can be part of the solution for a problem in cylindrical coordinates if the field variable (V, T, or C_A) is finite at the center ($r = 0$).*

An Abbreviated Table of Zero-Order Bessel Functions

r	$J_0(r)$	$Y_0(r)$	$I_0(r)$	$K_0(r)$
0.0	1	$-\infty$	1	∞
0.2	0.99	-1.0811	1.01	1.7527
0.4	0.9604	-0.606	1.0404	1.1145
0.6	0.912	-0.3085	1.092	0.7775
0.8	0.8463	-0.0868	1.1665	0.5653
1.0	0.7652	0.0883	1.2661	0.421
1.2	0.6711	0.2281	1.3937	0.3185
1.4	0.5669	0.3379	1.5534	0.2437
1.6	0.4554	0.4204	1.7500	0.188
1.8	0.34	0.4774	1.9896	0.1459
2.0	0.2239	0.5104	2.2796	0.1139
2.2	0.1104	0.5208	2.6291	0.0893
2.4	0.0025	0.5104	3.0493	0.0702
2.6	-0.0968	0.4813	3.5533	0.0554
2.8	-0.185	0.4359	4.1573	0.0438
3.0	-0.2601	0.3769	4.8808	0.0347
3.2	-0.3202	0.3071	5.7472	0.0276
3.4	-0.3643	0.2296	6.7848	0.022
3.6	-0.3918	0.1477	8.0277	0.0175
3.8	-0.4026	0.0645	9.5169	0.0139
4.0	-0.3971	-0.0169	11.302	0.0112
4.2	-0.3766	-0.0938	13.443	0.0089
4.4	-0.3423	-0.1633	16.010	0.0071
4.6	-0.2961	-0.2235	19.093	0.0057
4.8	-0.2404	-0.2723	22.794	0.0046
5.0	-0.1776	-0.3085	27.239	0.0037
5.2	-0.11029	-0.33125	32.584	0.00297
5.4	-0.04121	-0.34017	39.009	0.002385
5.6	0.02697	-0.33544	46.738	0.00192
5.8	0.0917	-0.317746	56.038	0.00154
6.0	0.15065	-0.28819	67.234	0.00124
6.2	0.20174	-0.24831	80.718	0.001
6.4	0.24331	-0.19995	96.962	0.00081
6.6	0.27404	-0.14523	116.54	0.00065
6.8	0.2931	-0.08643	140.14	0.00053
7.0	0.3001	-0.02595	168.59	0.00042
7.2	0.29507	0.03385	202.92	0.000343
7.4	0.2786	0.09068	244.34	0.000277
7.6	0.2516	0.1424	294.33	0.0002
7.8	0.2154	0.1872	354.69	0.000181
8.0	0.1717	0.2235	427.56	0.000146
8.2	0.1222	0.25012	515.59	0.000118
8.4	0.06916	0.26622	621.94	0.000096
8.6	0.01462	0.27146	750.5	0.000077
8.8	-0.0392	0.26587	905.8	0.000063
9.0	-0.0903	0.2498	1094	0.000051
9.2	-0.13675	0.22449	1321	0.000041
9.4	-0.17677	0.19074	1595	0.000033
9.6	-0.20898	0.15018	1927	0.0000271
9.8	-0.23277	0.10453	2329	0.0000219
10.0	-0.2459	0.05567	2816	0.0000178

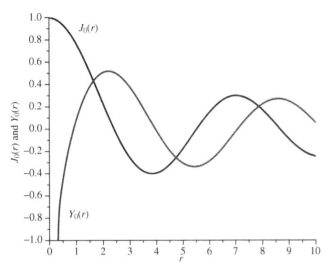

FIGURE 5.5. Bessel functions $J_0(r)$ and $Y_0(r)$ for r from 0 to 10.

$J_0(r)$ and $Y_0(r)$ are also shown graphically in Figure 5.5.

Now we are ready to proceed with our example. Since $K_0(0) = \infty$, we note $B = 0$. Using the surface concentration, $C_A(r = 1) = 1$, we can compute A:

$$A = \frac{1}{I_0\left(\sqrt{\dfrac{k_1 a}{D_{\text{eff}}}}\,R\right)}. \qquad (5.128)$$

From the previous table, we find $I_0(6) = 67.234$, and we can then calculate the following values for the concentration distribution in the interior of the catalyst pellet:

r	0.0	0.1	0.2	0.4	0.6	0.8	1.0
C_A	0.01487	0.01624	0.02073	0.04535	0.1194	0.3390	1.000

Naturally, it would be more efficient in terms of reactant conversion to use an *annular geometry* for the pellet (providing increased surface area per unit volume) so we should also look at the case where the reactant can flow both through the center and past the outer surface. We let the two radii be denoted by the subscripts, 1 and 2, respectively, such that $R_1 \le r \le R_2$, and we set the concentrations at both (inner and outer) surfaces to 1. In this case, K_0 must be retained in the solution. Again we let $\sqrt{(k_1 a)/D_{\text{eff}}} = 6$, just as before, and the constants of integration are found to be $A = 801.011$ and $B = 5.3 \times 10^{-5}$.

r	1.0	1.2	1.4	1.5	1.6	1.8	2.0
C_A	1.000	0.2856	0.1098	0.0987	0.1238	0.3254	1.000

Notice that we do not have symmetry about the midpoint of the annular region ($r = 1.5$). This is to be expected

because of the greater surface area available to the reactant at the outer surface, $r = 2$.

The need to differentiate Bessel functions arises frequently, particularly when a boundary condition involves a specified flux (e.g., Neumann or Robin's type). For J, Y, and K, we have

$$\frac{d}{dr}[Z_p(\alpha r)] = -\alpha Z_{p+1}(\alpha r) + \frac{p}{r} Z_p(\alpha r). \qquad (5.129)$$

Accordingly, we note that

$$\frac{d}{dr}[J_0(\beta r)] = -\beta J_1(\beta r), \quad \text{since } p = 0. \qquad (5.130)$$

For I_p, we have

$$\frac{d}{dr}[I_p(\alpha r)] = \alpha I_{p+1}(\alpha r) + \frac{p}{r} I_p(\alpha r). \qquad (5.131)$$

To illustrate, suppose we have to differentiate $J_0(\lambda r)$; by eq. (5.129), $(d/dr)[J_0(\lambda r)] = -\lambda J_1(\lambda r)$. Now assume that $\lambda = 1$ and that $r = 1.5$; if we use Figure 5.5 for a graphical estimate, we get about -0.51. We can reassure ourselves by looking up the exact value too, it is -0.558. Next, suppose we wanted to identify the maximum value for $Y_0(\lambda r)$; based on Figure 5.5, it appears to occur at about 2.1. We set

$$\frac{d}{dr}[Y_0(\lambda r))] = -\lambda Y_1(\lambda r) = 0$$

and find that $\lambda r = 2.19715$. Let us look at one more example: We have an annular catalytic pellet in which $C_A = A I_0(\beta r) + B K_0(\beta r)$, with inner and outer radii (R_1 and R_2) of 2 and 4, respectively. We take $\beta = 0.3$ and assume that the concentration at both exposed surfaces will be 1. Our interest is the driving force for diffusion in the interior of the annular solid: $dC_A/dr = A\beta I_1(\beta r) - B\beta K_1(\beta r)$. The boundary conditions are used to show $A = 0.7175$ and $B = 0.2784$, and we can now evaluate the derivative at any r-location of interest. We are going to illustrate this at the annular midpoint, $r = 3$, keeping in mind that the arithmetic center is *not* a surface of symmetry in this case; the result is $dC_A/dr = 0.0471$ for $r = 3$ (small, but *not* zero). There are many useful sources of information for Bessel's equation and Bessel functions, and a few of them are included in the references provided at the end of this chapter.

POWER SERIES SOLUTIONS OF ORDINARY DIFFERENTIAL EQUATIONS

We observed previously that for higher-order ODEs with variable coefficients, there is no guaranteed method of solution. In such cases, an approximate analytic solution may sometimes be useful, and we will review one approach here. The technique we describe in this section has been used by mathematicians for a very long time, and it has produced some important results. We will use the simple equation

$$\frac{d^2 y}{dx^2} + y = 0,$$ (5.132)

with $y(x = 0) = 0$ and $y'(x = 0) = 1$, for our initial exploration because we know the solution to be

$$y = C_1 \sin x + C_2 \cos x.$$ (5.133)

To satisfy the two boundary conditions, it is necessary that $y = \sin x$, of course. We are going to propose that the solution be represented by the series

$$a_0 + a_1 x + a_2 x^2 + a_3 x^3 + \cdots \quad \text{or} \quad \sum_{n=0}^{\infty} a_n x^n.$$ (5.134)

We note immediately that to conform to the boundary conditions, $a_0 = 0$ and $a_1 = 1$. Let us differentiate the series once, $\sum n a_n x^{n-1}$, and then again to obtain $\sum (n-1)(n) a_n x^{n-2}$. We take this last result back to our ODE, resulting in

$$\sum_{n=0}^{\infty} (n-1)(n) a_n x^{n-2} + \sum_{n=0}^{\infty} a_n x^n = 0.$$ (5.135)

We need to have the first summation written in terms of x^n for consolidation; therefore,

$$\sum_{n=0}^{\infty} [(n+1)(n+2) a_{n+2} + a_n] x^n = 0.$$ (5.136)

For this power series to be zero as indicated here, it is necessary for the coefficients to be zero and, therefore,

$$(n+1)(n+2) a_{n+2} + a_n = 0.$$ (5.137)

However, the boundary conditions require that $a_0 = 0$ and $a_1 = 1$, as we observed earlier. The first of these also requires $a_2 = 0$, and thus, all even a_ns must be zero. We now use eq. (5.137) to compute (by recursion) coefficients for $n = 3, 5, 7$. The result is

$$a_3 = -\frac{1}{(3)(2)}, \quad a_5 = \frac{1}{(5)(4)(3)(2)}, \quad a_7 = -\frac{1}{(7)(6)(5)(4)(3)(2)}.$$ (5.138)

Therefore, our series approximation for $y(x)$ looks like this:

$$y(x) \approx x - \frac{x^3}{3!} + \frac{x^5}{5!} - \frac{x^7}{7!} + \cdots.$$ (5.139)

How well can the truncated series represent the actual solution? If we limit n to values less than 27, we obtain the values shown in the following table:

x	$y(x)$ series	$y(x) = \sin(x)$
$\frac{1}{8}$	0.1246747	0.1246747
$\frac{1}{4}$	0.2474039	0.24740396
$\frac{1}{2}$	0.4794255	0.47942554
1	0.841471	0.84147098
2	0.9092975	0.90929743
4	−0.7568024	−0.756802495
8	0.9893362	0.98935825
9	0.4116477	0.41211848
10	−0.5542731	−0.54402111
11	−1.158563	−0.99999021
12	−2.469321	−0.53657292

These results make it clear that the truncated series does an excellent job of representing $y(x)$ for *small* values of x, as expected. Discrepancies do begin to appear as x exceeds about 4.

Let us explore a second example that offers a slightly different wrinkle. Consider the ODE

$$\frac{dy}{dx} = x^2 - 4x + y + 1 \quad \text{with} \quad y(x = 2) = 3.$$ (5.140)

It is effective for us to define a new independent variable, $z = x - 2$; consequently, $x^2 = z^2 + 4z + 4$ and $4x = 4z + 8$. We substitute into eq. (5.140) and obtain

$$\frac{dy}{dz} = z^2 + y - 3.$$ (5.141)

Now it is apparent that $y(z = 0) = 3$. Let us assume that the solution can be written as a series of the form

$$y = a_0 + a_1 z + a_2 z^2 + a_3 z^3 + \cdots,$$ (5.142)

and we see immediately that $a_0 = 3$, allowing us to write $y = 3 + \sum_{n=1}^{\infty} a_n z^n$. We differentiate, finding $dy/dz = \sum n a_n z^{n-1}$, and we take these expressions back to eq. (5.141) and consolidate:

$$\sum [((n+1) a_{n+1} - a_n) z^n] - z^2 = 0.$$ (5.143)

We group like powers, noting that the coefficients must combine to produce zero. For $n = 1$, $2a_2 - a_1 = 0$; for $n = 2$, $3a_3 - a_2 - 1 = 0$; for $n = 3$, $4a_4 - a_3 = 0$, and so on. From eq. (5.141), we note that $a_1 = 0$, and therefore $a_2 = 0$ as well. It is apparent that $a_3 = 1/3$, and by recursion, $a_4 = 1/12$ and $a_5 = 1/60$, and so on. We are in position to write down our series solution:

$$y = 3 + \tfrac{1}{3} z^3 + \tfrac{1}{12} z^4 + \tfrac{1}{60} z^5 + \cdots,$$ (5.144)

keeping in mind that $z = x - 2$. How well does this series represent the actual solution? The reader is encouraged to use the integrating factor approach to show that the analytic solution for eq. (5.141) is

$$y = -z^2 - 2(z + 1) + 3 + 2\exp(z). \qquad (5.145)$$

If we let $z = 2$, eq. (5.145) shows $y = 7.778$. If we truncate eq. (5.144) with the sixth-degree term, we get $y(2) = 7.71$. We can also solve this ODE numerically to confirm $y(2) = 7.778$.

The notion that we might want to seek power series solutions for ODEs may seem quaint to you. However, it is important that we remember that many very significant problems have been solved in this manner. The reader is encouraged to look at the application of the *method of Frobenius* to Bessel's differential equation and a useful description of the process is provided by Mickley et al. (1957) in Chapter 5.

REGULAR PERTURBATION

There are occasions when one simply must obtain an approximate analytic solution for a nonlinear ODE. In the previous section, we looked at the power series technique; now we turn our attention to regular perturbation. Perturbation is a powerful method that is especially useful when a *nonlinear* problem contains a parameter that is in some sense small. Let us consider steady-state heat transfer (conduction) in a slab of pure iron. The thermal conductivity of pure iron between $-200°C$ and $+700°C$ varies approximately as

$$k \cong a + bT = 76.333 - 0.0633T \text{ W/(m°C)}. \qquad (5.146)$$

Therefore, at $-200°C$, $k = 89\,\text{W/(m°C)}$, and at $+700°C$, $k = 32\,\text{W/(m°C)}$; from high-to-low temperature then, k increases by a factor of about 2.78. Such a variation will have a profound impact on the temperature distribution as we shall now see. Assume we have a slab of pure iron that extends from $y = 0$ to $y = 1$ m. The temperature at $y = 0$ is maintained for all time t at $-200°C$, and at $y = 1$ m, the constant temperature is $+700°C$. Under these steady conditions, the temperature distribution is governed by

$$\frac{d}{dy}\left[(a + bT)\frac{dT}{dy}\right] = 0 \qquad (5.147)$$

or

$$(a + bT)\frac{d^2T}{dy^2} + b\left(\frac{dT}{dy}\right)^2 = 0. \qquad (5.148)$$

One thing that we notice immediately about eq. (5.148) is that if the parameter, b, is *very* small, then the two non-linearities in the ODE are eliminated! We will assume that T can be represented as a sequence of unknown functions:

$$T = T_0 + bT_1 + b^2T_2 + b^3T_3 + \cdots. \qquad (5.149)$$

Our goal of course is to determine the unknown functions, T_0, T_1, and so on. We begin by writing out the derivatives we need:

$$\frac{dT}{dy} = \frac{dT_0}{dy} + b\frac{dT_1}{dy} + b^2\frac{dT_2}{dy} + \cdots \qquad (5.150)$$

and

$$\frac{d^2T}{dy^2} = \frac{d^2T_0}{dy^2} + b\frac{d^2T_1}{dy^2} + b^2\frac{d^2T_2}{dy^2} + \cdots. \qquad (5.151)$$

We now take eq. (5.149), eq. (5.150), and eq. (5.151) and substitute into eq. (5.148):

$$[a + b(T_0 + bT_1 + b^2T_2 + \cdots)]\left(\frac{d^2T_0}{dy^2} + b\frac{d^2T_1}{dy^2} + b^2\frac{d^2T_2}{dy^2} + \cdots\right)$$
$$+ b\left[\frac{dT_0}{dy} + b\frac{dT_1}{dy} + b^2\frac{dT_2}{dy} + \cdots\right]^2 \cong 0. \qquad (5.152)$$

Now let us assume b can take on any value, including a *very small* one; if b is vanishingly small, then we are left with

$$a\frac{d^2T_0}{dy^2} = 0 \quad \text{and, consequently,} \quad T_0 = C_1y + C_2. \qquad (5.153)$$

We treat the boundary conditions exactly the same way—we have -200 at $y = 0$ and $+700$ at $y = 1$. Therefore,

$$T_0 = 900y - 200. \qquad (5.154)$$

We differentiate eq. (5.154) as needed, and take those results back to eq. (5.152). Of course,

$$\frac{dT_0}{dy} = 900 \quad \text{and} \quad \frac{d^2T_0}{dy^2} = 0. \qquad (5.155)$$

If we divide by b and again allow b to become very small, then

$$a\frac{d^2T_1}{dy^2} = -(900)^2. \qquad (5.156)$$

Therefore,

$$T_1 = \frac{-(900)^2}{2a} y^2 + C_1 y + C_2. \qquad (5.157)$$

Now we return to eq. (5.149) to look at the boundary conditions, subtracting the results obtained with eq. (5.154) from T, and then dividing by b: Thus, $C_2 = 0$ and $C_1 = (900)^2/2a$. At this point, our approximate solution has the form

$$T \cong 900y - 200 + b\frac{(900)^2}{2a}(y - y^2),$$

and inserting the numerical values for a and b,

$$T \cong 335.851y^2 + 564.149y - 200. \qquad (5.158)$$

Our process guarantees that the boundary conditions are satisfied, but our real interest is whether or not this approximation will reproduce the expected curvature for intermediate values of y. Figure 5.6 provides a comparison of the exact solution with our truncated approximation.

Perturbation is a powerful technique that can produce very good approximate solutions under the right circumstances. Experience suggests, however, that if one must go past the third unknown function (past T_2 for the previous example), then the time invested for the likely return will probably be excessive. There are very good sources of information available for the applied scientist who must generate an analytic approximation to the solution of a nonlinear

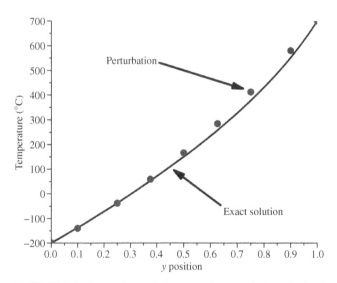

FIGURE 5.6. Comparison of the approximate solution obtained with regular perturbation with the exact solution for steady conduction in an iron slab with variable conductivity. Even though the perturbative solution was truncated, the agreement between the two is good enough for many practical purposes.

ODE; for example, this author has found Finlayson (1980) to be quite useful.

LINEARIZATION

Often the removal of nonlinear terms in ODEs will yield an equation that can be solved analytically. Of course, the analyst must decide if the approximation is adequate for the intended purpose. We will look at a couple of alternatives here and we will try to give the reader a sense of when such an approach may work and when it may not. Let us begin with the simple first-order ODE:

$$\frac{dh}{dt} = \tfrac{1}{2}h^{7/4}, \qquad (5.159)$$

with $h(t = 0) = 1$ and $0 \leq t \leq 2$. Equation (5.159) is easy to solve as it stands, so it will provide a very nice means of comparison. The analytic solution, which we will make use of later, is

$$h(t) = \left[1 - \frac{3}{8}t\right]^{-4/3}. \qquad (5.160)$$

Let us focus our attention on the right-hand side of eq. (5.159), and we will use a truncated Taylor series to approximate this function:

$$f(h) = h^{7/4} \cong f(h_0) + f'(h_0)(h - h_0). \qquad (5.161)$$

It is evident that our strategy is to place a tangent line on the $h^{7/4}$ curve at a particular value, h_0. It also seems likely that this will work well only if h does not deviate radically from h_0—the further we are away from that point, the worse the approximation will be. Since $f' = (7/4)h^{3/4}$, we can select an appropriate point and find the slope; we know from the problem statement that $h \geq 1$, so we will begin by trying $h_0 = 2$ such that $f'(h_0) = 2.9431$ and $f(h_0) = 3.3636$ Our approximation is therefore $h^{7/4} \cong 3.3636 + 2.9431(h - 2)$. First, we will see how well this linearized version of $f(h)$ represents the original function by constructing a little table.

h	1	1.5	2	2.5	3	4
$f(h)$	1.0000	2.0331	3.3636	4.9704	6.8385	11.3137
$f(h)$ linearized	0.4205	1.8921	3.3636	4.8352	6.3067	9.2498

The truncated Taylor series corresponds to the nonlinear function nicely *near* the selected point, $h = 2$; however, it begins to deviate dramatically for both smaller and larger values. These differences will impact the approximate solution of eq. (5.159) in a very significant way as we shall now see. The linearized differential equation is

$$\frac{dh}{dt} = 1.4716h - 1.2613 \qquad (5.162)$$

and the solution is

$$h = \frac{1}{1.4716}\left[C_1 \exp(1.4716t) + 1.2613\right]. \qquad (5.163)$$

The initial condition is $h(0) = 1$, so accordingly, $C_1 = 0.2103$. Now we can compare the two sets of results:

t	h Actual	h Linearized
0	1.00	1.00
0.25	1.1403	1.0636
0.5	1.3190	1.1554
0.75	1.5532	1.2880
1.0	1.8714	1.4796
1.25	2.3242	1.7565
1.5	3.0109	2.1564
1.75	4.1528	2.7342
2.0	6.3496	3.5689

What have we learned from this example? If the transient operation of the nonlinear system takes us far away from the expected value for h (which we decided might be 2), then we will have serious problems. On the other hand, if we are able to stay close to h_0, the linearized ODE may be completely acceptable. In terms of the table shown earlier, this probably means that we would need to constrain the independent variable, t, to $1 \leq t \leq 1.5$ if we wanted to get satisfactory results.

The ideas we employed in the previous example can be extended to higher-order problems too. Consider the nonlinear second-order system,

$$\frac{dx}{dt} = y - (x^2 + y^2)x \qquad (5.164)$$

and

$$\frac{dy}{dt} = -x - (x^2 + y^2)y. \qquad (5.165)$$

We will begin by letting $x_i = y_i = 3/4$, and solving the pair of equations, eq. (5.164) and eq. (5.165) numerically to obtain the system trajectory in the phase plane (merely a convenient way to visualize the system's dynamic behavior by cross plotting pairs of values for $x(t)$ and $y(t)$). The phase-plane construction shows that the system is stable and that both x and y oscillate with decaying amplitude. This results in an inward-directed spiral that is ultimately centered about the equilibrium point, $x = 0$ and $y = 0$, and the trajectory is provided in Figure 5.7.

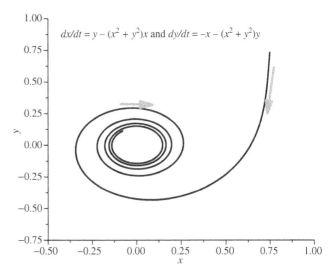

FIGURE 5.7. Phase plane portrait of the nonlinear second-order system revealing stable behavior with an inward-directed spiral.

Now let $u = x - x_0$ and $v = y - y_0$ such that $x = u + x_0$ and $y = v + y_0$. Our plan is to approximate the two nonlinear functions (just as we did for the first-order system in the initial example):

$$f(x, y) \cong f(x_0, y_0) + \left.\frac{\partial f}{\partial x}\right|_{x_0, y_0} u + \left.\frac{\partial f}{\partial y}\right|_{x_0, y_0} v \qquad (5.166)$$

and

$$g(x, y) \cong g(x_0, y_0) + \left.\frac{\partial g}{\partial x}\right|_{x_0, y_0} u + \left.\frac{\partial g}{\partial y}\right|_{x_0, y_0} v. \qquad (5.167)$$

The necessary partial derivatives correspond to the Jacobian matrix and the system can be represented in an alternative way:

$$\begin{aligned} \frac{du}{dt} \\ \frac{dv}{dt} \end{aligned} = \begin{bmatrix} -3x^2 - y^2 & 1 - 2xy \\ -1 - 2xy & -x^2 - 3y^2 \end{bmatrix} \begin{bmatrix} u \\ v \end{bmatrix} \qquad (5.168)$$

The equilibrium point is $(x_0, y_0) = (0, 0)$, so the Jacobian matrix is simply

$$\begin{bmatrix} 0 & 1 \\ -1 & 0 \end{bmatrix}$$

and consequently,

$$\frac{du}{dt} = v \quad \text{and} \quad \frac{dv}{dt} = -u. \qquad (5.169)$$

Equivalently, $(d^2u/dt^2) + u = 0$ such that $u = C_1 \sin t + C_2 \cos t$ (we obtain the identical result for v). This means that both u and v are purely periodic with *constant amplitude*. In the phase plane therefore, we will get a closed curve (a limit cycle) such that the system trajectory merely orbits around the equilibrium point. This is distinctly different from the nonlinear case where the oscillations exhibited decreasing amplitude, resulting in an inward-directed spiral as shown in Figure 5.7. Although the differences are worrisome in the absolute sense, it is important to note that, *qualitatively*, the original system and the linearized simplification exhibit similar oscillatory behaviors.

We conclude this discussion by looking at a different approach, one that has been used in fluid mechanics to eliminate nonlinear inertial terms from the Navier–Stokes equation (the interested reader should investigate *Oseen's correction*). Suppose we have the following differential equation:

$$\frac{d^2y}{dx^2} - y\frac{dy}{dx} = 2, \qquad (5.170)$$

with $y(x = 0) = 1/2$ and $y'(x = 0) = 0$ and let us assume we are interested in $0 \leq x \leq 1.75$. Our plan is too replace the $y\,(dy/dx)$ term with $y_0\,(dy/dx)$, where y_0 is an appropriately chosen constant; therefore,

$$\frac{d^2y}{dx^2} = 2 + y_0\frac{dy}{dx}. \qquad (5.171)$$

Equation (5.171) can be conveniently solved by reducing the order; for example, let $\phi = dy/dx$. The solution is (and the reader should verify this result)

$$y = \frac{2}{y_0}\left[\frac{1}{y_0}\exp(y_0 x) - x\right] + C_2. \qquad (5.172)$$

Since $y(x = 0) = \frac{1}{2}$, $C_2 = \frac{1}{2} - \left(2/y_0^2\right)$. How well our approximate solution corresponds to the exact numerical results depends on our choice for y_0, of course. The main difficulty is that with nonlinear differential equations we may not know much about $y(x)$ over the complete range of x, thus making a good choice for y_0 problematic. Here is a comparison of the actual computed and the linearized (approximate) results.

x	y (Exact)	y (Linearized with $y_0 = 2$)
0	0.5000	0.5000
0.3	0.5949	0.6111
0.5	0.7760	0.8591
0.7	1.0754	1.3276
0.9	1.5387	2.1248
1.2	2.8070	4.3116
1.5	6.3878	8.5428
1.75	40.8730	14.8077

Whether or not the examples of linearization applied to ODEs in this section will be satisfactory depends on the intended application. For cases requiring precision, these approaches will probably not be sufficient, but if we are merely interested in learning a little about the behavior of a difficult nonlinear problem (with a quick analytic approximation), then linearization may prove useful.

CONCLUSION

Toward the end of this chapter, we introduced a couple of techniques by which the analyst could obtain an approximate solution for an ODE, solution by power series and regular perturbation. Both can be used to advantage, but there are other more modern and perhaps more expedient ways to obtain analytic approximations. We will briefly introduce one such technique here, the *variational iteration method* (VIM), but we will defer a more complete discussion for Chapter 9, "Integro-Differential Equations," where VIM will prove to be extraordinarily powerful.

Consider the linear first-order ODE,

$$\frac{dy}{dx} + 2xy = 4x, \qquad (5.173)$$

with the analytic solution, $y = 2 + C_1 \exp(-x^2)$. We will choose $y(x = 0) = 4$ such that $C_1 = 2$ and assume we are interested in the behavior of $y(x)$ for $0 \leq x \leq 1$. To use VIM, we start with a simple estimate for y that satisfies the boundary condition and we will select $y_0 = x + 4$. Though this is not an optimal choice, we will still get to an acceptable approximation as we shall see. We now rearrange the ODE (eq. 5.173) and use it—integrated—to improve our initial guess:

$$y_{n+1} = y_n - \int_0^x [y_n' + 2sy_n - 4s]ds. \qquad (5.174)$$

Of course, $y_0' = 1$, and after we carry out the integration, we find

$$y_1 = -\frac{2}{3}x^3 - 2x^2 + 4, \qquad (5.175)$$

and then successively,

$$y_2 = \frac{4}{15}x^5 + x^4 - 2x^2 + 4 \qquad (5.176)$$

and

$$y_3 = -\frac{8}{105}x^7 - \frac{1}{3}x^6 + x^4 - 2x^2 + 4. \qquad (5.177)$$

Let us find out how well this sequence of approximations actually describes $y(x)$.

x	$y_{analytic}$	$y_0 = x + 4$	y_1	y_2	y_3
0	4	4	4	4	4
¼	3.8788	4.25	3.86458	3.87917	3.87882
½	3.5576	4.50	3.41667	3.57083	3.5567
¾	3.13957	4.75	2.59375	3.25469	3.12191
1	2.73576	5	1.33333	3.26667	2.59048

Even though we made a poor choice for our initial trial function, $y_0 = x + 4$, we have rapidly closed in on the values provided by the analytic solution. With just a few more iterations, we would have an exceptionally good match to $y(x)$. The student may wish to repeat this example by setting $y_0 = 4 - x$; how well do y_2 and y_3 agree with the analytic solution now? VIM is an incredibly useful tool for finding approximate solutions for difficult equations, and we will gain a much greater appreciation for its value in Chapter 9.

ODEs are common in applied mechanics as many force, mass, and energy balance problems lead to ODEs. The spectrum of application is broad, ranging from chemical kinetics to electronic circuits to mechanical flutter (aeroelasticity). It is obviously important that technical professionals be able to solve such problems with speed and accuracy. Our intent with this chapter is to provide the reader with a practical review of solution methods that have proven to be of value for frequently encountered ODEs.

PROBLEMS

5.1. Use the integrating factor approach to solve the following ODEs:

$$\frac{dy}{dx} + 3xy = 2x \quad \text{and} \quad x^2 \frac{dy}{dx} - xy = x^3 + 4.$$

5.2. Check the following differential equations to see if they are exact. If they are not exact, solve using the integrating factor approach:

$$x\frac{dy}{dx} = y - x^2 + 1$$

$$xy^3 \frac{dy}{dx} = x^4 + y^4.$$

5.3. A rifle bullet (130 grain WIN 270) leaves the barrel with an initial (muzzle) velocity of 3060 ft/s. Suppose the rifle is fired at an angle of 45° over level ground and that the only force acting on the bullet in flight is gravity. Find the maximum elevation achieved by the bullet and its maximum horizontal travel.

It may be obvious to you that the answers you obtained for the first part of this problem are quite unrealistic. Drag is very important! Fortunately, we have ballistic information available for the WIN 270 as follows:

Distance (yd)	0	100	200	300	400	500
Velocity (ft/s)	3060	2802	2559	2329	2110	1904

Notice that the rifle bullet loses about 38% of its initial velocity over the first 500 yd of travel. Devise a strategy to use these data to add drag force to your original differential equation and repeat your analysis. What is the maximum height achieved by the bullet under these more realistic conditions?

5.4. The example that accompanied Figure 5.2 was based on the differential equation

$$\frac{dy}{dx} = xy(y - 2).$$

We want to use the technique described in the text to explore the behavior of this nonlinear equation; connect a sequence of very short tangent lines obtained from the slope(s) at particular points (x_i, y_i) to develop a semiquantitative picture for this ODE. We will focus upon the first quadrant but only look at values of y *greater than* 2.

5.5. Consider the wheel–tire assembly on a vehicle equipped with IFS (independent front suspension). The assembly is subjected to a bump (perhaps a pothole) and the motion of the mass is constrained by a spring-shock absorber (or strut) combination. The resulting motion is described approximately by the second-order differential equation:

$$\tau^2 \frac{d^2z}{dt^2} + 2\tau\zeta \frac{dz}{dt} + z = F(t).$$

z corresponds to the vertical position of the assembly, τ is the time constant, and ζ is the damping coefficient. The challenge in suspension design, of course, is to keep the tire in contact with the pavement (for consistent traction) yet provide a tolerably smooth ride for the occupants. Suppose the time constant has the value 0.125 second and that the damping coefficient is 0.55; we then find

$$\frac{d^2z}{dt^2} + 8.8\frac{dz}{dt} + 64z = G.$$

If $F = 1$ for $t > 0$ (thus $G = 64$), what is the ensuing motion of the assembly? If the shock absorber is completely worn out, the damping coefficient will be smaller, say, 0.15.

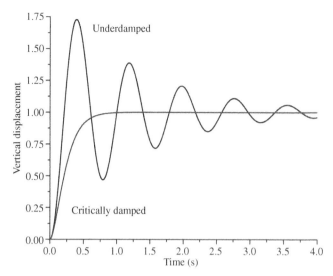

FIGURE 5.8. Suspension behavior for two cases, underdamped with $\zeta = 0.10$ and critically damped with $\zeta = 1.0$. The compromise inherent in suspension design is obvious: Very rapid response produces overshoot and oscillation.

What impact will that have upon the motion if the forcing function is exactly the same? Two examples are shown in Figure 5.8 to provide some guidance on what is expected from this ODE.

5.6. We need to analyze the performance of two identical continuous stirred-tank reactors (CSTRs) in series; this arrangement is being used to carry out the first-order, irreversible decomposition of species A. Both reactors are perfectly mixed and each has a volume of 500 gal. Reactant is introduced into the first reactor (and the initial, steady-state feed concentration is 2 mol/gal) at a volumetric flow rate of 25 gpm. The mass balance for species A in the first reactor takes the form

[Rate in] − [Rate out] − [Rate consumed] = [Accumulation]

$$\dot{V}C_{Ain} - \dot{V}C_{A1} - V_T k_1 C_{A1} = V_T \frac{dC_{A1}}{dt}.$$

We divide by the volumetric flow rate, $\dot{V}$, and rewrite the equation:

$$C_{Ain} - (1 + k_1 \tau)C_{A1} = \tau \frac{dC_{A1}}{dt}.$$

Let $k_1 = 0.033 \, \text{min}^{-1}$ and note that the mean residence time $\tau = 20 \, \text{min}$ $(V_T / \dot{V})$. At $t = 0$, the feed concentration to CSTR 1 doubles; it is instantaneously increased from 2 mol/gal to 4 mol/gal. Find the analytic solution for $C_{A1}(t)$ and then use that result in the mass balance (for A) on reactor 2. When will $C_{A2}(t)$ achieve 75% of its ultimate change?

5.7. A pipe carries a very hot fluid and circular fins (actually annular) are installed on the outer pipe wall to cast off unwanted thermal energy to the surroundings. Each fin extends from the outer pipe wall (at $r = R_1$) to the radial position, R_2. By assuming that the fin is thin and that it is made from a material with high thermal conductivity, the principal variation in temperature will occur in the r-direction (i.e., the temperature is nearly uniform in the transverse, or z-, direction). Assuming that thermal energy is lost from the fin's surface according to Newton's law of cooling, an energy balance reveals

$$\frac{d^2\theta}{dr^2} + \frac{1}{r}\frac{d\theta}{dr} - \frac{2h}{wk}\theta = 0,$$

where h is the heat transfer coefficient, k is the thermal conductivity of the fin, and w is the width of the fin in the z-direction. $\theta = (T - T_\infty)$, where T_∞ is the temperature of the surrounding air. Solve this ODE and find the temperature distribution in the fin given the following parametric values:

$$T(r = R_1) = 500°\text{F}, \quad T_\infty = 60°\text{F},$$

$$R_1 = 0.25 \, \text{ft}, \quad R_2 = 0.50 \, \text{ft}, \quad w = 0.0104 \, \text{ft},$$

$$h = 7.5 \, \text{Btu/(hr ft}^2 \, °\text{F)}, \quad k = 30 \, \text{Btu/(hr ft}°\text{F)}.$$

Assume that no thermal energy is lost through the end of the fin; that is,

$$-k\frac{dT}{dr}\bigg|_{r=R_2} = 0.$$

5.8. A cylindrical tank with a diameter of 12 ft and a depth of 10 ft is initially filled with water. At $t = 0$, a circular hole with a diameter of 2 in. is opened in the side wall at the bottom. The velocity through this orifice is approximately given by Torricelli's theorem, $V_0 = \sqrt{2gh}$; therefore, the initial velocity will be $V_0 = \sqrt{(2)(32.17)(10)} = 25.37$ ft/s. Since the orifice diameter is 2 in., the initial volumetric flow rate will be about 0.55 ft³/s, or 248 gpm. Find out when the tank will be 90% empty (i.e., when $h = 1$ ft). Then, in the second part of this problem, the tank is drained from its initial depth of 10 ft just as before, but this time, water is also added continuously at a rate of 55 gpm. When will the depth of water in the tank get within 3% of its ultimate value?

5.9. A particle with a diameter of 1 mm and a density of 1.45 g/cm³ is released at the surface of a flowing stream of water. The velocity of water in the stream varies from 130 cm/s at the surface ($y = 0$) to 0 cm/s at the bottom ($y = 175$ cm) according to $V = 130 - 0.004245 \, y^2$ (cm/s). Assume the particle is perfectly entrained by the moving water (its velocity in the stream direction always corresponds to the velocity of the water). If the only forces acting upon the particle in the y-direction are gravity and drag, determine how long it will take the particle to reach bottom and how far it will travel downstream during the settling process. The drag force is given by $F = AKf$, where A is the frontal area of the particle, K is the kinetic energy of the fluid (relative to the particle) per unit volume, and f is the drag coefficient:

$$A = \pi R^2, \quad K = \tfrac{1}{2}\rho v_y^2, \quad \text{and} \quad f \approx 0.5.$$

5.10. Consider the ODE

$$(1-x)\frac{dy}{dx} = 2x - y, \quad \text{with} \quad y(x=0) = 2.$$

Find a power series solution for this ODE and compare your result to the solution you obtain using an integrating factor.

5.11. Rosenhead (1940) and Landau and Lifshitz (1959) show how the radial component of the equation of fluid motion can be simplified for flow between converging walls. Provide a detailed outline of the process, showing how an elliptic integral of the first kind can be obtained from the analysis.

5.12. We want to use regular perturbation to find an approximate analytic solution for the example we explored at the end of the chapter, where

$$(a+bT)\frac{d^2T}{dy^2} + b\left(\frac{dT}{dy}\right)^2 = 0.$$

We will use the same terminal temperatures ($-200°$ and $+700°$); however, in this case, we will take $k = 87.7778 - 0.1111\,T$. This will cause a significant change in the profile, $T(y)$, which is illustrated in Figure 5.9.

In this case, it will be necessary for us to find T_0, T_1, and T_2. Is the resulting approximation an adequate representation for the profile, $T(y)$, shown in Figure 5.9?

5.13. *Please note: This problem is more difficult! You may find section 2 of chapter 16 in Gray et al. (1966) to be helpful.* Suppose we have a length of metal rod or wire positioned vertically. The bottom end is clamped in a vise; we want to determine the length of the wire segment that just results in instability (i.e., when the top of the wire is just ready to fall over). Let β be the rigidity of the wire or rod, and let W be the weight per unit length. The variable x is the vertical position above the clamped end and y is the lateral deviation from vertical (rigid) position. The total length of the wire is L. If the lateral deviation is not too large, then

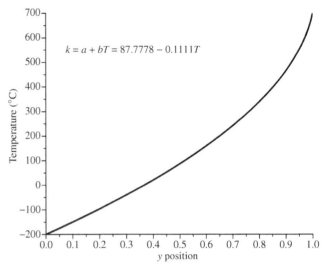

FIGURE 5.9. Exact solution for steady conduction in a slab of material for which $k = 87.7778 - 0.1111T$. The ends of the slab are maintained at $-200°$ and $+700°$ for all t.

$$\beta\frac{d^2y}{dx^2} \cong \int_x^L W(y'-y)dx'.$$

Differentiate this equation with respect to x and set $p = dy/dx$:

$$\frac{d^2p}{dx^2} + \frac{W}{\beta}(L-x)p = 0.$$

Now set $L - x = r^{2/3}$ and take $p = r^{1/3}\,z$. Show that these substitutions result in

$$\frac{d^2z}{dr^2} + \frac{1}{r}\frac{dz}{dr} + \left(\frac{4W}{9\beta} - \frac{1}{9r^2}\right)z = 0,$$

which is a form of Bessel's differential equation. At the top of the wire (the free end) where $x = L$ (or $r = 0$), we must have $dp/dx = 0$ or $r^{1/3}\,(dp/dr) = 0$. Find the critical length of the vertical wire.

5.14. The Leveque analysis applies to heat and mass transfer near a surface where the velocity distribution can—at least locally—be described by $v_x = ay$. If we confine our attention to the region very close to the wall, this linear approximation should be valid. For steady-state conditions then, the mass transfer process will be governed by

$$ay\frac{\partial C}{\partial x} = D\frac{\partial^2 C}{\partial y^2}.$$

By appropriate scaling of the independent variables x and y, namely,

$$\eta = y\left(\frac{a}{9Dx}\right)^{1/3},$$

this partial differential equation can be transformed into a second-order ODE:

$$\frac{d^2C}{d\eta^2} + 3\eta^2\frac{dC}{d\eta} = 0.$$

Prove that this transformation yields the result shown here and solve this ODE by first reducing the order; that is, let $\phi = dC/d\eta$. Then integrate twice to confirm that

$$C = \int_0^\eta A_1\exp(-\eta^3)d\eta + A_2.$$

Apply the appropriate boundary conditions: $C = C_0$ at the wall, where both y and $\eta = 0$, and $C = C_\infty$ for both y and $\eta = \infty$, to show that $C_2 = C_0$ and

$$C_1 = \frac{C_\infty - C_0}{\Gamma\left(\frac{4}{3}\right)}.$$

Please note that Γ is the *gamma function* defined by

$$\Gamma(n) = \int_0^\infty x^{n-1}\exp(-x)dx.$$

The recurrence formula is particularly useful, $\Gamma(n+1) = n\Gamma(n)$, and a short table of values for $\Gamma(n)$ is provided here to assist you:

n	$\Gamma(n)$
1.0	1.000
1.1	0.951
1.2	0.918
1.3	0.897
1.4	0.887
1.5	0.886
1.6	0.894
1.7	0.909
1.8	0.931
1.9	0.962
2.0	1.000

5.15. Two identical vented tanks, each 10 ft high and 6 ft in diameter, are interconnected by a straight length of pipe (with a valve in the middle) at the bottom (i.e., at $z = 0$). Tank 1 also has an opening equipped with a valve at the bottom to permit discharge into a floor drain. Tank 1 is completely full of water initially, and tank 2 is completely empty. At $t = 0$, *both* valves are opened (wide) and water begins to flow. Develop a model for this system by mass

balance and solve the ODEs so we can predict both $h_1(t)$ and $h_2(t)$. We will assume that the velocity in the interconnecting pipe can be described by $V_2 = b(h_1 - h_2)$. The velocity of the water leaving tank 1 destined for the floor drain is given by $V_0 = ch_1$. Assume that $b = c$ and that both are equal to 0.027 ft^3/s/ft. What would the maximum value of h_2 be? It would be somewhat more realistic to assume that the velocities were proportional to the square root of the head, for example, $V_0 = \sqrt{ch_1}$. If these changes are made, can the problem still be solved analytically?

5.16. A horizontal beam is being designed to carry a "skywalk" from one side of a shopping mall to the other, a span of L ft. The intent is to allow shoppers easy passage across a large, open, common area. It is anticipated that the maximum load to be carried will be W lb$_f$/ft (i.e., per foot of length, of course). We need to determine the maximum amount of deflection, and for the sake of customer confidence, it is not to exceed 0.5 in. The beam is rigidly clamped at both ends (at both $x = 0$ and $x = L$). The beam's deflection—if any occurs—will be in the negative y-direction. Assume that the appropriate model has the form

$$EI\frac{d^2y}{dx^2} = K + \frac{1}{2}WLx - \frac{1}{2}Wx^2,$$

where E is the modulus of elasticity and I is the moment of inertia. What is the maximum value of y?

5.17. The Chebyshev (Tschebysheff) differential equation is

$$(1-x^2)\frac{d^2y}{dx^2} - x\frac{dy}{dx} + n^2y = 0.$$

This ODE is of the Sturm–Liouville (SL) type and, of course, it does not have constant coefficients; problems of the SL type have the general form

$$\frac{d}{dx}\left[p(x)\frac{dy}{dx}\right] + [q(x) + \lambda r(x)]y = 0,$$

with $a \leq x \leq b$, $A_1y(a) + A_2y'(a) = 0$, and $B_1y(b) + B_2y'(b) = 0$. Solutions for an SL problem can only be found for certain values of the parameter, λ; hence, they are termed *eigenvalue* problems. Solutions for the Chebyshev ODE are given by Chebyshev polynomials and are written as $y = T_n(x)$. Consequently, if $n = 4$, then $y = T_4(x) = 8x^4 - 8x^2 + 1$. Set $n = 4$ such that $n^2 = 16$ in the Chebyshev ODE and then verify that $T_4(x)$ is a solution. Then let $n^2 = 64$ so that the solution is $T_8(x)$. *Without* looking up the T_8 polynomial, try to determine the form of the solution for this case (it would be comprised of *even* powers of x added to the constant 1). We note that the Chebyshev polynomials are orthogonal with respect to a weighting function,

$$\int_{-1}^{+1} T_n(x)T_m(x)\frac{1}{\sqrt{1-x^2}}\,dx = 0,$$

if $n \ne m$. Would you need additional information to arrive at a solution for the $n = 8$ case?

5.18. Laguerre's differential equation is

$$x\frac{d^2y}{dx^2} + (1-x)\frac{dy}{dx} + ny = 0.$$

This ODE arises in the study of harmonic oscillators in quantum mechanics and solutions (for integer values of n) are the Laguerre polynomials: $L_0 = 1$, $L_1 = 1 - x$, $L_2 = 1 - 2x + (1/2)\,x^2$, $L_3 = 1 - 3x + (3/2)\,x^2 - (1/6)\,x^3$, and so on. Laguerre polynomials are orthogonal with respect to the weight function, $\exp(-x)$, on the interval, $0 \le x < \infty$, so, for example,

$$\int_0^\infty e^{-x}(1-x)\left(1 - 2x + \frac{1}{2}x^2\right)dx = 0.$$

Investigate the *method of Frobenius* (you might find *Advanced Engineering Mathematics*, by Kreyszig (1972), very useful) and see if you can use it to verify the solution for $n = 2$.

REFERENCES

Abramowitz, M. and I. A. Stegun. *Handbook of Mathematical Functions*, Dover, New York (1972).

Ayres, F. *Theory and Problems of Differential Equations*, McGraw-Hill, New York (1952).

Carslaw, H. S. and J. C. Jaeger. *Conduction of Heat in Solids*, 2nd edition, Oxford University Press, Oxford (1959).

Davis, H. T. *Introduction to Nonlinear Differential and Integral Equations*, Dover Publications, New York (1962).

Dwight, H. B. *Tables of Integrals and Other Mathematical Data*, 3rd edition, Macmillan, New York (1957).

Finlayson, B. A. *Nonlinear Analysis in Chemical Engineering*, McGraw-Hill, New York (1980).

Gray, A., Mathews, G. B., and T. M. MacRobert. *A Treatise on Bessel Functions and Their Applications to Physics*, 2nd edition, Macmillan, New York (1931) and reprinted by Dover (1966).

Kreyszig, E. *Advanced Engineering Mathematics*, 3rd edition, John Wiley & Sons, New York (1972).

Landau, L. D. and E. M. Lifshitz. *Fluid Mechanics*, Pergamon Press, London (1959).

Mickley, H. S., Sherwood, T. K., and C. E. Reed. *Applied Mathematics in Chemical Engineering*, 2nd edition, McGraw-Hill, New York (1957).

Milne-Thomson, L. M. *Jacobian Elliptic Function Tables*, Dover Publications, New York (1950).

Rosenhead, L. The Steady Two-Dimensional Radial Flow of Viscous Fluid Between Two Inclined Planes. *Proceedings of the Royal Society of London*, Series A, 175:436–467 (1940).

Selby, S. M., editor. *CRC Handbook of Tables for Mathematics*, revised 4th edition, CRC Press, Cleveland, OH (1975).

6

NUMERICAL SOLUTION OF ORDINARY DIFFERENTIAL EQUATIONS

AN ILLUSTRATIVE EXAMPLE

In engineering and the applied sciences, transient mass and energy balances arise frequently, often leading to ordinary differential equations (ODEs). Suppose, for example, we have a jacketed process vessel in which an exothermic chemical reaction may occur. The entering (feed) stream has a temperature, T_{in}; the well-mixed contents have temperature, T; and the steam used to heat the vessel has temperature, T_s. A verbal statement of the appropriate energy balance might appear:

$$[\text{Rate in}] - [\text{Rate out}] + [\text{Rate supplied}]$$
$$+ [\text{Rate of production}] = [\text{Accumulation}].$$

And written out symbolically, we would expect something like this:

$$\dot{M}_{in}C_p(T_{in} - T_{ref}) - \dot{M}_{out}C_p(T - T_{ref}) + UA(T_s - T)$$
$$+ |\Delta H_{rxn}|(-r_A)V = \rho V C_p \frac{dT}{dt}. \quad (6.1)$$

The rate at which the reactant species, A, is consumed is r_A. We will let the mass flow rates in and out be the same, and we set the reference temperature equal to the inlet (feed) temperature and divide by $\dot{M}C_p$:

$$-(T - T_{in}) + \frac{UA}{\dot{M}C_p}(T_s - T) + \frac{|\Delta H_{rxn}|V}{\dot{M}C_p}(-r_A) = \tau \frac{dT}{dt}. \quad (6.2)$$

Please note that every term in the equation has the dimension of temperature. The characteristic time, τ, that appears on the right-hand side is the total mass in the vessel divided by the mass flow rate, and it is the time constant for this system. This is a first-order ODE that will generally be coupled to a comparable mass balance on the reactant species, A. If the heat of reaction is negligibly small, then this equation may be written as

$$aT - b = -\tau \frac{dT}{dt}. \quad (6.3)$$

The result is separable and easily integrated:

$$-\frac{dt}{\tau} = \frac{dT}{aT - b} \rightarrow -\frac{t}{\tau} + C_1 = \frac{1}{a}\ln(aT - b). \quad (6.4)$$

Consequently, we find

$$T = \frac{C_1}{a}\exp\left(-\frac{at}{\tau}\right) + \frac{b}{a}. \quad (6.5)$$

The constant of integration is determined from the initial condition: at $t = 0$, $T = T_0$; therefore, $C_1 = aT_0 - b$. Let us now suppose that $T_0 = 100$, $\tau = 4$, $a = 1$, and $b = 40$:

$$T = 60\exp(-t/4) + 40. \quad (6.6)$$

Applied Mathematics for Science and Engineering, First Edition. Larry A. Glasgow.
© 2014 John Wiley & Sons, Inc. Published 2014 by John Wiley & Sons, Inc.

We see an exponential decrease from the initial temperature (100) to the ultimate value of 40.

Time, t	Temperature, T
0	100
½	92.9498
1	86.7289
2	76.3918
4	62.0728
8	48.1201
16	41.0989
32	40.0201

While this particular differential equation was extremely simple to solve analytically, that will not always be the case. For example, we might have a set of simultaneous differential equations or a nonlinear ODE for which no analytic solution is known. Therefore, it is entirely appropriate for us to ask if a solution for our elementary example could also be obtained numerically. In the next section, we contemplate the simplest possible numerical approach and revisit the chemical reactor problem.

THE EULER METHOD

Recall that the definition of the first derivative is given by

$$\left.\frac{dy}{dx}\right|_x = \lim_{\Delta x \to 0} \frac{y(x + \Delta x) - y(x)}{\Delta x}. \qquad (6.7)$$

Suppose we render this definition discrete; that is, we let Δx assume some finite (but hopefully small) value and rearrange the result:

$$y(x + \Delta x) \cong \Delta x \left(\frac{dy}{dx}\right)_x + y(x). \qquad (6.8)$$

Therefore, given an initial (or beginning) value for $y(x)$ and an expression for dy/dx, we can simply forward march in the x-direction, computing new values for y as we go. Let us try this for the preceding energy balance example. We begin by isolating dT/dt:

$$\frac{dT}{dt} = -\frac{1}{\tau}(aT - b) = -\frac{1}{4}(T - 40). \qquad (6.9)$$

Next, let $\Delta t = 1/2$ and proceed:

Time, t	Temperature, T
0	100
½	92.5000
1	85.9375
2	75.1709
4	60.6165
8	47.0840
16	40.8364
32	40.0117

We should graphically compare these values computed with the Euler method to those obtained previously from the analytic solution; Figure 6.1 will illuminate the principal shortcoming of the technique.

Because the Euler method is a straight-line, piecewise approximation, it cannot accurately follow a function with curvature unless we make Δx (or Δt) *very* small. In some extreme cases, it might have to be prohibitively small, leading to cumulative roundoff error. Nevertheless, the Euler algorithm is often useful because it is easy to understand and easy to implement. We will illustrate this last point with a slightly more complicated example.

We applied the Euler method to a simple, first-order ODE and we obtained reasonable (though not perfect) results as evidenced by Figure 6.1. Suppose, however, we are interested in the second-order equation:

$$\frac{d^2 y}{dx^2} + \frac{dy}{dx} - 6y = 0, \quad \text{with } y(0) = 1 \text{ and } y'(0) = 0. \qquad (6.10)$$

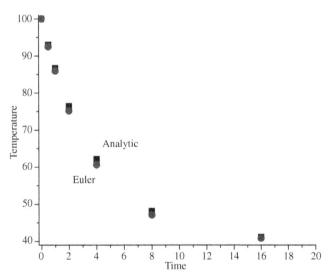

FIGURE 6.1. Comparison of the analytic and numerical (Euler method) solutions of the energy balance example. The discrepancies between the two are particularly apparent at $t = 2$, 4, and 8.

The corresponding analytic solution is $y = C_1 \exp(2x) + C_2 \exp(-3x)$, where $C_1 = 3/5$ and $C_2 = 2/5$. Now, we will find a numerical solution for this differential equation for values of x ranging from 0 to 0.5 using the Euler method. Let $y_1 = dy/dx$; therefore, $dy_1/dx = d^2y/dx^2 = -y_1 + 6y$. Of course, any nth order ODE can be written as n first-order equations. The essential part of the algorithm appears simply as

```
        x=0:dx=0.00390625:y=1:y1=0
20   REM *** find the new value for y
        y=dx*y1+y
     REM *** find the new value for y1
        y1=dx*(-y1+6*y)+y1
              x=x+dx
           PRINT x,y,y1
```

The initial values for x, y, and y' (which is y_1) allow us to find a new y that, in turn, is used to find a new y_1. In this case, we will begin by letting Δx be 1/32 (0.03125); we will carry out the calculation from $x = 0$ to $x = 1/4$. The analytic solution indicates that $y(x = 1/4) = 1.17818$.

Δx	$y(x = 1/4)$
1/32	1.158303
1/64	1.168356
1/128	1.173297
1/256	1.175746

These results show that by making Δx as small as 1/256 (or 0.00390625), we can get within 0.21% of the correct answer. For many practical calculations, this might be adequate; however, we can very easily encounter differential equations where the Euler scheme is not suitable. We will need a better tool, and one possibility is the modified Euler method (a self-starting, predictor–corrector approach).

Modified Euler Method

Let the differential equation of interest be

$$\frac{dy}{dx} = f(x, y), \qquad (6.11)$$

and we take $y(x = 0)$ to be a known value. A *predicted* value for $y(x = \Delta x)$ is found with the Euler method:

$$P = \Delta x f(x, y) + y(x = 0). \qquad (6.12)$$

This estimate for $y(x = \Delta x)$ is improved (corrected) by using the average slope:

$$C = \Delta x \left(\frac{y'(x = 0) + P(y')}{2} \right) + y(x = 0). \qquad (6.13)$$

The *corrected* value for y (which is C) can be taken back to $f(x, y)$ to get a better estimate of the new slope. This process can be repeated until a desired criterion is satisfied (until the change in $y(x = x + \Delta x)$ is sufficiently small). We can illustrate this process with an example that is easily solved analytically; suppose

$$\frac{dy}{dx} = 2y, \quad \text{with } y(x = 0) = 1. \qquad (6.14)$$

Of course, the solution is just $y = C_1 \exp(2x)$ with $C_1 = 1$. We set $\Delta x = 1/4$ and begin:

$$P = \tfrac{1}{4}(2) + 1 = 1.5$$

and, therefore,

$$P(y') = 2(1.5) = 3.$$

Accordingly,

$$C = \frac{1}{4}\left(\frac{2+3}{2} \right) + 1 = 1.625. \qquad (6.15)$$

Note that the correct value for $y(x = 1/4)$ is 1.64872. It is clear that our estimate for y has been improved significantly by the addition of the corrector step! Even so, far better methods have been developed, and we turn our attention to a technique that was originally proposed by Carl Runge.

RUNGE–KUTTA METHODS

Once again let us consider the differential equation

$$\frac{dy}{dx} = f(x, y). \qquad (6.16)$$

For the sake of making things a little more compact, we let $\Delta x = h$. A Runge–Kutta (RK) method is one in which we employ a series of calculations (to advance from x to $x + \Delta x$) of the form

$$k_0 = hf(x, y) \qquad (6.17)$$

$$k_1 = hf(x + h/2, y + k_0/2) \qquad (6.18)$$

$$k_2 = hf(x + h/2, y + k_1/2) \qquad (6.19)$$

$$k_3 = hf(x + h, y + k_2), \qquad (6.20)$$

with

$$y(x+h) = y(x) + \frac{1}{6}(k_0 + 2k_1 + 2k_2 + k_3). \quad (6.21)$$

The latter, in fact, is equivalent to a truncated Taylor series expansion for $y(x)$; see chapter 6 in James et al. (1977) for elaboration. Note that four evaluations for the ks are necessary to compute the new value for y; this is an illustration of the *fourth-order* RK algorithm, which has long been a standard tool for science and engineering. It is a forward-marching, *self-starting* technique that is easy to implement. All one needs is an initial value for $y(x)$. We can explore this process with the following differential equation:

$$\frac{dy}{dx} = f(x, y) = x^2 + y, \quad \text{with } y(x=1) = 1. \quad (6.22)$$

We choose $h = 1/2$ and find

$$k_0 = (0.5)(2) = 1 \quad (6.23)$$

$$k_1 = (0.5)(3.0625) = 1.53125 \quad (6.24)$$

$$k_2 = (0.5)(3.3281) = 1.66406 \quad (6.25)$$

$$k_3 = (0.5)(2.25 + 2.66406) = 2.45703. \quad (6.26)$$

Therefore, we find $y(x = 1.5) = 2.6413$. How does this compare with the analytic solution?

We can use the integrating factor technique to solve this equation, resulting in

$$y = -x^2 - 2(x+1) + C_1 / e^{-x}, \quad (6.27)$$

where $C_1 = 2.20728$. Therefore, $y(x = 1.5) = 2.64234$; the RK approach with $h = 1/2$ (*a single step*) produced an error of about 0.039%.

The procedure we just examined is a standard fourth-order RK method, and as we indicated, it is used extensively throughout the world for scientific calculations. Of course, one can construct and use RK routines of any order. We will demonstrate this by solving the same ODE using a *third-order* RK procedure. The necessary steps are shown as follows:

```
        #COMPILE EXE
#DIM ALL
    REM *** Example of 3rd order Runge-Kutta
        solution of an ODE
    GLOBAL H,XI,YI,XF,X,Y,K1,K2,K3,YNEW,XNEW,
        FXY,XX AS SINGLE
```

```
FUNCTION PBMAIN
    H=0.10:XI=1:YI=1:XF=1.5
    X=XI:Y=YI
80 REM *** CONTINUE
    GOSUB 200
        K1=H*FXY
        X=XI+H/2:Y=YI+K1/2
    GOSUB 200
        K2=H*FXY
        X=XI+H:Y=YI-K1+2*K2
    GOSUB 200
        K3=H*FXY
        YNEW=YI+1/6*(K1+4*K2+K3)
        XNEW=XI+H
        PRINT XNEW,YNEW
        XI=XNEW:YI=YNEW
    IF XI>XF THEN 190 ELSE 80
190 REM *** CONTINUE
        INPUT "Shall we continue?";XX
        IF XX>0 THEN 195
195 END
200 REM *** here is the differential equation
        FXY=X^2+Y
        RETURN
END FUNCTION
```

Once again, we will integrate out to $x = 1.5$, but this time, we will start with a step size of 0.10 then cut h in half repeatedly to see the effect on the computation.

h	$y(x = 1.5)$
0.10	2.644123
0.05	2.642912
0.025	2.642494
0.0125	2.642373
0.00625	2.642340

Note that the last value (using $h = 0.00625$) is essentially identical to the value obtained from the analytic solution. Of course, this third-order RK procedure requires only *three* k-function evaluations per step; if h is appropriately small, we can get quite satisfactory results for many ODEs.

Now we will examine a second-order ODE using an elementary example from mechanics where we utilize a force balance, setting $ma = \Sigma F$. Suppose we have a mass (m) that is suspended from a horizontal surface; it is attached to the surface with a spring and a viscous dashpot (a shock absorber). Position is represented by the variable y. A simple model for this situation is

$$m\frac{d^2y}{dt^2} + C\frac{dy}{dt} + Ky = F(t). \quad (6.28)$$

We divide by Hooke's constant (K), resulting in

$$\frac{m}{K}\frac{d^2y}{dt^2} + \frac{C}{K}\frac{dy}{dt} + y = \frac{F(t)}{K}. \qquad (6.29)$$

This equation is now in standard form, which can be written equivalently as

$$\tau^2\frac{d^2y}{dt^2} + 2\zeta\tau\frac{dy}{dt} + y = X(t). \qquad (6.30)$$

τ is the time constant for the system, ζ is the damping coefficient, and $X(t)$ is the forcing function. Let $\tau = 1$ and $\zeta = 0.3$ (underdamped); we will drive the system by giving the mass some initial displacement, and we can expect oscillatory behavior to result. *Please note the effect that the damping coefficient has on the solution of this differential equation; that is, look at* $b^2 - 4ac$, *or* $4\tau^2(\zeta^2 - 1)$ *alternatively.* Typical program logic for a fourth-order RK procedure is shown as follows:

```
        #COMPILE EXE
#DIM ALL
    REM ** 4th-order Runge-Kutta scheme for 2nd order
        ODE.
    GLOBAL xi,yi,y1i,h,xt,xbeg,pk1,x,y,y1,pk2,pk3,pk4,j,
        f,D,tau,Xforc,cc AS SINGLE
FUNCTION PBMAIN
    XI=0:YI=1.00:Y1I=0.0:H=.1:XT=18:tau=
        1:D=0.9
    OPEN "c:Dampp9.dat" FOR OUTPUT AS #1
    XBEG=XI:J=0:X=XI:Y=YI:Y1=Y1I
120 REM *** continue
    GOSUB 570
        PK1=H*F
        X=XI+H/2
        Y=YI+H/2*Y1I
        Y1=Y1I+PK1/2
    GOSUB 570
        PK2=H*F
        Y=Y+H/4*PK1
        Y1=Y1I+PK2/2
    GOSUB 570
        PK3=H*F
        X=XI+H
        Y=YI+H*Y1I+H/2*PK2
        Y1=Y1I+PK3
    GOSUB 570
        PK4=H*F
        J=J+1
        Y=YI+H*Y1I+H/6*(PK1+PK2+PK3)
        Y1=Y1I+(PK1+2*PK2+2*PK3+PK4)/6
            PRINT x,y,y1
            WRITE#1,x,y,y1
```

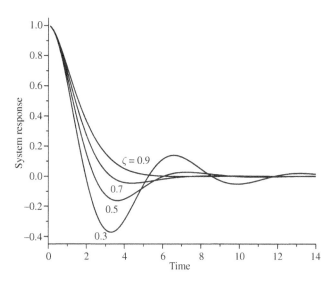

FIGURE 6.2. Effect of the damping coefficient, ζ, on the dynamic behavior of the initially displaced mass. Results are obtained from the fourth-order Runge–Kutta method with fixed step size. The significantly underdamped case, $\zeta = 0.3$, exhibits a large overshoot and a very oscillatory response.

```
    XI=X:YI=Y:Y1I=Y1
        IF X>XT THEN 380 ELSE 120
380 REM *** continue
        INPUT "Shall we continue?";CC
        IF cc>0 THEN 500
500 REM *** continue
        CLOSE:END
570 F=-2*D/tau*y1-y/tau^2+Xforc/tau^2
580 RETURN

END FUNCTION
```

Results obtained from this program are shown in Figure 6.2 for some values of the damping coefficient ranging from 0.3 to 0.9. Note how increasing the value for ζ makes the solution less oscillatory but more sluggish. This has important implications in examples from mechanics (such as an automobile suspension), but it is also significant to many other fields like automatic process control where a decaying oscillatory response to a forcing function (like a step function or an impulse) is quite common.

RK methods have been used for a very long time, and it is important that the reader understand that newer, more powerful techniques have been developed for solving ODEs. Indeed, Press et al. (1989) have offered the slightly severe view that "Runge–Kutta is what you use when you don't know any better ... or you have a trivial problem where computational efficiency is of no concern."

They advocate strongly for the Bulirsch–Stoer method, which we will discuss later in this chapter.

SIMULTANEOUS ORDINARY DIFFERENTIAL EQUATIONS

Often we must solve sets of simultaneous ODEs. This may come about from reducing an nth order ODE to an n first-order ODEs as we pointed out previously, or it could result from a problem in which several related phenomena are occurring at the same time. As an example of the latter, consider a series of chemical reactions carried out in a batch reactor:

$$A \to B \to C. \qquad (6.31)$$

A situation can occur in series reactions in which we might wish to maximize production of the intermediate product, B. Naturally, if this reaction sequence is allowed to proceed for a long period of time, we will get mostly C. The mass balances for this system are written as

$$\frac{d[A]}{dt} = -k_1[A], \qquad (6.32)$$

$$\frac{d[B]}{dt} = k_1[A] - k_2[B], \qquad (6.33)$$

and

$$\frac{d[C]}{dt} = k_2[B]. \qquad (6.34)$$

We assume that we begin the process with only species A present in the reactor. Note that if the rate constants are such that $k_2 \gg k_1$, it will be extremely difficult to obtain much B from the process. Consequently, we will focus our attention on the case in which the rate constants are comparable. We employ a fourth-order RK scheme designed specifically to handle sets of simultaneous ODEs. We let the initial concentration of A be 1, and set the reaction rate constants $k_1 = 0.85$ and $k_2 = 0.55$ (both with dimensions of reciprocal time). The results are shown in Figure 6.3; note that the maximum concentration of the desirable intermediate, B, occurs at approximately $t = 1.45$.

Some Potential Difficulties Illustrated

The focus of this section is simultaneous ODEs, and in the preceding example, we saw computational results for a series of chemical reactions carried out in a batch process. The model was elementary and the dynamic response of the system was easy to anticipate even with no prior exposure to problems in chemical kinetics. However, we may be required to seek solutions for more difficult cases—for example, ones that exhibit sharp fluctuations. In the example we are about to explore, we will look at oscillatory behavior associated with cellular cycles.

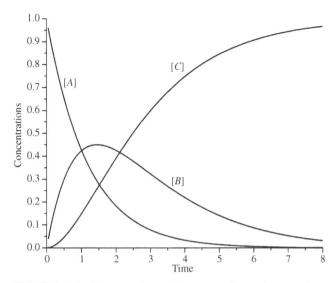

FIGURE 6.3. Concentrations of reactants for series reactions carried out in a batch reactor. The desirable intermediate, B, achieves a maximum concentration of about 0.45 at $t \approx 1.45$.

Recent work published by Ferrell et al. (2011) sought to understand the dynamic behavior of cell cycles driven by a protein circuit in which the activation of cyclin-dependent protein kinase (CDK1, represented by C) drives a cell into mitosis and the activation of the anaphase-promoting complex (APC, represented by A) leads the cell back out. Ferrell et al. note that proteins such as polo-like kinase 1 (Plk1, represented by P) may also play a role in the activation of A, and thus they directed their attention to a three-component system governed by the ODEs:

$$\frac{dC}{dt} = \frac{1}{10} - \frac{3CA^8}{\left(\frac{1}{2}\right)^8 + A^8}, \qquad (6.35)$$

$$\frac{dP}{dt} = 3(1-P)\frac{C^8}{\left(\frac{1}{2}\right)^8 + C^8} - P, \qquad (6.36)$$

and

$$\frac{dA}{dt} = 3(1-A)\frac{P^8}{\left(\frac{1}{2}\right)^8 + P^8} - A. \qquad (6.37)$$

These three ODEs are to be solved numerically with the initial conditions (for $t = 0$) $C = P = A = 0$. The problem is clearly nonlinear and if the analyst is unfamiliar with such systems, he/she may not have any idea what to expect. This is where the potential difficulty lies as we shall see. Suppose we decide to use the Euler method due to the ease with which it can be applied here. We start with a Δt of 0.0125 time units and plot the results in Figure 6.4.

The sharp changes in slope produced by this model are cause for concern. Had we had selected a larger Δt, say, $\Delta t = 0.4$ instead of 0.0125, the resulting activity of

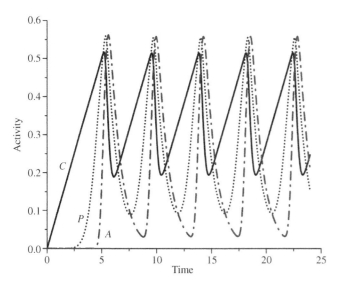

FIGURE 6.4. Oscillations in the cellular activity of CDK1 (*C*), APC (*A*), and Plk1 (*P*). These results are in *qualitative* agreement with Ferrell et al.

C (CDK1) would then have a *negative* (unphysical) value at $t = 20$ and the maximum value for *A* (APC) at $t = 19.6$ would be about 0.76. Neither is correct, of course. This example underscores the fact that we need to be vigilant when dealing with an unfamiliar system of nonlinear differential equations and that we should not be reluctant to employ a better solution technique for such problems even if the algorithm is a little more complicated. Indeed, if we selected the pedestrian fourth-order RK procedure, we could prepare a little comparison of the effect of step size on the calculations:

h	$C(t = 20)$	$A(t = 19.6)$
0.8	0.4402	0.019
0.4	0.2439	0.3021
0.2	0.2487	0.3004
0.1	0.2413	0.3174
0.01	0.2409	0.3186

Thus, it is clear that abrupt changes in activity produced by this model can be problematic for the numerical procedures that we have discussed so far. In the material to follow, we will examine some better tools for problems of this type. However, we should also observe that the dynamic behavior of systems similar to (this example of) cellular oscillation can often be very effectively revealed with *phase-space analysis*. In phase space, an oscillatory signal of sustained amplitude produces a limit cycle (a closed-loop trajectory). We will explore this technique later in this chapter and discover its value, particularly in the investigation of nonlinear systems. We will also have an opportunity in a student exercise at the end of the chapter to explore oscillating chemical reactions for the classic two-phase reactor problem of

Schmitz and Amundson (1963). This chemical reactor problem bears some (dynamic) similarity to the previous cellular oscillation example.

LIMITATIONS OF FIXED STEP-SIZE ALGORITHMS

The methods we have discussed so far in this chapter have at least one feature in common: All employ a fixed step size, that is, a constant value for h. This can produce an incorrect result for certain equation types; more importantly, the discrepancy can be very difficult to detect unless one has some prior knowledge of that particular problem. To illustrate this point, consider the elementary differential equation

$$\frac{dy}{dx} = xy^{3/2}, \quad \text{with } y(0) = 2. \tag{6.38}$$

Now suppose we want to compute the value of y for $x = 1.65$. The analytic solution for this problem can be found easily:

$$y = \left\{ -\frac{1}{2} \left(\frac{x^2}{2} - 1.41421 \right) \right\}^{-2}. \tag{6.39}$$

Consequently, $y(x = 1.65) = 1426.15$. Let us now use the *third-order* RK scheme described (and used) previously. We begin with $h = 0.05$; the computation produces $y(x = 1.65) = 1211.9$, which corresponds to an error of about 15%. If we were to proceed without prior knowledge of the problem or without a computed error estimate, this discrepancy would almost certainly remain undetected. And if we were to base a system design or perhaps a complex system model on such a result, we might blunder into a disaster. Fortunately, there are very simple means we can use to avoid such errors, and we will first describe here an approach developed by Bailey (1969). The underlying idea is straightforward: If the change in y (for $x = x + h$) is "small," we double the step size, and if the change in y is "large," we cut h in half. Therefore, if we encounter a region in which y is growing very rapidly (which is exactly what happens in our example), h can (and will) become *very* small. Note that there is a certain arbitrariness in play here—the analyst decides what changes are "large" or "small." The logic to be inserted into the RK code is just

```
REM *** BAILEY METHOD FOR VARIABLE
   STEP SIZE
      DELY=ABS(YI-YOLD)
         IF DELY<0.0005 THEN 180
      TESTY=ABS(DELY/YI)
         IF TESTY<0.002 THEN H=2*H
         IF TESTY>0.02 THEN H=H/2
REM *** CONTINUE
```

This simple change to the third-order RK procedure yields $y(x = 1.65) = 1434.24$, corresponding to an error of about 0.57%, which would often be tolerable for practical calculations. As an alternative, one might think that simply using a fixed step-size procedure would work if h were suitably small. We can test this hypothesis on eq. (6.38) with the fourth-order RK algorithm. We will start with $h = 0.05$ and use *double precision*:

h	$y(x = 1.65)$
0.05	1222.466
0.01	1423.938
0.005	1425.794
0.001	1425.961
0.0005	1425.961+

This is an important lesson: For some types of ODEs, fixed step-size methods such as the RK scheme used here, or Mathcad's *rkfixed*, will be at best computationally expensive and possibly quite ineffective.

In addition to the step-halving approach utilized earlier, there are other strategies that can be employed when *adaptive* step-size control is required. One very appealing alternative is the Runge–Kutta–Fehlberg (RKF) scheme, which is sometimes referred to as an *embedded* RK procedure since both fourth- and fifth-order estimates come from the same sequence of calculations. The essential idea is to compute both fourth- and fifth-order estimates then compare them to obtain an estimate of the local error. That error is then used to make a step-size adjustment if needed. The particular algorithm we will employ comes from Cash and Karp (1990). Once again, we will take

$$\frac{dy}{dx} = xy^{3/2}, \quad \text{with } y(x = 0) = 2, \quad (6.40)$$

where we are interested in $y(x = 1.65)$.

The required sequence of calculations follows:

$$k_1 = f(x_i, y_i) \quad (6.41)$$

$$k_2 = f\left(x_i + \frac{1}{5}h, y_i + \frac{1}{5}k_1 h\right) \quad (6.42)$$

$$k_3 = f\left(x_i + \frac{3}{10}h, y_i + \frac{3}{40}k_1 h + \frac{9}{40}k_2 h\right) \quad (6.43)$$

$$k_4 = f\left(x_i + \frac{3}{5}h, y_i + \frac{3}{10}k_1 h - \frac{9}{10}k_2 h + \frac{6}{5}k_3 h\right) \quad (6.44)$$

$$k_5 = f\left(x_i + h, y_i - \frac{11}{54}k_1 h + \frac{5}{2}k_2 h - \frac{70}{27}k_3 h + \frac{35}{27}k_4 h\right) \quad (6.45)$$

$$k_6 = f\left(x_i + \frac{7}{8}h, y_i + \frac{1631}{55,296}k_1 h + \frac{175}{512}k_2 h + \frac{575}{13,824}k_3 h + \frac{44,275}{110,592}k_4 h + \frac{253}{4096}k_5 h\right). \quad (6.46)$$

The fourth-order estimate is computed from

$$y_{i+1} = y_i + \left(\frac{37}{378}k_1 + \frac{250}{621}k_3 + \frac{125}{594}k_4 + \frac{512}{1771}k_6\right)h, \quad (6.47)$$

and the fifth-order estimate comes from

$$y_{i+1} = y_i + \left(\frac{2825}{27,648}k_1 + \frac{18,575}{48,384}k_3 + \frac{13,525}{55,296}k_4 + \frac{277}{14,336}k_5 + \frac{1}{4}k_6\right)h. \quad (6.48)$$

We will now use this technique for our example equation using constant step-size ($h = 0.05$), determining the local error from the difference between the two estimates at each step. These computations reveal that the error begins to grow objectionably when x exceeds about 1.2 (see Figure 6.5).

Now that we understand how the estimated error is behaving in this problem, we can make suitable changes to h. One possibility, suggested by Press et al. and discussed by Chapra and Canale (2002), is

$$h_{\text{revised}} = h\left|\frac{\varepsilon_{\text{crit}}}{\varepsilon}\right|^{0.25}, \quad (6.49)$$

where $\varepsilon_{\text{crit}}$ is the threshold (desired) accuracy level. Suppose we take $\varepsilon \sim$ *error*, and further, assume "typical" values for both εs:

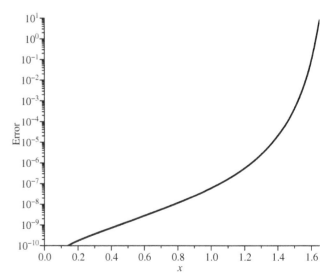

FIGURE 6.5. A measure of the local error for the equation $dy/dx = xy^{3/2}$, obtained from the Runge–Kutta–Fehlberg computation (fourth and fifth orders) using a fixed step size with $h = 0.05$.

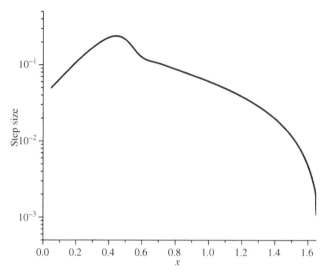

FIGURE 6.6. Behavior of the step size, h, using the Runge–Kutta–Fehlberg scheme with step-size adaptation and an initial h value of 0.05. With this technique, the computed value for $y(x = 1.65)$ is between 1426 and 1427, within about 0.03% of the correct value.

$$h_{\text{revised}} = h \left| \frac{10^{-7}}{10^{-6}} \right|^{0.25} = 0.5623h. \qquad (6.50)$$

Now we are in a position to return to our RKF routine with the addition of adaptive step-size control. We will track the impact of this modification on h as we approach $x = 1.65$ and provide the result in Figure 6.6.

The RKF procedure offers us a very effective way to implement adaptive step-size control, and when applied to nonlinear systems where little is known about the behavior of the dependent variable(s), it can save us a lot of grief. The principal advantage of RKF is that both fourth- and fifth-order estimates are obtained from the same sequence of calculations, providing us with a built-in estimate of error.

RICHARDSON EXTRAPOLATION

Suppose we take the two Taylor series expansions:

$$f(x+h) = f(x) + hf'(x) + \frac{h^2}{2}f''(x) + \frac{h^3}{6}f'''(x) + \cdots \qquad (6.51)$$

and

$$f(x-h) = f(x) - hf'(x) + \frac{h^2}{2}f''(x) - \frac{h^3}{6}f'''(x) + \cdots. \qquad (6.52)$$

We subtract the latter from the former:

$$f(x+h) - f(x-h) = 2hf'(x) + \frac{h^3}{3}f'''(x) + \cdots. \qquad (6.53)$$

Now we rearrange the equation to solve for the derivative, $f'(x)$:

$$f'(x) = \frac{f(x+h) - f(x-h)}{2h} - \frac{h^2}{6}f'''(x) - \frac{h^4}{120}f'''''(x) - \cdots. \qquad (6.54)$$

You may recognize that this expression—when truncated—is the second-order, central difference approximation for the first derivative. We rewrite this equation with a small modification, letting the difference approximation be represented by $\phi(h)$:

$$f'(x) = \phi(h) - \frac{h^2}{6}f'''(x) - \frac{h^4}{120}f'''''(x) - \cdots. \qquad (6.55)$$

Now we cut the interval in half; that is, we replace h with $h/2$:

$$f'(x) = \phi(h/2) - \frac{h^2}{24}f'''(x) - \frac{h^4}{1920}f'''''(x) - \cdots. \qquad (6.56)$$

We multiply this equation by 4 and then subtract the previous expression from it, isolating $f'(x)$; the truncated result is

$$f'(x) \cong \frac{4}{3}\phi\left(\frac{h}{2}\right) - \frac{1}{3}\phi(h). \qquad (6.57)$$

By adding one more function evaluation, we have significantly improved the quality of our estimate of $f'(x)$. Instead of the neglected term being of the order of h^2, it is now of order h^4! Let us see how well this works with a simple example: Take $f(x) = x^2 \sin x$, such that $f'(x) = 2x \sin x + x^2 \cos x$. We will use the latter to evaluate $f'(x = 1/2)$, which is 0.698821. Now we take $h = 1/8$ and evaluate the derivative using the second-order central difference expression $(f(5/8) - f(3/8))/(1/4)$; the result is 0.708188, which is about 1.3% too large. We repeat but add the additional evaluation at $h/2$ (1/16), finding $f'(1/2) = 0.698828$, which is about 0.001% too large. At the cost of one additional function evaluation, we have dramatically reduced the error in our estimate of the derivative. This is an extremely powerful procedure: We made two calculations (one using h and one using $h/2$), and then extrapolated to $h = 0$, obtaining an extremely accurate estimate for the derivative at $x = 1/2$. This is an example of the *Richardson extrapolation*, and it has been characterized by Press et al. as "turning straw into gold." It is so powerful that it has become a critical element in some newer ODE solvers, and we will see an example later in this chapter.

MULTISTEP METHODS

You should recall that forward-marching techniques (like Euler's method) take the slope of the function to be constant over some finite interval. We cannot expect this to work very well unless h is small—perhaps *very* small. But suppose we were able to evaluate the slope at multiple points; if we could fit a polynomial to these values, we could extrapolate this function. By accounting for the change in slope over the interval in this way, we might significantly improve our results. This is the idea behind the Adams' *four-point* formula (also known as the Adams–Bashforth method). The algorithm for this technique is

$$y_{n+1} = y_n + \frac{h}{24}(55y_n' - 59y_{n-1}' + 37y_{n-2}' - 9y_{n-3}'). \quad (6.58)$$

In the notation being used here, y_n' is simply $y'(x_n)$ or $(dy/dx)_{x=x_n}$. Note that three points ($n-1$, $n-2$, and $n-3$) *lie to the left* of the interval; the Adams–Bashforth method is not self-starting, so another technique must be used to get under way. The reader interested in the derivation of this formula (which is obtained easily through use of the Gregory–Newton backward interpolation) may consult Southworth and Deleeuw (1965). Now, suppose we have the differential equation

$$\frac{dy}{dx} = (x - x^2)y, \quad \text{with } y(0) = 2. \quad (6.59)$$

We wish to know the value of $y(x = 1)$; we begin by using the Euler method with $h = 0.2$, which results in $y(1) = 2.33944$ (the analytic result is 2.36272). Now let us apply the Adams–Bashforth method using Euler's technique to get started. But this time, we begin the calculation with $h = 0.002$ so that the initial values for the slopes (y_n') will be more accurate. In this case, the Adams–Bashforth method yields $y(1) = 2.36274$, corresponding to an error of about 0.0008%. The technique can be further improved by the addition of a corrector step (which can be applied repeatedly) to the predictor that we have used here. The corrector is

$$y_{n+1} = y_n + \frac{h}{24}(9y_{n+1}' + 19y_n' - 5y_{n-1}' + y_{n-2}'). \quad (6.60)$$

SPLIT BOUNDARY CONDITIONS

A situation that occurs regularly in problems concerning momentum, heat, and mass transfer in boundary layers is split boundary conditions; for example, some variables (or derivatives) might be known at $x = 0$ but others at some different value of the independent variable. Consider two-dimensional flow past a flat surface. The model developed by Prandtl and Blasius consists of the third-order non-linear ODE:

$$\frac{d^3 f}{d\eta^3} + \frac{1}{2} f \frac{d^2 f}{d\eta^2} = 0, \quad (6.61)$$

where $\eta = y(V/vx)^{1/2}$, $v_x = Vf'$, and $v_y = 1/2\sqrt{(\nu V/x)} \times (\eta f' - f)$. The plate surface corresponds to $\eta = 0$, where both v_x and v_y must be zero. Consequently, both $f(0)$ and $f'(0) = 0$. However, the required third boundary condition comes from the fact that v_x must correspond to the external potential flow outside the boundary layer (where the velocity is V): As $\eta \to \infty$, $v_x \to V$; since $f' = v_x/V$, it is clear that the needed condition is $f'(\eta \to \infty) = 1$. Now consider application of the RK method to this equation. Because RK is self-starting, we must have values for $f(0)$, $f'(0)$, and $f''(0)$ to solve the boundary-value problem, eq. (6.61). Of course, we can guess the value of the second derivative, integrate across the boundary layer, find the result for f', and then adjust $f''(0)$ accordingly. And the reader may want to use this approach to find that $f''(0) = 0.33206$. The results for this problem are given in Figure 6.7 to facilitate verification. The inefficient process employed here is equivalent to converting a boundary-value problem into *multiple* initial value problems; the process will take some time, and naturally one might wonder if there is a line of attack that would diminish our workload.

Hamming (1973) provided a useful example that illustrates one possible approach for split boundary conditions

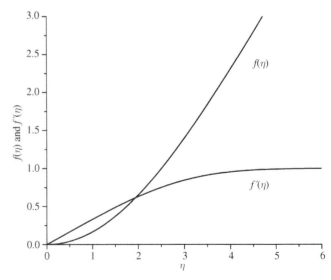

FIGURE 6.7. Solution of the Blasius equation for flow in the boundary layer on a flat plate. Note how $f'(\eta)$ approaches 1.0 asymptotically as η becomes large.

using a finite-difference approximation for the second derivative. Suppose we have the following second-order ODE:

$$\frac{d^2 y}{dx^2} = y + x, \quad \text{with } y(0) = 0 \text{ and } y(1) = 1. \quad (6.62)$$

We use the second-order central difference approximation for the left-hand side, resulting in

$$\frac{y_{n+1} - 2y_n + y_{n-1}}{h^2} \cong y_n + x_n. \quad (6.63)$$

Now suppose we break the interval (0–1) into four pieces—this gives us five nodal points with $n = 0$ and $n = 4$ corresponding to the ends of the interval. We apply the approximation at the three interior nodes, obtaining

$$\frac{y_2 - 2y_1 + y_0}{(1/4)^2} = y_1 + \frac{1}{4} \quad (6.64)$$

$$\frac{y_3 - 2y_2 + y_1}{(1/4)^2} = y_2 + \frac{1}{2} \quad (6.65)$$

$$\frac{y_4 - 2y_3 + y_2}{(1/4)^2} = y_3 + \frac{3}{4}. \quad (6.66)$$

Of course, $y_0 = 0$ and $y_4 = 1$; we have three equations and three unknown nodal values. It is easy to show that the interior nodal values (y_1, y_2, and y_3) are 0.180229, 0.387348, and 0.649927, respectively. How do these compare with the analytic solution (which is $y = 0.850918 \, (e^{+x} - e^{-x}) - x$)? The three corresponding values for y_n are 0.179905, 0.386819, and 0.649448. The results are encouraging and even more accurate results could be obtained by simply increasing the number of interior nodal points. But the equation we used for this example (eq. 6.62) certainly lent itself to the process we carried out, so we might want to think about some alternatives that are more broadly applicable. To underscore the point, the reader may want to consider how "easily" a finite-difference technique could be applied to the Blasius equation (eq. 6.61).

Shooting methods have been developed precisely to deal with this situation; let us consider the more general differential equation,

$$\frac{d^2 y}{dx^2} = f\left(x, y, \frac{dy}{dx}\right), \quad (6.67)$$

along with the boundary values $y(x = a) = y_a$ and $y(x = b) = y_b$. Now we contemplate the corresponding *initial value* problem:

$$\frac{d^2 y}{dx^2} = f\left(x, y, \frac{dy}{dx}\right), \quad \text{with } y(x = a) = ya$$
$$\text{and } dy/dx = s \text{ at } x = a. \quad (6.68)$$

It is clear that we would find the solution we want for the *boundary-value* problem if we could identify a zero for the function:

$$F = y(b, s) - y_b. \quad (6.69)$$

Naturally, each value selected for s will yield a corresponding F. Identifying the particular value we want can be done in many different ways including an iterative scheme using Newton's method of tangents (or Newton–Raphson). The superscripts refer to the iterate number in eq. (6.70):

$$s^{n+1} = s^n - \frac{F(s^n)}{F'(s^n)}. \quad (6.70)$$

Stoer and Bulirsch (1993) point out that the derivative of F appearing in the denominator is often replaced by a first-order forward difference approximation. They add that Δs must be chosen carefully—a value either too large or too small may either produce a derivative that is not sufficiently accurate or possibly lead to convergence problems. Moreover, as Hanna and Sandall (1995) note, it is clear that we now have two sources of error: We have error associated with the estimate of s and we have error associated with the numerical solution of the ODE itself.

We can better appreciate the shooting method through an example. Suppose we have the following second-order ODE:

$$\frac{d^2 y}{dx^2} + 0.06667 \frac{dy}{dx} + 0.11111 y = 0, \quad (6.71)$$

with $y(0) = 2$ and $y(3) = 10/3$. We plan to use an RK method, but we do not know $(dy/dx)_{x=0}$. Let our initial guess for s be 0.4 (with $\Delta s = 0.1$) and proceed:

$y'(0)$	$y(3)$	$F = y(3) - 10/3$
0.4	2.05325	−1.28008
0.5	2.28208	−1.05125

These values allow us to obtain a new estimate for s:

$$s^{n+1} = 0.4 - \frac{-1.28008}{2.2883} = 0.959402. \quad (6.72)$$

Now we return to our RK procedure but this time with $y'(0) = 0.959402$. The result is $y(3) = 3.333318$, which is pretty close to the correct value of 10/3. We could then refine our estimates if greater precision was required:

$y'(0)$	$y(3)$	$F = y(3) - 10/3$
0.959	3.332397	−0.000936
0.960	3.334686	+0.001353

The improved estimate for s is 0.959409, which yields $y(3) = 3.333333$. We have now seen the ease with which a shooting method can be employed; this will be far more efficient than solving *many* initial value problems in an effort to identify $y'(0)$ or $y''(0)$.

FINITE-DIFFERENCE METHODS

We demonstrated in the previous section how finite-difference approximations could be conveniently used to solve certain types of ODEs with split boundary conditions. We would like to further explore this topic, providing some amplification and treating a problem of some practical importance. Let us consider the steady-state conduction in a homogeneous medium in which thermal energy is absorbed (possibly by an endothermic chemical reaction). We will also assume that the energy absorption is directly proportional to local temperature; therefore,

$$\frac{d^2T}{dx^2} - \beta T = 0. \qquad (6.73)$$

The ends of the material are maintained at different temperatures: $T(x = 0) = T_0$, and $T(x = L) = T_L$. The continuum is discretized; that is, we place n equally spaced nodes in the interior, resulting in $n + 1$ intervals of length h (Δx), along with $n + 2$ nodes (including the boundaries). If we use a second-order central difference for the second derivative, our *algebraic* approximation for the differential equation appears as

$$\frac{T_{n+1} - 2T_n + T_{n-1}}{h^2} \cong \beta T_n. \qquad (6.74)$$

It is important to remember that we have introduced *truncation* error with this approximation; the Taylor series expansions used to construct this finite difference have been truncated and, accordingly, we have introduced an error that is of the same order as the neglected terms. Note that the coefficients for the three nodal points are

$n - 1$	n	$n + 1$
1	$-h^2\beta - 2$	1

When we apply this pattern to all of the interior nodes (the *interior* nodes begin with $n = 1$), we find

$n = 0$	$n = 1$	$n = 2$	$n = 3$	$n = 4\ldots$
1	$-h^2\beta - 2$	1	0	0
0	1	$-h^2\beta - 2$	1	0
0	0	1	$-h^2\beta - 2$	1
0	0	0	1	$-h^2\beta - 2$
0	0	0	0	$1\ldots$etc.

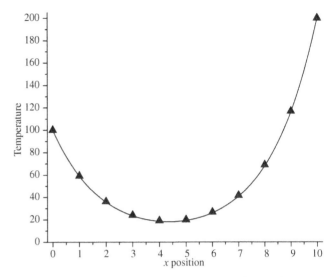

FIGURE 6.8. Comparison of the analytic solution (solid curve) for steady-state conduction in a slab with absorption of thermal energy with the finite-difference computation (filled triangles) using just nine interior nodes.

Please remember that the first column ($n = 0$) corresponds to the *left-hand boundary where the temperature is specified*. We see that the coefficient matrix is sparse—it contains many zeroes. Furthermore, it follows the *tridiagonal* form; this structure arises so frequently in applied mathematics that very efficient procedures for solving such sets of equations have been developed.

Let us give our conduction problem some quantitative definition and apply this technique to its solution: We have a slab of material 10 cm wide, with one edge ($x = 0$) maintained at $100°$ and the other at $200°$. Let $\beta = 0.3 \text{ cm}^{-2}$ and let $h = 1 \text{ cm}$ so we have nine interior nodes (we must solve a set of nine simultaneous algebraic equations). The analytic solution is just

$$T = 99.1655\exp(-\sqrt{0.3}x) + 0.8345\exp(+\sqrt{0.3}x), \qquad (6.75)$$

which allows for easy comparison. The results for the finite-difference computation are provided in Figure 6.8, along with the analytic solution; the largest discrepancy between the two is a little worse than 3%. We could easily cut the interval in half (to 0.5 cm) if we wanted improved accuracy. Indeed, if we do this, the worst-case deviation is reduced to less than about 0.9%.

STIFF DIFFERENTIAL EQUATIONS

A serious complication can arise in the numerical solution of ODEs when the equation(s) is *stiff*. When we say that an ODE (or a system of ODEs) is stiff, we are recognizing that the equations have widely varying characteristic values—they may differ by several orders of magnitude. In such cases, the stability of the calculation may become an issue. To examine the full range of system behavior, we may be forced to use a *very small* step size, yet the important (ultimate) outcome may take a *very long time* to develop. This is exactly the case for the following hypothetical system where the characteristic times differ by three orders of magnitude:

$$\frac{dy_1}{dt} = -y_1 \quad \text{and} \quad \frac{dy_2}{dt} = -1000y_2, \qquad (6.76)$$

where both $y_1(0) = 1$ and $y_2(0) = 1$.

Of course, both have elementary exponential solutions, making the difficulty quite apparent. Such a parametric "disparity" can be very difficult to reconcile. Furthermore, the techniques we have discussed so far in this chapter may not work very well or perhaps not at all. We will illustrate this point with the following example. Suppose we wish to solve the following test equation given by Hanna and Sandall (1995):

$$\frac{dy}{dx} = 50(x^2 - y), \quad \text{with } y(0) = 0. \qquad (6.77)$$

Our objective is to find y for $x = 2$; we select the third-order RK method we introduced previously and we will begin by using $h = 0.1$, and then successively halving h:

h	$y(x = 2)$
0.1	-2.78×10^{27}
0.05	-1.515×10^{35}
0.025	-1.783×10^{33}
0.0125	-1.601
0.00625	3.9208
0.003125	3.92075
0.0015625	3.92072
0.0010	$3.9208+$

The exact value for $y(2)$ is 3.9208. Though we were able to obtain something very close to that value by making h sufficiently small, the behavior we discovered above is disconcerting! It would be nice to be able to anticipate this problem and to assess its severity. One way to approach this difficulty is to define a "stiffness ratio" as follows:

$$S = |\lambda|(x_f - x_i), \qquad (6.78)$$

where x_f and x_i are the end and initial points of the integration, respectively. We can obtain a *local* estimate for λ from

$$\lambda = \frac{dy / dx}{y}. \qquad (6.79)$$

In this example, $dy/dx = 50(x^2 - y)$ and for $x = 2$, $y = 3.9208$; thus,

$$\lambda \cong \frac{50(4 - 3.9208)}{3.9208} = 1.01, \qquad (6.80)$$

resulting in $S = 2$. There are two important points to remember here: First, this is a very modest value for S, and second, for many nonlinear problems of interest, the value of S will vary considerably between the initial- and endpoints of the integration. In such cases, a *single-point* evaluation of S may not be very informative.

Gear's method is one of the best-known techniques for dealing with systems of stiff differential equations. In fact, Gear devised an example problem (unsteady, batch, chemical kinetics) that has been used extensively to test ODE solvers; it consists of three simultaneous equations:

$$\frac{dy_1}{dt} = -0.013y_1 - 1000y_1y_3 \qquad (6.81)$$

$$\frac{dy_2}{dt} = -2500y_2y_3 \qquad (6.82)$$

$$\frac{dy_3}{dt} = -0.013y_1 - 1000y_1y_3 - 2500y_2y_3, \qquad (6.83)$$

with $y_1(0) = 1$, $y_2(0) = 1$, and $y_3(0) = 0$. The range of integration is from $t = 0$ to $t = 50$. Suppose we begin our exploration of this example using the fourth-order RK procedure that we described previously. We begin by setting $h = 0.002$; to reach $t = 50$, about 100,000 k-function evaluations would be necessary. However, we never get to that point since the calculation self-destructs almost immediately. Even with $h = 0.001$, we find negative (unphysical) values for y_3. For the fourth-order RK scheme (with fixed step size) to work, it is necessary that h be less than about 0.0005. The results from such a computation are shown in Figure 6.9.

Gear (1971) developed a fourth-order backward differentiation formula (BDF) method specifically to deal with stiff ODEs, and it has been widely applied in this context. BDF methods will be discussed in detail in the next section.

Backward Differentiation Formula (BDF) Methods

BDF methods have been widely used for the solution of stiff differential equations, particularly since the publication of Gear's book in 1971. We begin our discussion of these techniques with an elementary illustration. Suppose that the differential equation to be solved is

$$\frac{dy}{dx} = f(x, y). \qquad (6.84)$$

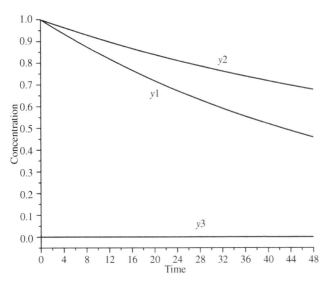

h	$y(x = 8)$
0.5	134.696
0.25	271.482
0.125	466.908
0.0625	431.449
0.03125	109.204
0.015625	69.176
0.007813	60.645
0.003906	57.370
0.001953	55.908
0.000977	55.211
0.000488	54.876
0.000244	54.652
0.000122	54.540

FIGURE 6.9. Computed results for Gear's example (chemical kinetics problem). The concentration for the third species, y_3, is very small but *not* zero.

We formulate an Euler algorithm (but one that is *implicit*):

$$y_{i+1} = y_i + h\frac{dy_{i+1}}{dx} = y_i + hf(x_{i+1}, y_{i+1}). \quad (6.85)$$

What distinguishes this technique from those we have seen previously is that the functional value that we wish to calculate (y_{i+1}) appears on *both sides of the equation*. Moreover, many of the problems for which this method is appropriate will be nonlinear; therefore, we will almost certainly need to solve nonlinear algebraic equations. But the enhanced stability of such methods for stiff ODEs will make the extra pain worthwhile.

Let us apply this technique to a specific example. Suppose we have

$$\frac{dy}{dx} = 0.01xy^2, \quad \text{with } y(x = 1) = 3. \quad (6.86)$$

We want to know how y behaves as $x \rightarrow 8$. The analytic solution is simply

$$y = \frac{1}{0.338333 - 0.005x^2}, \quad (6.87)$$

so $y(x = 8) = 54.5464$. We will now use the implicit Euler method starting with $h = 1/2$ to see how well the technique performs for this example (eq. 6.86); we must keep in mind, though, that this is a first-order algorithm and accuracy may be poor.

By making h sufficiently small, we were able to obtain a reasonable result. However, it is clear that a higher-order BDF method is needed here. The generalized algorithm for order n can be written as

$$a_0 y_{i+1} + a_1 y_i + \cdots + a_n y_{i-n+1} = hf(x_{i+1}, y_{i+1}). \quad (6.88)$$

The coefficients required for this equation have been compiled by Lee and Schiesser (2004), and their table is presented in modified form as follows:

n	a_0	a_1	a_2	a_3	a_4
1	1	−1			
2	3/2	−2	1/2		
3	11/6	−3	3/2	−1/3	
4	25/12	−4	3	−4/3	1/4
5	137/60	−5	−10/3	5/4	−1/5

Therefore, if we wished to employ the third-order BDF method, we would use

$$\frac{11}{6}y_{i+1} - 3y_i + \frac{3}{2}y_{i-1} - \frac{1}{3}y_{i-2} = hf(x_{i+1}, y_{i+1}). \quad (6.89)$$

Since this algorithm requires two previous values for y, it is not self-starting and some initial calculations will have to be made with another method.

BULIRSCH–STOER METHOD

We previously introduced the Richardson extrapolation and we saw what a powerful tool it could be. Bulirsch and Stoer recognized that it might be used as the basis for an extremely efficient method for solving ODEs; the technique they developed has garnered some very enthusiastic advocates (see, e.g., Press et al., 1989, p. 563). For problems in which it is essential that the computational effort be minimized, Bulirsch–Stoer is definitely worth consideration.

The principal idea is a simple one: Suppose we have a differential equation $dy/dx = f(x, y)$ we wish to solve from some initial $x = 0$, to $x = L$. The function y is computed using a midpoint method across a *large* interval initially using two steps. The calculation is carried out again, but this time using four steps. We now have a function $f(h)$ from the two different values of h that can be fit to an analytic form. An extrapolation is then attempted to $h = 0$ (corresponding to an infinite number of steps); if the estimated error is deemed satisfactory, we proceed. If not, we increase the number of steps (to 6, then 8, then 12, etc.), and after each increase, we try to the extrapolation to $h = 0$ again. The usual sequence of step number is 2, 4, 6, 8, 12, 16, 24, 32, 48, 64, 96, and so on. Press et al. note that a more *expected sequence* where we double the number of steps with each trial will cause h to become too small too rapidly.

We will treat an elementary example to demonstrate the Bulirsch–Stoer method. Consider the differential equation

$$\frac{dy}{dx} = xy^2, \quad \text{with } y(x = 0) = 2. \tag{6.90}$$

We are interested in $y(x = 0.975)$; since the analytic solution is $y = 2/(1 - x^2)$, $y(0.975) = 40.5063$. Let us use the modified midpoint method (which is sometimes referred to as Gragg's method) with $h = H/n_s$, where H is the interval we wish to cover (which is 0.975) and n_s is even (we start with $n_s = 4$):

$$y_{ep} = 6.159404.$$

We repeat the process, but with eight steps rather than four:

$$y_{ep} = 10.258197.$$

Now we use the two endpoint (*ep*) estimates for a polynomial extrapolation:

$$y(x = 0.975) \cong \frac{4y_8 - y_4}{3}$$
$$= \frac{(4)(10.258197) - (6.159404)}{3} = 11.6245. \tag{6.91}$$

It is clear that we are not there yet. We continue, but this time with 16, 32, 64, 128,... steps (we will double the number of steps each time to make use of the preceeding polynomial extrapolation). The resulting succession of estimates takes the form

18.4863, 26.9801, 34.5736, 38.8291,
40.2004, 40.4649, and 40.5014.

Now notice how the error for our estimate has diminished through this sequence:

54.36%, 33.39%, 14.64%, 4.14%,
0.755%, 0.102%, and 0.012%.

The very last estimate is from the 512 and 1024 step pair. As you can see, this gets us within about 0.012% of the correct value. We would need about 2000 steps with a third-order RK scheme to get within the same error window.

The Bulirsch–Stoer method is also available with Mathcad™ (*bulstoer*) and we illustrate its application using eq. (6.90). Note that there are seven parameters in the parameter list (unlike *rkfixed* which has only five).

Mathcad™ Implementation of Bulirsch–Stoer Method for eq. (6.90)

Solve $dy/dx = xy^2$ with $y(x = 0) = 2$.

$$y_0 := 2$$

$$D(x, y) := x(y_0)^2$$

$$Z := \text{bulstoer}(y, 0, 0.975, 0.00001, D, 10, 0.000001)$$

$$Z = \begin{pmatrix} 0 & 2 \\ 9.75 \times 10^{-3} & 2 \\ 0.107 & 2.023 \\ 0.61 & 3.183 \\ 0.705 & 3.973 \\ 0.817 & 6.006 \\ 0.927 & 14.29 \\ 0.975 & 40.503 \end{pmatrix}$$

The fourth parameter controls the accuracy of the solution, and in this case, we have set its value to 1×10^{-5}; note that the value reported for $y(x = 0.975)$ is 40.503, which corresponds to an error of about 0.008%.

PHASE SPACE

Among the problems that arise when an analyst must work with a nonlinear differential equation (or model) is the fact that he/she may not have any idea what form the dynamic behavior of the system will take. This makes it difficult to interpret a numerical result, and extremely difficult to detect errors when they occur. Furthermore, an output stream of numbers appearing on screen (or written to a file) does not provide much feedback—it might be nearly impossible to detect a periodicity, or a lack of periodic behavior from such output. This is exactly the kind of situation where a phase-space analysis can be useful. Our strategy is to construct a *system trajectory* by cross plotting dependent variables and

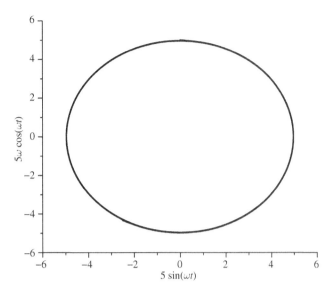

FIGURE 6.10. The system trajectory for this sinusoid is a closed path or a limit cycle. The amplitude of the oscillation is neither increasing nor decreasing. Motion on this limit cycle is clockwise.

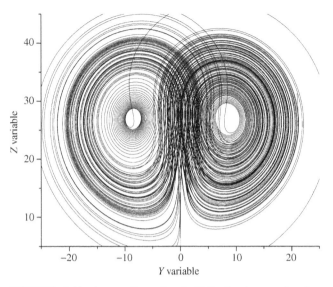

FIGURE 6.11. A two-dimensional (Y-Z plane) portrait of a *strange attractor* from the Lorenz model with $r = 28$.

removing time from the visualization of results. For systems of one, two, or three dimensions, this is straightforward as we shall see.

Consider the simple sinusoid $y(t) = A \sin \omega t$, with the derivative $y'(t) = \omega A \cos \omega t$. Of course, if we plot $y(t)$, we know exactly what to expect. But suppose instead that we plot corresponding values of $y(t)$ and the derivative $y'(t)$ as time evolves; we will see a *closed* system trajectory in the form of what is called a *limit cycle*. We will select $A = 5$ and $\omega = 1$ for this illustration.

If the amplitude of the sinusoid was increasing with time (unstable behavior), then the form we would see in Figure 6.10 would be an outward-directed spiral. If the sinusoid was decaying, for example, if $y(t) = A \exp(-\beta t) \sin(\omega t)$, then we would expect an inward-directed spiral that would approach 0, 0 on the phase plane as time became large. Of course, what we really want to know is what phase-space analysis can do for us in the context of challenging nonlinear differential equations.

In 1963, Edward Lorenz published an extremely important paper entitled "Deterministic Nonperiodic Flow" in the *Journal of Atmospheric Sciences*. Lorenz set out to develop the simplest possible model for atmospheric phenomena, accounting for the intensity of convective motion (X), the temperature difference between rising and falling currents (Y), and deviation of the vertical temperature profile from linearity (Z). The resulting set of ODEs can be written as

$$\frac{dX}{dt} = \text{Pr}(Y - X), \quad \frac{dY}{dt} = -XZ + rX - Y,$$

$$\text{and} \quad \frac{dZ}{dt} = XY - bZ. \tag{6.92a,b,c}$$

We will assume the following values for the parameters that appear in eq. (6.92a, b, c): $\text{Pr} = 10$, $r = 28$, and $b = 8/3$. For initial conditions (X, Y, Z), we select (0, 1, 0), then we will obtain the projected system trajectory (on the Y-Z plane) by numerical solution of the differential equations. The two-dimensional cut (from the three-dimensional system) shown in Figure 6.11 is a "portrait" of a *strange attractor*. For a simple mechanical system that oscillates with decaying amplitude, the phase-space trajectory (two-dimensional) as we observed previously will be an inward spiral—this is characteristic of dissipative systems. The point in phase space to which the trajectory is drawn is called an "attractor." If a frictionless system oscillates with constant amplitude, the phase-space portrait will be an ellipse (a limit cycle as we saw in Figure 6.10); such systems are said to be conservative because the phase "volume" remains constant.

What the two-dimensional cut provided in Figure 6.11 *cannot* reveal is that no point in phase space is *ever* revisited. Thus, the Lorenz model—though fully deterministic—is nonperiodic. The implications are staggering and it would be far more difficult to comprehend the behavior of this system if we were trying to do so from a table of numbers, say, $Y(t)$. This is a case where phase-space analysis is invaluable.

We will conclude our discussion of the utility of phase space with a final example adopted from LaSalle's contribution in *Proceedings of Symposia in Applied Mathematics, Volume XIII, Hydrodynamic Instability* (1962). Consider the deceptively simple (but nonlinear) ODE:

$$\frac{d^2 y}{dt^2} + a \frac{dy}{dt} + 2y + 3y^2 = 0, \tag{6.93}$$

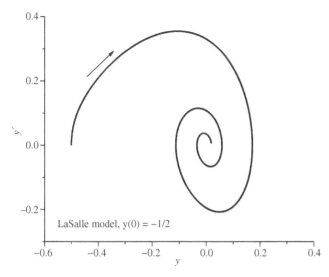

FIGURE 6.12. Phase plane for the LaSalle model with $y(0) = -1/2$. Note that the system trajectory is approaching the equilibrium point located at $y = 0$.

with $a = 1/2$. We will solve this ODE numerically beginning with $y(0) = -1/2$ and $y'(0) = 0$. This system is stable and the phase-space portrait reveals an inward spiral as shown in Figure 6.12.

The LaSalle model (eq. 6.93) becomes especially interesting when we change the initial condition to explore the threshold of instability; for example, suppose we first take $y(0) = +0.562866$ and then $y(0) = +0.562867$ (we maintain the zero initial value for the derivative both times) and solve eq. (6.90) for large values for t. In these cases, the phase-plane portraits are very different from that depicted in Figure 6.12; in fact, the dynamic behavior in the second case is unstable. We will explore this model further in a student exercise at the end of the chapter.

SUMMARY

We have said very little about the error inherent in the numerical solutions of differential equations. This is because the general topic is too broad and too complex for the space available here. But there are some elementary observations we should make. For example, we need to draw the reader's attention to the difference between round-off error and truncation error. Roundoff error is hardware-dependent; that is, it is a consequence of the precision of the computing device being used. For example, a calculator might report 2/3 as 0.666667. Truncation error, on the other hand, is the result of neglected terms in the construction of the algorithm. For example, if we truncate a Taylor series expansion, the resulting error will be of the order of the first neglected term.

And there is another aspect to this discussion that is crucial: numerical stability (sometimes referred to as *weak* stability in the literature). We have primarily talked about choosing the step size in terms of the accuracy of the resulting solution. And often—particularly for nondemanding problems—that is entirely appropriate. In the case of certain ODEs, however, step size is chosen not for accuracy but to ensure *stability* of the computation. It is this feature of the numerical solution of ODEs that we wish to discuss briefly here. For additional detail, consult Lee and Schiesser (2004), and for greater mathematical rigor, see Hairer and Wanner (1996).

Consider the elementary differential equation:

$$\frac{dy}{dx} = -\beta y. \tag{6.94}$$

Of course the solution has an exponential form, $y/y_0 = \exp(-\beta x)$. Now suppose we wished to use the Euler method for this equation beginning from $x = 0$; the first step would look like this:

$$y(\Delta x) \cong -\Delta x \beta y(0) + y(0) = y(0)(1 - \Delta x \beta), \tag{6.95}$$

and the second step,

$$y(2\Delta x) \cong -\Delta x \beta y(\Delta x) + y(\Delta x) = y(\Delta x)(1 - \Delta x \beta). \tag{6.96}$$

Note that the right-hand side of the latter equation can be written equivalently as

$$y(2\Delta x) = y(0)(1 - \Delta x \beta)^2, \tag{6.97}$$

and in general, for n-steps,

$$y(n\Delta x) = y(0)(1 - \Delta x \beta)^n. \tag{6.98}$$

If Δx is selected such that the product, $\Delta x \beta = 2$, then the procedure results in an oscillation between $-y(0)$ and $+y(0)$. Even worse, if $\Delta x \beta > 2$, the oscillations grow in amplitude; for example, for $\Delta x \beta = 4$, we get the sequence $-3y(0)$, $+9y(0)$, $-27y(0)$, and so on. On the other hand, if $\Delta x \beta < 1$, we at least get qualitatively correct behavior where successive steps reveal diminishing y, which is in accord with the *decaying* exponential solution. Setting $\Delta x \beta = 4/5$ produces the sequence 1/5, 1/25, 1/125, 1/625,..., each multiplied by $y(0)$, of course. It is clear, therefore, that the explicit Euler method will require that $|\Delta x \beta| < 2$, although for the sake of accuracy, the product $\Delta x \beta$ will usually be much smaller than this.

In the case of explicit RK methods, a stability assessment can be carried out in a similar manner. We take the model equation (sometimes referred to as the Dahlquist test equation):

$$\frac{dy}{dt} = \lambda y, \quad \text{with } y(0) = y_0, \qquad (6.99)$$

and let $z = h\lambda$. The solution for the test equation is exponential, and for a single step forward, we write it as $\exp(\lambda t) = e^z$. Therefore, a stability function can be written (for the fourth-order RK procedure in this case) as

$$1 + \frac{z}{1!} + \frac{z^2}{2!} + \frac{z^3}{3!} + \frac{z^4}{4!}. \qquad (6.100)$$

For stability of the numerical calculation, it is necessary that $|y_{i+1}/y_i| \leq 1$, so the stability function given earlier *is set equal to* 1. This determines the limiting value for z (or $h\lambda$), which is -2.785 for the fourth-order RK. Lambert (1973) shows that, for the third-order RK procedure, the limiting value is -2.51. The implication, of course, is that the explicit RK methods are only *slightly more stable* than the Euler technique; that is, the *limit* for the magnitude of $h\lambda$ is only a bit larger. Complete stability domains are given graphically in Hairer and Wanner (1996), which permit easy comparison of different solution techniques. Because the elementary explicit techniques have limited stability domains, they are not suitable for stiff problems; in these cases, one must turn to an implicit method (like BDF) as we noted previously.

Although BDF methods are not self-starting, their enhanced stability has made them very popular for the solution of stiff ODEs. Lee and Schiesser (2004) summarize the stability properties of BDF methods in chapter 1 of their book, and they note that BDF algorithms through the sixth order are unconditionally stable along the negative real axis—clearly superior in this regard to the Euler and RK techniques considered earlier. Lee and Schiesseer also point out that there are many very high-quality codes available for the solution of ODEs, both within commercial software packages and from the public domain. For the latter, a good starting point is the Netlib Repository at http://www.netlib.org/index.html.

PROBLEMS

6.1. Show that the analytic solution for the ODE, $dy/dx = x - (y/x)$ is $y = (x^2/3) + (C/x)$. Then, given that $y(x = 2) = 2$, use the modified Euler method and find $y(x = 4)$.

6.2. Consider the second-order differential equation,

$$16\frac{d^2y}{dt^2} + 5\frac{dy}{dt} + y = x(t).$$

Initially, both y and dy/dt are zero. If $x(t) = 1/(1 + t)$, what is the maximum value attained by y, and when does that occur?

6.3. A tank with a capacity of 1900 ga (the tank diameter is 7 ft) is initially half-full. Water enters the tank at a constant rate of 85 gpm; it also drains from the bottom of the tank through a round orifice. The velocity through the orifice is given (approximately) by Torricelli's theorem: $V_0 = \sqrt{2gh}$, where h is the depth of water above the hole. The area of that orifice is 0.01 ft. Will the tank overflow? Is so, when? If not, when will the maximum depth be attained?

6.4. We want to examine the dynamic behavior of two populations in conflict, coyotes and rabbits (this is a subset of Volterra's problem but with heredity neglected). We will initiate the calculation with 100 rabbits (n_2) and three coyotes (n_1). Assume the governing equations are

$$\frac{dn_1}{dt} = an_1n_2 - bn_1 \quad \text{and} \quad \frac{dn_2}{dt} = cn_2{}^2 - dn_1n_2.$$

Solve the two simultaneous equations and prepare a phase-space portrait by cross plotting n_1 and n_2. The values for the constants a, b, c, and d are 0.3, 0.2, 0.25, and 0.7, respectively. Is extinction of a species possible in this problem? If heredity factors are added to this problem, exactly how will the nature of the problem be changed?

6.5. In a misguided attempt to celebrate the New Year, Eric fires his 9-mm pistol vertically into the air. The 115-grain bullet leaves the barrel with an initial velocity of 1225 ft/s. Assume the drag acting on the projectile is given by $F = AKf$, where $A = \pi R^2$, $K = \frac{1}{2}\rho V^2$, and the drag coefficient, f, is assumed to be constant at 0.5. When will the bullet strike the ground, and what will the velocity be at impact? Is it possible that the returning projectile could be lethal?

6.6. The driven pendulum has been the focus of intensive investigation because of the possibility of chaotic behavior. Begin your analysis with the elementary, linearized damped pendulum with the equation of motion:

$$\frac{d^2\theta}{dt^2} + \frac{d\theta}{dt} + \theta = 0.$$

Solve this equation (giving the pendulum an initial disturbance) and prepare a plot of the system trajectory by cross plotting θ with its derivative. How does this system behave dynamically? Now, assume the three governing equations for the *driven pendulum* have the form

$$\frac{d\omega}{dt} = -\frac{\omega}{q} - \sin\theta + g\cos\phi$$

$$\frac{d\theta}{dt} = \omega$$

$$\frac{d\phi}{dt} = \omega_D.$$

We want to solve this set of equations using $q = 2$, $\omega_D = 2/3$, and $g = 1.0$, and once again we would like to construct the system trajectory (using ω and θ) as we suspect that chaotic behavior may be possible. If you need further help with this problem, see Baker and Gollub (1990).

6.7. Begin this problem by consulting the work of Durham et al. (1964) entitled *Study of Methods for the Numerical Solution of ODEs*. On page 104, they describe a model for a restricted three-body problem:

$$\frac{d^2 x}{dt^2} = x + 2\frac{dy}{dt} - \mu' \frac{x+\mu}{((x+\mu)^2 + y^2)^{3/2}} - \mu \frac{x-\mu'}{((x-\mu')^2 + y^2)^{3/2}}$$

and

$$\frac{d^2 y}{dt^2} = y - 2\frac{dx}{dt} - \mu' \frac{y}{((x+\mu)^2 + y^2)^{3/2}} - \mu \frac{y}{((x-\mu')^2 + y^2)^{3/2}},$$

where $\mu' = 1 - \mu$. We want to solve this fourth-order system. The initial conditions are $x(0) = 0.994$, $y(0) = 0$, $dx/dt = 0$ for $t = 0$, and $dy/dt = -2.03173262955733683566$ for $t = 0$. The parameter μ is 0.012277471. The period for the motion is 11.12434033726608513507, and the time step will need to be very small (and probably variable).

6.8. Hiemenz stagnation flow is governed by the nonlinear third-order ODE:

$$\nu f''' = f'^2 - ff'' - a^2,$$

where ν is the kinematic viscosity of the fluid and a is the strength of the potential flow approaching the flat surface. Assume the fluid is water and that $a = 5$. Find a solution for this problem given that $f(0) = 0$, $f'(0) = 0$, and $f'(\infty) = a$.

6.9. One form of Bessel's differential equation is

$$\frac{d^2 y}{dx^2} + \frac{1}{x}\frac{dy}{dx} - k^2 y = 0.$$

This might describe, for example, the disappearance of a reactant species in an infinitely long cylinder (catalyst pellet). It is obvious that there is a regular singular point at $x = 0$. The analytic solution for this problem is

$$y = C_1 I_0(kx) + C_2 K_0(kx).$$

We want to solve this equation numerically, integrating from $x = 0$ to $x = 1$ given that the concentration is finite at

the center and equal to a surface value (say, 1) at $x = 1$. Take $k = 2.325$.

6.10. Consider the differential equation, $dy/dx = 3x + y^2$. We know that $y(x = 0) = 1$, and we want to find an estimate of $y(x = 0.1)$. In particular, Ayres (1952) gives a value for this problem of 1.12725, and we want to know how accurate this value is.

6.11. A steel cable with a per-unit-length weight of W (pounds per foot) hangs between two supports. The deflection is described by the differential equation

$$\frac{d^2 y}{dx^2} = \frac{W}{h}\sqrt{1 + \left(\frac{dy}{dx}\right)^2}.$$

The shape assumed by a suspended cable is a *catenary*, which can be described with the hyperbolic cosine (cosh). At each support, of course, $y(x = -L$ and $x = L) = 0$. Solve the differential equation numerically, assuming $W/h = 2$ and $L = 10$.

6.12. The Leveque equation describes the temperature distribution for an inlet flow past a heated wall. The x-axis corresponds to the wall and the y-coordinate extends into the fluid phase. By assuming that the velocity distribution in proximity to the stationary wall is linear, $v_x = cy$, the following differential equation can be developed:

$$\frac{d^2 T}{d\eta^2} + 3\eta^2 \frac{dT}{d\eta} = 0,$$

where

$$\eta = y\left(\frac{c}{9\alpha x}\right)^{1/3}.$$

Of course, α is the thermal diffusivity of the fluid (assume that it is water, $\alpha \cong 0.00155\,\text{cm}^2/\text{s}$). Given that $T(\eta = 0) = T_w$, the temperature of the heated wall, and that $T(\eta \to \infty) = T_\infty$, solve this second-order equation numerically and compare your results with the analytic solution for $\eta = 0.3$, 0.5, 0.7, 0.9, and 1.10. Let $T_w = 50°$ and $T_\infty = 25°$.

6.13. The Rayleigh–Plesset equation describes the oscillatory behavior of the interface of a disturbed spherical gas bubble immersed in a liquid:

$$\frac{P_i - P_\infty}{\rho} = R\frac{d^2 R}{dt^2} + \frac{3}{2}\left(\frac{dR}{dt}\right)^2 + \frac{4\nu}{R}\frac{dR}{dt} + \frac{2\sigma}{\rho R}.$$

R, of course, is the bubble radius. This differential equation is notoriously stiff and the term that includes the kinematic viscosity (ν) is often small and therefore frequently neglected.

We are interested in the case in which the spherical gas bubble is subjected to an instantaneous increase in external pressure. This should result in compression and rebound—repeatedly. Use Borotnikova and Soloukhin (1964) as a guide and solve the Rayleigh–Plesset equation for this case.

6.14. Use the Bulirsch–Stoer method to solve the differential equation

$$\frac{dy}{dx} = x^2 y^3,$$

with $y(x = 0) = 3$, and find estimates for $y(x = 0.5475)$.

6.15. Phase-space analysis is a powerful tool for exploring the dynamic behavior of complex systems (with a limited number of dimensions). It will allow us to assess stability and to identify any attractors should they exist. We wish to consider the nonlinear LaSalle model:

$$\frac{d^2 y}{dt^2} + a\frac{dy}{dt} + 2y + 3y^2 = 0,$$

with $a = 1/2$. There is an obvious equilibrium point at $y = 0$, obtained by setting the time derivatives equal to zero. There is another located at $y = -2/3$. Solve this equation numerically using different initial values for $y(t = 0)$, letting $y'(t = 0) = 0$ in every case. Draw a sufficient number of trajectories on the phase plane such that the delineation between stable and unstable regions is apparent. We are particularly interested in trajectories near the second equilibrium point (at $y = -2/3$). Can you find any trajectory that actually terminates at $y = -2/3$? How would you describe that point? What is a saddle point? A composite example appears in Figure 6.13 to guide you in your work (y is plotted on the horizontal axis and y' on the vertical axis).

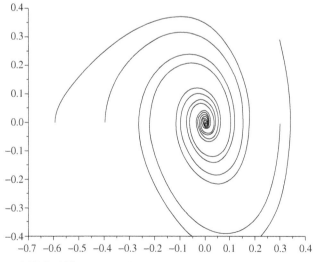

FIGURE 6.13. A few trajectories plotted for the LaSalle model.

6.16. The Chebyshev ODE is

$$(1 - x^2)\frac{d^2 y}{dx^2} - x\frac{dy}{dx} + n^2 y = 0.$$

We are interested in the case where $n = 5$ such that the last term on the left-hand side is $25y$. We know that $y(x = -1) = -1$ and $y(x = +1) = +1$. We want to solve this equation *numerically* on the interval $-1 \leq x \leq +1$ and we will pretend that *we do not know* that the analytic solution is $T_5(x)$. One of the requirements of this problem is that we need to obtain a very accurate estimate for $dy/dx|_{x=-1}$. Use the method of your choice and prepare a plot of your results for $y(x)$.

6.17. Schmitz and Amundson (1963) studied the behavior of a reacting system consisting of two immiscible liquid phases (α and β) that are fed continuously to a stirred-tank reactor (CSTR). A first-order, irreversible chemical reaction occurs in both phases with the rate constants k_α and k_β. The temperatures of the two phases are equal and the components are distributed according to an equilibrium law. Because the heat generation function (for the exothermic case) is sigmoidal in shape with respect to temperature (T), and the heat removal rate is linear with T, it is possible that the two intersect three times (i.e., there could be a maximum of three "steady-state" temperatures). Naturally, this condition would only occur if the slope and intercept of the heat removal line were injudiciously chosen (a reactor designer would certainly try to avoid such a regime for safety reasons). Schmitz and Amundson solved three simultaneous (nonlinear) ODEs that formed the model for this system (their equations 18, 19, and 20). The three equations describe the dynamic behavior of the reduced total concentration, the reduced temperature, and the volume fraction for the α-phase. The parametric values of interest to us are provided as case d, table 1, page 280, in the Schmitz–Amundson paper; they are reproduced here:

$$k_\alpha = \exp(23.1 - 63.5/T')$$

$$k_\beta = \exp(22.1 - 51.5/T')$$

$$v_\beta / v_\alpha = C_\beta / C_\alpha = 1$$

$$K_A = 0.2$$

$$\varepsilon = 1$$

$$A_{\beta 0} / A_{\alpha 0} = 0.2$$

$$Uv_\alpha / C_\alpha V = 6.37 \times 10^{-3}$$

$$\tau = \exp(6)$$

$$T'_{o0} = T'_{\beta0} = 1.85$$

$$\rho_s = 1.$$

Solve the Schmitz–Amundson model numerically, recognizing that the oscillatory behavior expected for this case may require a procedure with variable step size (at a minimum). Prepare a phase-space portrait of the system by cross plotting reduced mole fraction and reduced temperature. Verify the limit-cycle behavior presented in the original paper.

6.18. In Chapter 1, we contemplated a problem in which a recently acquired video revealed the flight characteristics of a falcon stalking and attacking a flying crow. This is a classic curve of pursuit problem but in three dimensions, of course. In a planar (x, y) problem, *if* the prey travels a straight line in the y-direction, starting from $x = a$, then the trajectory of the pursuer is described by the differential equation

$$1 + \left(\frac{dy}{dx}\right)^2 = k^2(a-x)^2\left(\frac{d^2y}{dx^2}\right)^2,$$

where k is the ratio of velocities (velocity of the pursuer)/(velocity of the prey). Solve this equation numerically and plot the path taken by the pursuer given that $k = 1.5$ and $a = 10$. The prey starts from the x-axis and the pursuer starts from the origin. What happens to the solution if $k = 1$?

6.19. Suppose an SR-71 (the Lockheed reconnaissance aircraft that won the Collier Trophy in 1963) is flying a photographic intel mission over hostile territory. The "Blackbird" flies a straight course at 2045 mph at a constant altitude of 89,500 ft. As the plane passes directly overhead, an antiaircraft installation fires a ground-to-air missile that accelerates rapidly to Mach 5 (take this to be 3600 mph); the missile maintains that speed until its fuel is exhausted (after 38 seconds). Can an intercept occur? If so, how far downrange does it occur? If not, how close does the missile get to the SR-71? This is another *curve of pursuit* problem (see Problem 6.18 and also Davis, 1962).

6.20. Previously, we looked at the behavior of the driven pendulum modeled as a third-order system (Problem 6.6). One of the earliest studies of nonlinear oscillators was the work of Van der Pol (1926). In dimensionless form, Van der Pol's model is

$$\frac{d^2\theta}{dt^2} - (\varepsilon - \theta^2)\frac{d\theta}{dt} + \theta = 0,$$

and he arrived at this form by assuming that the dissipative (friction) term would be a function of the amplitude of the

oscillation. We want to solve this equation numerically using two different starting points: $(\theta, \theta') = (-0.1, 0)$ and $(-0.1, 2.8)$. Determine the phase-plane portrait of the Van der Pol system by cross plotting θ with θ', and use both $\varepsilon = 0.4$ and $\varepsilon = 0.1$. Do all of the trajectories end up at the same limit cycle?

REFERENCES

Ayres, F. *Differential Equations*, Schaum's Outline Series, McGraw-Hill, New York (1952).

Bailey, H. E. Numerical Integration of the Equations Governing the One-Dimensional Flow of a Chemically Reactive Gas. *Physics of Fluids*, 12:2292 (1969).

Baker, G. L. and J. P. Gollub. *Chaotic Dynamics: An Introduction*, Cambridge University Press, Cambridge (1990).

Borotnikova, M. I. and E. I. Soloukhin. A Calculation of the Pulsations of Gas Bubbles in an Incompressible Liquid Subject to a Periodically Varying Pressure. *Soviet Physics—Acoustics*, 10:28 (1964).

Cash, J. R. and A. H. Karp. A Variable Order Runge-Kutta Method for Initial Value Problems with Rapidly Varying Right-Hand Sides. *ACM Transactions on Mathematical Software*, 16:201 (1990).

Chapra, S. C. and R. P. Canale. *Numerical Methods for Engineers*, 4th edition, McGraw-Hill, Boston (2002).

Davis, H. T. *Introduction to Nonlinear Differential and Integral Equations* (Chapter 5, Section 7), Dover Publications, New York (1962).

Durham H. L., Francis, O. B., Gallaher, L. J., Hale, H. G., and I. E. Perlin. *Study of Methods for the Numerical Solution of ODEs* (NASA-CR-57430) Huntsville, AL (1964).

Ferrell, J. E., Tsai, T. Y. C., and Q. Yang. Modeling the Cell Cycle: Why Do Certain Circuits Oscillate? *Cell*, 144:874 (2011).

Gear, C. W. *Numerical Initial Value Problems in Ordinary Differential Equations*, Prentice-Hall, Englewood Cliffs, NJ (1971).

Hairer, E. and G. Wanner. *Solving Ordinary Differential Equations II*, 2nd revised edition, Springer, Berlin (1996).

Hamming, R. W. *Numerical Methods for Scientists and Engineers*, 2nd edition, McGraw-Hill, New York (1973).

Hanna, O. T. and O. C. Sandall. *Computational Methods in Chemical Engineering*, Prentice Hall PTR, Upper Saddle River, NJ (1995).

James, M. L., Smith, G. M., and J. C. Wolford. *Applied Numerical Methods for Digital Computation*, 2nd edition, Harper & Row, New York (1977).

Lambert, J. D. *Computational Methods in Ordinary Differential Equations*, John Wiley & Sons, London (1973).

LaSalle, J. P. Asymptotic Stability Criteria. *Proceedings of Symposia in Applied Mathematics, Volume XIII, Hydrodynamic Instability*. American Mathematical Society, Providence, RI (1962).

Lee, H. J. and W. E. Schiesser. *Ordinary and Partial Differential Equation Routines in C, C++, Fortran, Java, Maple, and MATLAB*, Chapman and Hall, Boca Raton, FL (2004).

Lorenz, E. Deterministic Nonperiodic Flow. *Journal of the Atmospheric Sciences*, 20:130 (1963).

Press, W. H., Flannery, B. P., Teukolsky, S. A., and W. T. Vetterling. *Numerical Recipes: The Art of Scientific Computing*, Cambridge University Press, Cambridge (1989).

Schmitz, R. A. and N. R. Amundson. An Analysis of Chemical Reactor Stability and Control-Va: Two-Phase Systems in Physical Equilibrium. *Chemical Engineering Science*, 18:265 (1963).

Southworth, R. W. and S. L. Deleeuw. *Digital Computation and Numerical Methods*, McGraw-Hill, New York (1965).

Stoer, J. and R. Bulirsch. *Introduction to Numerical Analysis*, 2nd edition, Springer-Verlag, New York (1993).

Van der Pol, B. On Relaxation Oscillations. *Philosophical Magazine*, 2:978 (1926).

.

7

ANALYTIC SOLUTION OF PARTIAL DIFFERENTIAL EQUATIONS

INTRODUCTION

Many of the phenomena that are of interest to us in engineering and the applied sciences are modeled with partial differential equations (PDEs). Fluid flow, heat transfer, and mass transfer are prime examples, but problems in gravitation, electrostatics, and quantum theory all give rise to similar equations. The purpose of this chapter is to provide the reader with some basic skills, enabling him/her to find analytic solutions for many commonly encountered PDEs.

Several valuable references will be provided as we move through this material, but at the outset, we want to point out that there are two uniquely important monographs devoted to the analytic solution of PDEs: *The Mathematics of Diffusion*, Second Edition, by Crank (1975), and *Conduction of Heat in Solids*, Second Edition, by Carslaw and Jaeger (1959). These two books are known to nearly every worker in applied mathematics. Both are incredibly useful as guides to the solution of practical problems where diffusional (molecular) transport processes are dominant. Practitioners in this field are often heard to say, "I found a similar problem in Crank" or "I verified my solution with Carslaw and Jaeger." Anyone wishing to become adept with the subject matter of this chapter simply must own both of these books.

CLASSIFICATION OF PARTIAL DIFFERENTIAL EQUATIONS AND BOUNDARY CONDITIONS

We have to be able to recognize and classify PDEs to attack them successfully; a book such as Powers (1979) can be a valuable ally in this effort. Consider the generalized second-order PDE where ϕ is the dependent variable and x and y are arbitrary independent variables:

$$A\frac{\partial^2\phi}{\partial x^2} + B\frac{\partial^2\phi}{\partial x\partial y} + C\frac{\partial^2\phi}{\partial y^2} + D\frac{\partial\phi}{\partial x} + E\frac{\partial\phi}{\partial y} + F\phi + G = 0. \tag{7.1}$$

A, B, C, D, E, F, and G can be functions of x and y but *not* of ϕ. This linear PDE can be classified as follows:

$$B^2 - 4AC < 0 \quad elliptic$$

$$B^2 - 4AC = 0 \quad parabolic$$

$$B^2 - 4AC > 0 \quad hyperbolic$$

For illustration, we look at the "heat" equation (one-dimensional transient conduction):

$$\frac{\partial T}{\partial t} = \alpha\frac{\partial^2 T}{\partial y^2}. \tag{7.2}$$

You can see that $A = \alpha$, $B = 0$, and $C = 0$; the equation is parabolic. Compare this with the governing (Laplace) equation for two-dimensional potential flow (ψ is the stream function):

$$\frac{\partial^2\psi}{\partial x^2} + \frac{\partial^2\psi}{\partial y^2} = 0. \tag{7.3}$$

Applied Mathematics for Science and Engineering, First Edition. Larry A. Glasgow.
© 2014 John Wiley & Sons, Inc. Published 2014 by John Wiley & Sons, Inc.

In this case, $A = 1$ and $C = 1$, while $B = 0$; the equation is elliptic. Next, we consider a vibrating string (the wave equation):

$$\frac{\partial^2 u}{\partial t^2} = s^2 \frac{\partial^2 u}{\partial y^2}. \qquad (7.4)$$

Note that $A = 1$ and $C = -s^2$; therefore, $-4AC > 0$ and eq. (7.4) is hyperbolic. In applied mathematics, transient problems with molecular transport only (heat or diffusion equations) will have parabolic character. Equilibrium problems such as steady-state diffusion, conduction, or viscous flow in a duct will be elliptic in nature (phenomena governed by Laplace- or Poisson-type PDEs). We will see numerous examples of both in this chapter. Hyperbolic equations are common in quantum mechanics and high-speed, compressible flows; for example, inviscid supersonic flow about an airfoil. The Navier–Stokes equations that have been the focus of much attention by physicists and mathematicians over the last 160 years are of mixed character.

The three common types of boundary conditions used in applied mathematics are Dirichlet, Neumann, and Robin's. For Dirichlet boundary conditions (or conditions of the first kind) the field variable is specified at the boundary. Two examples follow: In a conduction problem, the temperature at a surface might be fixed (at $y = 0$, $T = T_0$); alternatively, in a viscous fluid-flow problem, the velocity at a stationary duct wall would be zero (for a Newtonian fluid). A condition of the first kind can also be written as a function of time, at $y = 0$, $T = f(t)$.

For Neumann conditions (or boundary conditions of the second kind), the flux is specified; for example, for a conduction problem with an insulated wall located at $y = 0$, $(\partial T/\partial y)_{y=0} = 0$. Of course, this gradient could also be written as a function of time.

A Robin's-type boundary condition (or condition of the third kind) results from equating the fluxes; for example, consider the solid–fluid interface in a heat transfer problem. On the solid side, heat is transferred by conduction (Fourier's law), but on the fluid side of the interface, we might have mixed heat transfer processes approximately described by Newton's "law" of cooling:

$$-k \left(\frac{\partial T}{\partial y} \right)_{y=0} = h(T_0 - T_f). \qquad (7.5)$$

We hasten to add that the heat transfer coefficient, h, that appears in eq. (7.5) is an empirical quantity. The numerical value of h is known only for a small number of cases, usually those in which molecular transport is dominant. Thus, the use of a Robin's-type boundary condition usually means that an additional unknown has been brought into the problem.

An analogous relationship can be used for mass transfer at interfaces:

$$-D_{AB} \left(\frac{\partial C_A}{\partial y} \right)_{y=0} = K(C_{A0} - C_{A\infty}).$$

And again, the mass transfer coefficient, K, is unknown and would generally have to be estimated.

A critical observation with regard to these boundary conditions is that all three kinds are *linear* with respect to the dependent variable.

FOURIER SERIES

In his prize-winning work submitted to the Paris Academy in 1811, and later published within *Theorie Analytique de la Chaleur* in 1822, Fourier claimed that an arbitrary function, $f(x)$, could be represented by the trigonometric series,

$$f(x) = a_0 + (a_1 \cos x + b_1 \sin x) + (a_2 \cos 2x + b_2 \sin 2x) + \cdots . \qquad (7.6)$$

The constants that appear in this series are given by

$$a_0 = \frac{1}{2\pi} \int_{-\pi}^{+\pi} f(x)dx, \qquad (7.7)$$

$$a_n = \frac{1}{\pi} \int_{-\pi}^{+\pi} f(x)\cos nx dx, \qquad (7.8)$$

and

$$b_n = \frac{1}{\pi} \int_{-\pi}^{+\pi} f(x)\sin nx dx. \qquad (7.9)$$

The general idea had surfaced earlier; prior to Fourier's work (in fact, in the eighteenth century) a number of prominent mathematicians worked on the vibrating string problem. Carslaw (1950) records that Bernoulli obtained a solution (for a string starting from rest) in the form of a trigonometric series. Euler responded to this work by noting that if Bernoulli was correct, then an arbitrary function of a single variable could be represented by an infinite series of sines (of integer multiples of the independent variable). Euler did not believe this was possible; he observed that sine was both a periodic function and one that was odd. If the function that was being represented did not have the same characteristics, how could it be obtained from sine?

Let us assume we are interested in a function, $f(x)$, for $(-\pi \le x \le +\pi)$:

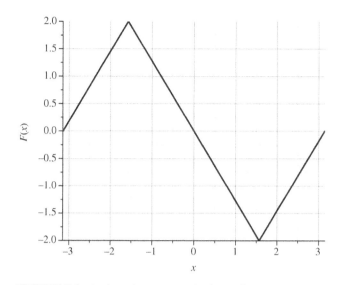

FIGURE 7.1. A triangular wave on the interval $(-\pi \leq x \leq +\pi)$.

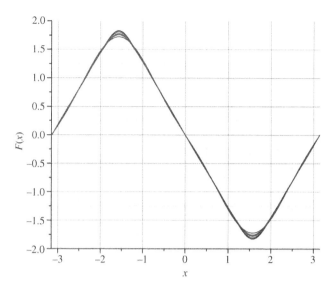

FIGURE 7.2. Representation of the triangular wave with Fourier's series technique. Many of the essential characteristics of $f(x)$ are reproduced reasonably well with just three terms in the series.

$$f(x) = \frac{4}{\pi}x + 4 \quad \text{for} - \pi \leq x \leq -\pi/2, \qquad (7.10)$$

$$f(x) = -\frac{4}{\pi}x \quad \text{for} - \pi/2 \leq x \leq +\pi/2, \qquad (7.11)$$

and

$$f(x) = \frac{4}{\pi}x - 4 \quad \text{for} + \pi/2 \leq x \leq +\pi. \qquad (7.12)$$

The function's behavior (a triangular wave) is illustrated in Figure 7.1.

We would like to know if this function can be represented in the manner suggested by Fourier, and if so, how do the coefficients actually behave? Figure 7.2 shows the approximations for $f(x)$ using 3, 5, 10, 20, and 90 terms in the series. Though minor discrepancies are apparent, the results are similar to the function illustrated in Figure 7.1.

If this is your first exposure to Fourier series and their application to boundary-value problems, then Spiegel's (1974) book can be an extremely useful tool and learning guide. The basic idea with Fourier series is that we use superposition to construct a representation of a periodic function using combinations of the oscillating functions, *sine* and *cosine*. Since many boundary-value problems require us to expand a function into trigonometric series, we can expect Fourier series to prove extremely useful. If you are skeptical about using sine and cosine in this way, you are in good company. Korner (1989) notes that both Laplace and Lagrange initially had doubts about Fourier's development; part of their concern was a consequence of Fourier's lack of rigor.

Consider a function, $f(x)$, defined over an interval, $-L < x < +L$. The Fourier series corresponding to this function is, by definition,

$$f(x) = \frac{A_0}{2} + \sum_{n=1}^{\infty} A_n \cos\frac{n\pi x}{L} + B_n \sin\frac{n\pi x}{L}. \qquad (7.13)$$

One concern that students new to Fourier series typically have is exactly how this expression will prove to be of value. After all, if this equation is to be used to reconstruct the function, $f(x)$, then we might need to know a very large number of A_ns and B_ns. The effort required appears formidable—until we think about orthogonality of the functions sine and cosine. Specifically, consider that

$$\int_{-\pi}^{\pi} \sin nx\, dx = 0 \qquad (7.14)$$

$$\int_{-\pi}^{\pi} \sin nx \cos mx\, dx = 0 \qquad (7.15)$$

$$\int_{-\pi}^{\pi} \cos nx \cos mx\, dx = 0 \text{ if } n \neq m, \quad \text{and } \pi \text{ if } n = m \qquad (7.16)$$

$$\int_{-\pi}^{\pi} \cos nx \cos mx\, dx = 0 \text{ if } n \neq \text{m}, \quad \text{and } \pi \text{ if } n = m \neq 0. \qquad (7.17)$$

These relationships suggest the following approach: Multiply the expression for $f(x)$ by $\sin(mx)\,dx$ and integrate from $-\pi$ to $+\pi$, so that

$$\int_{-\pi}^{\pi} f(x)\sin mx\,dx = \int_{-\pi}^{\pi} \frac{A_0}{2}\sin mx\,dx$$
$$+ \int_{-\pi}^{\pi}(A_n\cos nx + B_n\sin nx)\sin mx\,dx.$$

$$(7.18)$$

It follows immediately that for a function $f(x)$ of period 2π,

$$B_n = \frac{1}{\pi}\int_{-\pi}^{\pi} f(x)\sin nx\,dx. \qquad (7.19)$$

More generally, for a function $f(x)$ defined over the interval, $-L < x < +L$, we have

$$B_n = \frac{1}{L}\int_{-L}^{+L} f(x)\sin\frac{n\pi x}{L}\,dx. \qquad (7.20)$$

Of course, the A_ns can be determined analogously.

In many of the problems of interest to us, the series solutions we obtain may only involve either sine or cosine terms. We refer to such cases as half-range Fourier sine (or cosine) series, and often our attention in such problems is focused on just half of the interval, that is, from $x = 0$ to $x = +L$. In case of the half-range series,

$$\text{Cosine: } A_n = \frac{2}{L}\int_{0}^{L} f(x)\cos\frac{n\pi x}{L}\,dx \qquad (7.21)$$

and

$$\text{Sine: } B_n = \frac{2}{L}\int_{0}^{L} f(x)\sin\frac{n\pi x}{L}\,dx. \qquad (7.22)$$

An obvious question of concern to us is whether an arbitrary function that is piecewise continuous over some interval 0 to L can be represented successfully in this way (by "successfully," we mean that we can obtain sufficient accuracy using a reasonable number of terms). Recall that in Figure 7.2, we saw that a triangular waveform could be very easily represented with just a few Fourier series terms. Now let us consider a function formed by two straight lines, represented by $f(x) = 3x$ from $x = 0$ to $x = 3$, and then $f(x) = 18 - 3x$ from $x = 3$ to $x = 6$. Since this function is defined only from $x = 0$ to $x = L$, and since the form requires an odd function representation,

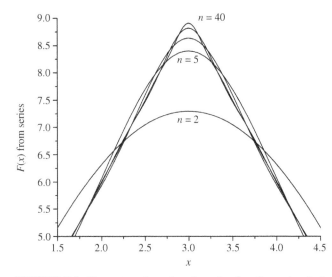

FIGURE 7.3. Reconstruction of a triangular function using 2, 5, 10, 20, and 40 terms in the Fourier sine series. Please note that both axes have been abridged to better show the differences between the results.

$$f(x) = \sum_{n=1}^{\infty} B_n\sin\frac{n\pi x}{L}. \qquad (7.23)$$

The necessary coefficients are determined directly from the integral(s):

$$B_n = \frac{2}{L}\left[\int_{0}^{3} 3x\sin\frac{n\pi x}{L}\,dx + \int_{3}^{6}(18-3x)\sin\frac{n\pi x}{L}\,dx\right]. \qquad (7.24)$$

We will compute the B_ns and then see how well the series represents $f(x)$ using 2, 5, 10, 20, and 40 terms successively. The first seven B_ns are 7.2951, -2.6408×10^{-5}, -0.8106, 2.3685×10^{-15}, 0.2918, -8.8027×10^{-6}, and -0.1489, and we see from the numerical computation that the even terms are actually zero.

As you can see from Figure 7.3, the approximation obtained with $n = 40$ is very good, with $f(x = 3)$ only 1% below the correct value (8.9088 as opposed to $f(3) = 9.0$).

A Preview of the Utility of Fourier Series

We want to explore a problem that will make it very clear why Fourier's work is so useful to us in our efforts to solve PDEs and we will preface this example with an observation made by Lord Kelvin: "Fourier's theorem is not only one of the most beautiful results of modern analysis, but it is said to furnish an indispensable instrument in the treatment of nearly every recondite question in modern physics." Let us see why Kelvin was so enthusiastic.

We will assume that we have a slab of homogeneous material that extends in the x-direction such that $0 \le x \le 3$. This slab has an initial temperature distribution given by

$$T = f(x) = 6x - 2x^2, \qquad (7.25)$$

which yields a temperature of zero (°C) at both ends and $4.5°$ at the center ($x = 1.5$). At $t = 0$, the temperature at both ends is instantaneously raised to 10°C. The evolution of temperature in the slab is governed by a parabolic PDE,

$$\frac{\partial T}{\partial t} = \alpha \frac{\partial^2 T}{\partial x^2}, \qquad (7.26)$$

which we will modify by defining a new dependent variable, $\theta = T - 10$. Of course, this means that θ will be zero at both ends, which proves to be very convenient. α that appears in eq. (7.26) is the thermal diffusivity, $\alpha = k/pC_p$. A solution for this problem has the form

$$\theta = C_1 \exp(-\alpha \lambda^2 t)[A \sin \lambda x + B \cos \lambda x]. \qquad (7.27)$$

It is easy to show that, to satisfy the boundary conditions, we must have $B = 0$ and $\lambda_n = n\pi/3$. Consequently, the solution we seek has the form

$$\theta = \sum_{n=1}^{\infty} A_n \exp\left(-\alpha \frac{n^2 \pi^2}{9} t\right) \sin \frac{n\pi x}{3}. \qquad (7.28)$$

Our initial condition ($t = 0$) corresponds to

$$6x - 2x^2 - 10 = \sum_{n=1}^{\infty} A_n \sin \frac{n\pi x}{3}. \qquad (7.29)$$

Of course, this is a Fourier sine series and we know that the unknown coefficients are determined by integration:

$$A_n = \frac{2}{3} \int_0^3 (6x - 2x^2 - 10) \sin \frac{n\pi x}{3} dx. \qquad (7.30)$$

Notice that the solution that we developed, eq. (7.28), has a lot of exponential damping when t becomes *large*. This means that for very small times, we should anticipate that the infinite series may converge *very slowly*! To further explore this "worst-case" scenario, we will find a few A_ns by integration (for even ns, the coefficients are zero).

n	A_n
1	−8.08812
3	−4.07208
5	−2.50927
7	−1.80530
9	−1.40824
11	−1.15388
13	−0.97716
15	−0.84729
17	−0.74784
19	−0.66925
21	−0.60559
23	−0.55296
25	−0.50874
27	−0.47106
29	−0.43856

We see immediately that these coefficients are diminishing rather slowly—which is not a good sign for convergence of the infinite series. Since we have all of the pieces in place, we will compute the initial temperature distribution using increasing values of n (we start with $n = 50$ and go up to $n = 700$).

This example, illustrated in Figure 7.4, has taught us an important lesson: Convergence of the Fourier series to the initial temperature distribution is abysmally slow—we needed hundreds of terms to get a good approximation. However, we must remember that, for modestly larger times, the exponential damping in the infinite series will greatly improve convergence; so much so that for intermediate ts we might only need *one or two terms* to get satisfactory

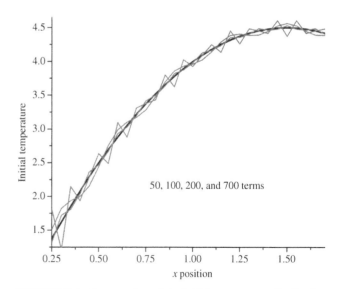

50, 100, 200, and 700 terms

FIGURE 7.4. Computation of the initial temperature distribution in the slab using 50, 100, 200, and 700 terms. Both axes have been truncated to better reveal the behavior of the Fourier series (which is especially bad near the ends of the slab for small n).

results. We will illustrate this behavior by using the solution (eq. 7.28) to calculate the temperature at the center ($x = 1.5$) for times ranging from 1/32 (0.03125) to 10. Recall that the initial temperature at $x = 1.5$ was 4.5 as given by eq. (7.25). We will arbitrarily choose a thermal diffusivity that corresponds to a material such as lead, $\alpha = 1/5$.

Time (s)	Temperature for $x = 1.5$	Number of Terms Required in Series
1/32	4.475	19
1/16	4.450	15
1/8	4.400	12
1/4	4.300	10
1/2	4.116	7
1	4.060	5
2	4.862	4
4	6.638	2
6	7.831	2
8	8.601	2
10	9.098	1

Note that the temperature in the center of the slab *decreases* at first; this is a consequence of the shape of the initial temperature distribution and the fact that thermal energy is transferred downhill, in the direction of decreasing temperature. These results show that by the time we get to $t = 4$ seconds, we only need two terms in the series to get an acceptable value for T.

THE PRODUCT METHOD (SEPARATION OF VARIABLES)

The product method is a technique by which certain PDEs can be solved analytically. As the name implies, the method is based on equating the dependent variable to a product of functions of the independent variables. If the separation is successful, the result will be ordinary differential equations (ODEs) for which familiar methods of solution may be employed. There are, however, important restrictions on the applicability of this technique with respect to the form of the differential equation, the shape of the boundary, and the nature of the boundary conditions.

It is obvious that the PDE itself must be separable; that is, it must be linear and it must not have any cross derivatives. For example, the equation

$$A\left(\frac{\partial \phi}{\partial x}\right)^2 + B\frac{\partial^2 \phi}{\partial x \partial y} + \cdots \qquad (7.31)$$

would violate both of these restrictions. Furthermore, the equation must be homogeneous, or of a form that can be rendered homogeneous through suitable transformation. Thus, a Poisson-type PDE might be handled successfully if

the constant can be removed through a change in the dependent variable. We should also note that it is not necessary that an equation have constant coefficients. Consider the wave (hyperbolic) equation:

$$\frac{\partial^2 \phi}{\partial t^2} - s^2(x)\frac{\partial^2 \phi}{\partial x^2} = 0. \qquad (7.32)$$

We propose $\phi = f(t)g(x)$, which results in $f''g = s^2 fg''$, and then $f''/f = s^2(g''/g) = -\lambda^2$.

So the two ODEs are simply

$$f'' + \lambda^2 f = 0 \quad \text{and} \quad s^2(x)g'' + \lambda^2 g = 0. \qquad (7.33)$$

For the problem types discussed earlier, the boundary conditions must correspond to constant values of x and y. Thus, $\phi(x = 0) = \phi_0$ or $\phi(y = B) = \phi_B$ would be satisfactory, but $\phi = \phi_0$ for $y/x = 2$ would not. Also, boundary conditions applied at, say, $x = B$, cannot include any partial derivatives involving y. Similarly, a boundary condition written for $y = A$ cannot include partial derivatives with respect to x.

The preceding discussion may make it seem as though the applicability of the product method is severely limited, and it is certainly true that there are many PDEs of interest to us that cannot be solved with this technique. However, we should not be too hasty to discount separation. There are many important problems in applied mathematics that can be solved in this manner, and we will find examples in fluid flow, heat transfer, diffusion, and wave phenomena, among others. Moreover, we have more than two centuries of work in this branch of mathematics to draw on, so usefully complete examples abound.

Before we begin to examine applications of this technique, a final word of caution: The solutions we are about to construct (and use) give the impression of being exact. As Weinberger (1965) notes, we truncate these infinite series solutions as a practical matter. This means that only a finite number of terms will be used to construct a Fourier series, for example, and therefore we will be generating an *approximation* to the solution. In many cases, it will be a very close approximation, but keep in mind that our results will not be *exactly correct*.

Parabolic Equations

The parabolic PDEs that we are most likely to see will involve transient molecular transport (by viscous friction, conduction, or diffusion) in one or more spatial directions. The general procedure we will employ will be similar in every case: We perform the separation, solve the resulting ODEs, use the boundary conditions to simplify the result and identify the constant of separation, and finally, use the initial condition (with Fourier theorem or orthogonality) to identify the proper values for the leading coefficient in the

TABLE 7.1. Illustration of Infinite Series Convergence for Small ts

Term No.	$t = 0.001$	$t = 0.005$	$t = 0.025$	$t = 0.125$	$t = 0.625$
1	1.271981	1.266969	1.242205	1.12546	0.6870893
3	0.851322	0.8609938	0.9023096	0.9856378	0.6854422
5	1.099763	1.086086	1.039727	0.9972914	0.6854423
7	0.926459	0.9432634	0.9854355	0.9968604	0.6854423
9	1.05706	1.038121	1.004608	0.9968669	0.6854423
11	0.954341	0.9744126	0.9987616	0.9968669	0.6854423
13	1.037236	1.01695	1.000275	0.9968669	0.6854423
15	0.969256	0.9889856	0.9999457	0.9968669	0.6854423
17	1.025566	1.006978	1.000006	0.9968669	0.6854423
19	0.97864	0.9956936	0.9999966	0.9968669	0.6854423
21	1.017874	1.002573	0.9999977	0.9968669	0.6854423
23	0.985031	0.9985044	0.9999976	0.9968669	0.6854423
25	1.012515	1.000835	0.9999976	0.9968669	0.6854423
27	0.98955	0.9995433	0.9999976	0.9968669	0.6854423
29	1.008694	1.000235	0.9999976	0.9968669	0.6854423
31	0.992785	0.9998772	0.9999976	0.9968669	0.6854423
33	1.005956	1.000056	0.9999976	0.9968669	0.6854423
35	0.995097	0.9999698	0.9999976	0.9968669	0.6854423
37	1.004008	1.00001	0.9999976	0.9968669	0.6854423
39	0.996732	0.9999919	0.9999976	0.9968669	0.6854423
41	1.002642	0.9999996	0.9999976	0.9968669	0.6854423
43	0.997868	0.9999964	0.9999976	0.9968669	0.6854423

infinite series. The process is straightforward and usually quite transparent. There are, however, a few subtle issues that can complicate our analysis. We will try to illustrate some of these in the examples that follow.

Let us begin by examining transient conduction in a *finite* slab of material, for which $\partial T/\partial t = \alpha(\partial^2 T/\partial y^2)$; let this object extend from $y = 0$ to $y = 1$. We can have either a uniform initial temperature or a temperature distribution that can be written as a function of y. At $t = 0$, both faces are instantaneously heated to some new temperature, T_s. Define a dimensionless temperature, $\theta = (T - T_s)/(T_i - T_s)$, and let $\theta = f(y)g(t)$. The product method yields

$$g' = -\alpha\lambda^2 g \quad \text{and} \quad f'' + \lambda^2 f = 0. \quad (7.34)$$

As expected, we get

$$g = C_1 \exp(-\alpha\lambda^2 t) \quad \text{and} \quad f = A\sin\lambda y + B\cos\lambda y. \quad (7.35)$$

Since B must be zero, and $\sin(\lambda) = 0$, we very quickly find

$$\theta = \sum_{n=1}^{\infty} A_n \exp(-\alpha\lambda_n^2 t)\sin\lambda_n y, \quad \text{where } \lambda_n = n\pi. \quad (7.36)$$

If we have a uniform initial temperature, T_i, then application of the initial condition results in

$$1 = \sum_{n=1}^{\infty} A_n \sin\lambda_n y, \quad (7.37)$$

a half-range Fourier sine series. By definition,

$$A_n = \frac{2}{L}\int_0^L f(y)\sin\frac{n\pi y}{L}dy, \quad (7.38)$$

but for our case, $L = 1$, and the function, $f(y)$, is also 1. The preceding integral is zero for even n and equal to $4/(n\pi)$ for $n = 1, 3, 5, \ldots$. With this example, we have a good opportunity to examine the convergence of the infinite series solution. Let $y = 1/2$, $\alpha = 0.1$, and let t range from 0.001 to 0.625 by repeated factors of 5. We shall examine the series for ns from 1 to 43 (see Table 7.1). Note that for small ts, the series does not converge quickly. However, for $t = 0.125$, we need only five terms, and at $t = 0.625$, only three. The results should not be surprising. For very small ts, the temperature profile is virtually half a cycle of a square wave.

Now suppose we have a Neumann condition at one boundary; in particular, let us assume we have an insulated boundary located at $y = L$, such that $(\partial T/\partial y)_{y=L} = 0$. The reader may wish to show, using eq. (7.36) as our starting point, that $\cos\lambda_n L = 0$. Therefore, $\lambda_n = \pi/2L$, $3\pi/2L$, $5\pi/2L$, and so on. Once again, we have discovered a Fourier series problem since the separation constants are integer multiples of pi (π).

We also want to address the issues that arise when one of the boundary conditions for our slab of material is of the Robin's type (the third kind). All of the initial steps in the problem are the same as in the previous case, but we make one adjustment by defining a new dependent variable,

$\theta = T - T_\infty$, where T_∞ is the temperature of the surroundings at a large distance. The situation we are describing would correspond to the case where one end of the slab (at $y = L$) loses heat to the surroundings, and we assume that this process can be described by Newton's law of cooling. Since the first part of this problem is common to what we saw previously, we can start with

$$\theta = A \exp(-\alpha\lambda^2 t)\sin\lambda y. \tag{7.39}$$

But this time, at the end of the slab, where $y = L$, we equate the fluxes:

$$-k\left.\frac{\partial\theta}{\partial y}\right|_{y=L} = h(T_{y=L} - T_\infty). \tag{7.40}$$

Consequently, we find

$$-kA\lambda\exp(-\alpha\lambda^2 t)\cos\lambda L = hA\exp(-\alpha\lambda^2 t)\sin\lambda L, \tag{7.41}$$

and this is equivalent to

$$-\frac{k}{hL}\lambda L = \tan\lambda L,$$

which we write as

$$\lambda L\cot\lambda L + \frac{hL}{k} = 0. \tag{7.42}$$

This transcendental equation arises frequently in applied mathematics, and you may recognize the dimensionless grouping, hL/k, as the Biot number (or modulus). We can find a few roots for this equation in Carslaw and Jaeger (1959) and a table of values (with eight roots) is provided here for convenience. Equation (7.42) has been rewritten as $x\cot(x) + C = 0$, and the negative values of C come about for a sphere that loses heat to the surroundings through a Robin's-type boundary condition.

$C = -1$	-0.8	-0.6	-0.4	-0.2	0
0.0000	0.7593	1.0528	1.2644	1.4320	1.5708
4.4934	4.5379	4.5822	4.6261	4.6696	4.7124
7.7253	7.7511	7.7770	7.8028	7.8284	7.8540
10.9041	10.9225	10.9408	10.9591	10.9774	10.9956
14.0662	14.0804	14.0946	14.1088	14.1230	14.1372
17.2208	17.2324	17.2440	17.2556	17.2672	17.2788
20.3713	20.3811	20.3909	20.4007	20.4106	20.4204
23.5195	23.5280	23.5365	23.5450	23.5535	23.5619

0.2	0.4	0.6	0.8	1.0	2.0
1.6887	1.7906	1.8798	1.9586	2.0288	2.2889
4.7544	4.7956	4.8358	4.8750	4.9132	5.0870
7.8794	7.9045	7.9295	7.9542	7.9787	8.0962
11.0137	11.0318	11.0498	11.0677	11.0855	11.1727
14.1513	14.1654	14.1795	14.1935	14.2074	14.2764
17.2903	17.3019	17.3134	17.3249	17.3364	17.3932
20.4301	20.4399	20.4497	20.4594	20.4692	20.5175
23.5704	23.5789	23.5874	23.5958	23.6043	23.6463

4	6	8	10	20	40
2.5704	2.7165	2.8044	2.8628	2.9930	3.0651
5.3540	5.5378	5.6669	5.7606	5.9921	6.1311
8.3029	8.4703	8.6031	8.7083	9.0018	9.1987
11.3348	11.4773	11.5993	11.7027	12.0250	12.2688
14.4080	14.5288	14.6374	14.7335	15.0625	15.3417
17.5034	17.6072	17.7032	17.7908	18.1136	18.4180
20.6120	20.7024	20.7877	20.8672	21.1772	21.4980
23.7289	23.8088	23.8851	23.9574	24.2516	24.5817

Now, suppose for our example $hL/k = 1/2$, the first eight values for the product λL are

$$1.8366, 4.8158, 7.9171, 11.0409, 14.1724,$$
$$17.3076, 20.4448, \text{ and } 23.5831.$$

Notice the *spacing* between the consecutive pairs: 2.9792, 3.1013, 3.1238, 3.1315, and so on. It is now clear that values of the constant of separation, λ, are *not* integer multiples of *pi*, and this means that this case is *not a Fourier series problem*. So, although our solution has the form

$$\theta = \sum_{n=1}^{\infty} A_n\exp(-\alpha\lambda_n^2 t)\sin\lambda_n y, \tag{7.43}$$

the coefficients (A_ns) must be determined using orthogonality as we will now demonstrate. At $t = 0$, $\theta = \theta_0$, so

$$\theta_0 = \sum_{n=1}^{\infty} A_n\sin\lambda_n y. \tag{7.44}$$

We multiply both sides of the equation by $\sin\lambda_m y\,dy$ and integrate from $y = 0$ to $y = L$, making use of the fact that

$$\int_0^L \sin\lambda_n y\sin\lambda_m y\,dy = 0 \quad \text{unless } n = m. \tag{7.45}$$

Therefore, the needed coefficients are obtained from the quotient of integrals:

$$A_n = \frac{\displaystyle\int_0^L \theta_0\sin\lambda_n y\,dy}{\displaystyle\int_0^L \sin^2\lambda_n y\,dy} = \frac{-\dfrac{\theta_0}{\lambda_n}(\cos\lambda_n L - 1)}{\dfrac{L}{2} - \dfrac{1}{4\lambda_n}\sin 2\lambda_n L}. \tag{7.46}$$

It is important that we be able to deal with analogous problems for "infinite" ($L >> d$) cylinders too. Suppose we begin with a cylinder that has some initial temperature (or initial temperature distribution). At $t = 0$, the surface at $r = R$ is instantly cooled or heated; we wish to determine how the temperature of the medium for $0 < r < R$ responds to this change. For this case, the governing PDE is

$$\frac{\partial T}{\partial t} = \alpha \left[\frac{\partial^2 T}{\partial r^2} + \frac{1}{r} \frac{\partial T}{\partial r} \right]. \qquad (7.47)$$

This fits the criteria that we had established for separation, so we propose $T = f(r)g(t)$; this results in

$$\frac{g'}{\alpha g} = \frac{f'' + \frac{1}{r} f'}{f} = -\lambda^2. \qquad (7.48)$$

The two ODEs (the second one is a form of Bessel's differential equation) with their solutions are shown here:

$$g' = -\alpha \lambda^2 g \quad \text{and} \quad f'' + \frac{1}{r} f' + \lambda^2 f = 0, \qquad (7.49)$$

resulting in

$$g = C_1 \exp(-\alpha \lambda^2 t) \quad \text{and} \quad f = A J_0(\lambda r) + B Y_0(\lambda r). \qquad (7.50)$$

For all problems of this type that involve a solid cylindrical medium that extends from $r = 0$ to $r = R$, we can immediately simplify (since T must be finite at the center and since $Y_0(0) = -\infty$ as we saw in Chapter 5, $B = 0$):

$$T = A \exp(-\alpha \lambda^2 t) J_0(\lambda r). \qquad (7.51)$$

As noted previously, we assume that the *surface* temperature is changed instantaneously to a new value, T_s. Furthermore, we define a new dependent variable, $\theta = T - T_s$, which means that $\theta(r = R, t) = 0$. This will be satisfied as long as $J_0(\lambda R) = 0$; this equation has an infinite number of roots and the first few λRs are 2.40483, 5.52008, 8.65373, 11.79153, 14.93092, and so on. Therefore,

$$\theta = \sum_{n=1}^{\infty} A_n \exp(-\alpha \lambda_n^2 t) J_0(\lambda_n r). \qquad (7.52)$$

We are ready to apply the initial condition for the interior of the cylinder; typically, that would be $T(r, t = 0) = $ constant, or $T(r, t = 0) = f(r)$. We start with the constant case and utilize orthogonality (readers unfamiliar with this process for Bessel functions may find section 7.5 of Carslaw and Jaeger, or chapter 10 in Spiegel, 1971, to be quite helpful): We begin with the initial condition

$$\theta_0 = \sum_{n=1}^{\infty} A_n J_0(\lambda_n r), \qquad (7.53)$$

and multiply both sides by $r J_0(\lambda_m r) dr$, then integrate from $r = 0$ to R. Please note the inclusion of the weighting function, r. Consequently,

$$A_n = \frac{\displaystyle\int_0^R \theta_0 r J_0(\lambda_n r) dr}{\displaystyle\int_0^R r J_0^2(\lambda_n r) dr}. \qquad (7.54)$$

The integral in the denominator is known to be $(R^2/2) J_1^2(\lambda_n R)$, so therefore,

$$A_n = \frac{2\theta_0}{R^2 J_1^2(\lambda_n R)} \int_0^R r J_0(\lambda_n r) dr. \qquad (7.55)$$

Completing the problem, we find

$$\theta = \frac{2\theta_0}{R} \sum_{n=1}^{\infty} \exp(-\alpha \lambda_n^2 t) \frac{J_0(\lambda_n r)}{\lambda_n J_1(\lambda_n r)}. \qquad (7.56)$$

We should look at a specific example to better understand how well this will work for us. Suppose we have a long acrylonitrile butadiene styrene (ABS) plastic rod exactly 2 cm in diameter. The thermal diffusivity (α) for ABS is about 0.00108 cm^2/s, and we will assume our interest is for $r = 1/2$ cm and $t = 50$ seconds (we have purposefully selected a time that is neither short nor long). We obtain

$$\frac{\theta}{\theta_0} = \frac{2}{R} [0.39267 + 0.017297 - 0.002659 \cdots],$$

and since $R = 1$, this corresponds to $\theta/\theta_0 \cong 0.79$. The reader should also be aware that the solution of this infinitely long cylinder problem is so important in practical situations that it has been presented graphically throughout the literature of heat and mass transfer, including Carslaw and Jaeger (1959) and Glasgow (2010). A similar graph is reproduced here as Figure 7.5, and this graph indicates that $\theta/\theta_0 \cong 0.81$ for the ABS rod example.

A related problem, but one that is a bit more difficult, arises when a boundary condition of the third kind must be used at the cylinder surface where $r = R$. Although the discussion that follows concerns heat transfer to a cylinder, keep in mind that the analysis is the same for the equivalent mass transfer problem. Suppose we have heat transfer from

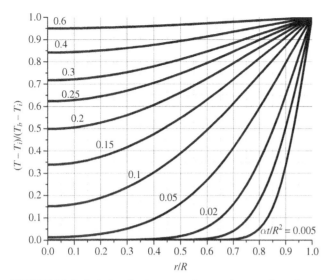

Biot, hR/k	$\lambda_1 R$	$\lambda_2 R$	$\lambda_3 R$	$\lambda_4 R$	$\lambda_5 R$
0.01	0.1412	3.8343	7.0170	10.1745	13.3244
0.02	0.1995	3.8369	7.0184	10.1754	13.3252
0.05	0.3401	3.8443	7.0225	10.1784	13.3274
0.1	0.4417	3.8577	7.0298	10.1833	13.3312
0.2	0.6170	3.8835	7.0440	10.1931	13.3387
0.5	0.9408	3.9594	7.0864	10.2225	13.3611
1	1.2558	4.0795	7.1558	10.2710	13.3984
2	1.5994	4.2910	7.2884	10.3658	13.4719
5	1.9898	4.7131	7.6177	10.6223	13.6786
10	2.1795	5.0332	7.9569	10.9363	13.9580
20	2.2880	5.2568	8.2534	11.2677	14.2983
100	2.3809	5.4652	8.5678	11.6747	14.7834

FIGURE 7.5. Solution for heat transfer to a long, solid cylinder when the surface acquires a new temperature instantaneously. Note that the centerline temperature attains 50% of the total change when $\alpha t / R^2 = 0.2$.

the fluid phase to a long, solid cylinder with relatively large conductivity. Because k (or α) is large, much of the thermal energy that arrives at the interface will be conducted readily into the interior of the cylinder. Consequently, the surface temperature will not acquire the fluid temperature *instantaneously*. It is appropriate to think of this behavior in terms of the relative resistances: A small Biot number indicates that the bulk of the resistance to heat transfer lies in the fluid phase. The initial steps for this problem are the same as before, so

$$T = A \exp(-\alpha \lambda^2 t) J_0(\lambda r). \qquad (7.57)$$

However, at $r = R$, a Robin's-type boundary condition must be used:

$$-k \frac{\partial T}{\partial r}\bigg|_{r=R} = h\left(T|_{r=R} - T_\infty\right). \qquad (7.58)$$

It is convenient to define $\theta = T - T_\infty$, so that when we introduce eq. (7.57) into eq. (7.58),

$$\lambda R J_1(\lambda R) - \frac{hR}{k} J_0(\lambda R) = 0. \qquad (7.59)$$

Naturally, the roots of this transcendental equation depend on the value of the Biot number. In a typical application of the kind we are discussing, hR/k might be about 0.7 for which the $\lambda_n R$s are 1.0873, 4.0085, 7.1143, 10.2419, 13.3761, 16.5131, and so on. The roots for this transcendental eq. (7.59) are needed frequently in applied mathematics, so it may be useful to provide an abbreviated table here:

Of course, we now know that the solution for the modified problem must be written as

$$\theta = \sum_{n=1}^{\infty} A_n \exp(-\alpha \lambda_n^2 t) J_0(\lambda_n r), \qquad (7.60)$$

and the coefficients (A_ns) are determined by orthogonality as before, with one important difference: The values for the separation constant come from the transcendental eq. (7.59) rather than from the zeros of J_0, so the required integrations produce a different result. Again, this is described clearly in chapter 7 in Carslaw and Jaeger (1959) and also in chapter 10 in Spiegel; the consequence for this case is

$$A_n = \frac{2\lambda_n R J_1(\lambda_n R)}{\left[\dfrac{h^2 R^2}{k^2} + \lambda_n^2 R^2\right] J_0^2(\lambda_n R)}. \qquad (7.61)$$

Let us take $k = 0.09$, $R = 1.27$, and $h = 0.0496$ (all centimeter-gram-second units), such that $hR/k = 0.7$. Then, using eq. (7.59), we find $\lambda_1 = 0.8561$, and by eq. (7.61), $A_1 = 1.1522$ and $A_2 = -0.21197$.

To this point, we have said nothing about parabolic equations arising in spherical geometries. Let us now consider transient conduction in a spherical entity; thermal energy is transferred only in the r-direction:

$$\rho C_p \frac{\partial T}{\partial t} = k\left[\frac{1}{r^2}\frac{\partial}{\partial r}\left(r^2 \frac{\partial T}{\partial r}\right)\right]. \qquad (7.62)$$

We rewrite the equation as

$$\frac{\partial T}{\partial t} = \alpha\left[\frac{\partial^2 T}{\partial r^2} + \frac{2}{r}\frac{\partial T}{\partial r}\right], \qquad (7.63)$$

and let $T = \theta/r$. This variable change results in the familiar "slab" equation,

$$\frac{\partial \theta}{\partial t} = \alpha \frac{\partial^2 \theta}{\partial r^2}, \qquad (7.64)$$

for which we know

$$\theta = C_1 \exp(-\alpha\lambda^2 t)[A\sin\lambda r + B\cos\lambda r]. \quad (7.65a)$$

Naturally, we need only to divide by r to return to $T(r, t)$. This means that a great many problems of this type in spherical coordinates can be solved by simply adapting appropriate results from solutions worked out for slab problems.

We will illustrate a typical solution procedure for spheres with a detailed example: A solid sphere of radius 3 has a uniform initial temperature of $30°$. At $t = 0$, the surface of the sphere is instantaneously heated to $80°$. Of course, we already know that $T = (C_1/r)\exp(-\alpha\lambda^2 t)[A\sin\lambda r + B\cos\lambda r]$. Since T must be finite at the center, we require $B = 0$. If we now define a dimensionless temperature, $\theta = (T - 80)/(30 - 80)$, then $\theta = 0$ for $r = R$, and

$$\theta = \sum_{n=1}^{\infty} \frac{A_n}{r}\exp(-\alpha\lambda_n^2 t)\sin\lambda_n r. \quad (7.65b)$$

The constants of separation are integer multiples of pi (π), $\lambda_n = n\pi/R$, and we use the initial condition to find $1 = \sum_{n=1}^{\infty}(A_n/r)\sin\lambda_n r$. This is a Fourier series problem, so we can determine the leading coefficients by

$$A_n = \frac{2}{R}\int_0^R r\sin\lambda_n r\, dr = -\frac{2R}{n\pi}\cos n\pi.$$

We note that $\cos(n\pi)$ will be -1 for odd n and $+1$ for even n. A complete solution is now at hand, and we will replace R with its value, 3:

$$\frac{T-80}{30-80} = -6\sum_{n=1}^{\infty}\frac{\cos n\pi}{n\pi}\exp\left(-\alpha\frac{n^2\pi^2}{9}t\right)\frac{\sin\dfrac{n\pi r}{3}}{r}. \quad (7.65c)$$

We will set $\alpha = 0.005$, $r = 1.5$, vary t from 1 to 1000, and use 50 terms in the series:

t	1	10	100	400	1000
$T(r/R = 0.5)$	29.99999	30.00021	43.36091	72.89833	79.7354

Because this is such an important practical problem, the solution is often presented graphically in the form shown in Figure 7.6. Please observe that the dimensionless temperature used in the figure is different from the definition we used for eq. (7.65b); it is convenient to have $\theta \to 0$ as t becomes large in the analytic solution.

We can use this graphical presentation to confirm our analytic results. For $t = 400$, we find $\alpha t/R^2 = 0.222$, which means that $(T - 30)/80 - 30 \cong 0.855$, yielding a temperature of $72.7°$ (very close to the result we obtained in the table above using the infinite series solution).

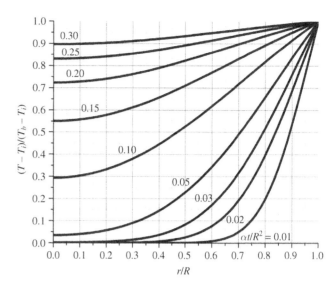

FIGURE 7.6. Solution for a sphere, initially at a uniform temperature, T_i. At $t = 0$, the temperature at the surface of the sphere is instantaneously elevated to T_b. We have selected $r/R = 0.5$ for our example, which corresponds to the vertical line in the middle of the figure.

We also need to point out that Fourier series solutions can be extended to parabolic problems with *multiple* spatial variables. Suppose we have a slab of material that extends from $x = 0$ to $x = L$ and from $y = 0$ to $y = H$ that has some initial distribution of temperature in the interior, $T(x, y)$. At $t = 0$, all four edges are instantaneously changed to a new temperature which we take to be zero for convenience. The governing equation is

$$\frac{\partial T}{\partial t} = \alpha\left[\frac{\partial^2 T}{\partial x^2} + \frac{\partial^2 T}{\partial y^2}\right], \quad (7.66)$$

and we let $T = f(x)g(y)h(t)$. After dividing by αfgh, we find

$$\frac{h'}{\alpha h} = \frac{f''}{f} + \frac{g''}{g} = -\lambda^2. \quad (7.67)$$

Of course, we immediately see that $h = C_1\exp(-\alpha\lambda^2 t)$, and that

$$\frac{f''}{f} = -\lambda^2 - \frac{g''}{g}, \quad (7.68)$$

which gives us a function of x on the left and a function of y on the right. We use the familiar argument and conclude that both sides must be equal to a constant:

$$\frac{f''}{f} = -\lambda^2 - \frac{g''}{g} = -\eta^2. \quad (7.69)$$

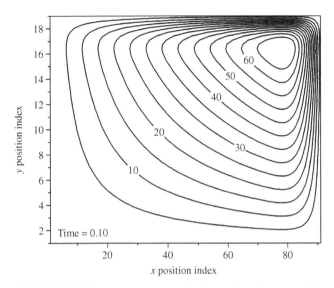

FIGURE 7.7. Typical temperature distribution in a slab at $t = 0.10$ given an initial distribution of $T(x, y, t = 0) = 100xy$; at $t = 0$, all four edges are reduced to $T = 0$.

Accordingly, we find

$$f = A\sin\eta x + B\cos\eta x \quad \text{and}$$
$$g = C\sin\sqrt{\lambda^2 - \eta^2}\,y + D\cos\sqrt{\lambda^2 - \eta^2}\,y. \quad (7.70)$$

We arrange the dependent variable, T, such that $T = 0$ for both $x = 0$ and $y = 0$; therefore, $B = D = 0$ leaving us with

$$T = A\exp(-\alpha\lambda^2 t)\sin\eta x\sin\sqrt{\lambda^2 - \eta^2}\,y. \quad (7.71)$$

When $x = L$, $T = 0$, so $\sin(\eta L) = 0$, resulting in $\eta = m\pi/L$. Similarly, when $y = H$, $T = 0$, so $\sin\sqrt{\lambda^2 - \eta^2}\,H = 0$, requiring that $\sqrt{\lambda^2 - \eta^2} = n\pi/H$. The solution for our problem can be written as

$$T = \sum_{m=1}^{\infty}\sum_{n=1}^{\infty} A_{mn}\exp(-\alpha\lambda^2 t)\sin(\eta x)\sin\sqrt{\lambda^2 - \eta^2}\,y, \quad (7.72)$$

with λ and η determined as shown previously. Now we will assume that the slab has an initial distribution of temperature, $T(x, y, t = 0) = T_0 xy$ (with $T_0 = 100$), and let $L = H = 1$. This means that the maximum *initial* temperature is in the upper right-hand corner, and for small ts, we should see something similar to the contours shown in Figure 7.7.

The coefficients for our double Fourier series are determined from

$$A_{mn} = 400\int_0^1\int_0^1 xy\sin m\pi x\sin n\pi y\,dx\,dy. \quad (7.73)$$

A few of the computed coefficients are provided as follows to allow the student to further explore this situation:

(m, n) =					
(1, 1)	40.5283	(1, 2)	−20.268	(1, 3)	13.5125
(2, 2)	10.1321	(2, 3)	−6.7571	(2, 4)	5.0673
(3, 3)	4.5031	(3, 4)	−3.3780		
(4, 4)	2.5330	(4, 5)	−2.0268		
(5, 5)	1.6211	(5, 6)	−1.3513		
(6, 6)	1.1258				

We want to offer a final word regarding the solution of parabolic PDEs: You may recall that, at the beginning of this chapter, we mentioned the possibility of time-varying boundary conditions. For example, we can envision a heat (or mass) transfer problem in which a surface condition (or flux) changes with time; as an illustration, we might think about the diurnal variation of solar radiation on an outside surface. Another possibility is that the concentration of solute in a solvent might increase from zero and approach some maximum value (maybe its solubility) asymptotically. Such problems, when posed correctly, can be solved with *Duhamel's theorem*; the solution is constructed from the *fundamental* solution obtained for *constant* surface conditions. An illustration for the interested reader will be provided (along with a numerical analysis) in the next chapter; a very useful discussion also appears in Carslaw and Jaeger (1959) in section 1.14.

Elliptic Equations

Elliptic (or often, *potential*) equations apply to equilibrium phenomena. Familiar situations include steady-state conduction (of thermal energy) in a slab and viscous flow in a duct—these are examples of the Laplace and Poisson equations, respectively. We should keep in mind, however, that there are many other applications for *potential* equations including gravitation, electrostatics, and ideal (inviscid) fluid flow. Let us illustrate some of the issues we will encounter while using molecular conduction as our example phenomenon.

We have a two-dimensional slab of material with three sides maintained at some fixed temperature and the upper (top) surface maintained at a different (elevated) temperature (this is exactly the same problem as *slow* viscous flow in a rectangular duct in which motion is driven by an upper surface sliding in the z-, or axial, direction). Since the dependent variable (T) is specified everywhere on the boundary, this is an example of a *Dirichlet problem*. We will place the origin at the lower left-hand corner. The slab extends from $x = 0$ to $x = L$ and from $y = 0$ to $y = H$. The governing equation in this case is

$$\frac{\partial^2 T}{\partial x^2} + \frac{\partial^2 T}{\partial y^2} = 0. \quad (7.74)$$

As has been our practice, we take $T = f(x)g(y)$. This leads directly to two ODEs:

$$f'' + \lambda^2 f = 0 \quad \text{and} \quad g'' - \lambda^2 g = 0. \quad (7.75)$$

The solutions for the two differential equations are

$$f = A \sin \lambda x + B \cos \lambda x \quad (7.76)$$

and

$$g = C \sinh \lambda y + D \cosh \lambda y. \quad (7.77)$$

We now define the dependent variable T as the difference between T and the temperature of the two sides and the bottom; this gives us $T = 0$ on the left, the right, and the bottom. Since we placed the origin in the lower-left corner, we can use odd functions to build our solution. Thus,

$$T = A \sin \lambda x \sinh \lambda y. \quad (7.78)$$

Note how the boundary conditions for $x = 0$ and $y = 0$ are satisfied. Of course, T must also be zero when $x = L$; consequently, it is necessary for $\sin(\lambda L) = 0$, which means $\lambda = n\pi/L$. The solution we seek is therefore

$$T = \sum_{n=1}^{\infty} A_n \sin \frac{n\pi x}{L} \sinh \frac{n\pi y}{L}. \quad (7.79)$$

It remains for us to identify the A_ns, and we use the boundary condition at the top for this purpose. Let us assume that $T = 100°$ for $y = H$; of course, this leads to the Fourier sine series:

$$\frac{100}{\sinh \dfrac{n\pi H}{L}} = \sum_{n=1}^{\infty} A_n \sin \frac{n\pi x}{L}, \quad (7.80)$$

and by definition,

$$A_n = \frac{200}{L \sinh \dfrac{n\pi H}{L}} \int_0^L \sin \frac{n\pi x}{L} dx. \quad (7.81)$$

The integral is -2 for odd n (which must be multiplied by $-L/n\pi$) and 0 for even n. Therefore,

$$T = \frac{400}{\pi} \sum_{n=1,3,5,\ldots}^{\infty} \frac{\sin \dfrac{n\pi x}{L} \sinh \dfrac{n\pi y}{L}}{n \sinh \dfrac{n\pi H}{L}}. \quad (7.82)$$

A contour plot of this result, given $L = 10$ and $H = 10$, follows in Figure 7.8.

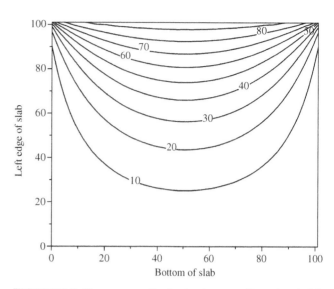

FIGURE 7.8. Temperature distribution in a two-dimensional slab with the top maintained at $100°$ and the other three sides at $T = 0°$.

Now, let us consider how the solution will be impacted if the right-hand boundary has a Neumann boundary condition; for example, it may be insulated such that $(\partial T/\partial x)_{x=L} = 0$. The preliminary steps are exactly the same, up to the point where

$$T = A \sin \lambda x \sinh \lambda y. \quad (7.83)$$

The Neumann condition for the right-hand boundary will require that

$$0 = A\lambda \cos \lambda L \sinh \lambda y, \quad (7.84)$$

and therefore, $\lambda = \pi/2L$, $3\pi/2L$, $5\pi/2L$, and so on. Applying the condition at the top of the slab leads to a Fourier sine series just as before; the A_ns are determined by definition and the integration results in

$$T = \frac{400}{\pi} \sum_{n=0,1,2,3,\ldots}^{\infty} \frac{\sin \dfrac{(1+2n)\pi x}{2L} \sinh \dfrac{(1+2n)\pi y}{2L}}{(1+2n) \sinh \dfrac{(1+2n)\pi H}{2L}}. \quad (7.85)$$

Of course, in this case, we have blocked heat transfer through the right-hand side, so we can expect the temperature contours to be perpendicular to that edge; this is illustrated in Figure 7.9.

In our previous examples of solutions for the Laplace equation for steady conduction in a two-dimensional slab, we placed the origin at the lower left-hand corner. However, there can be definite advantages to placing it at the center; for example, in cases with symmetry, the solution can be built from even functions. And it could also facilitate adapting a solution to a problem in a spherical geometry. Consider a slab extending from $x = -L$ to $x = +L$, and from $y = -H$

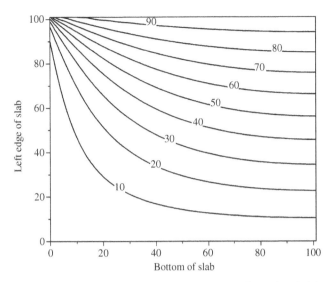

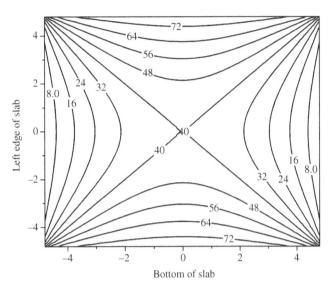

FIGURE 7.9. Temperature distribution in a two-dimensional slab with the right-hand side insulated and the top edge maintained at 100°.

FIGURE 7.10. Temperature distribution in a slab with symmetric (left-right and top-bottom) Dirichlet boundary conditions: 80° top and bottom, and 0° for the left-hand and right-hand sides.

to $y = +H$. We maintain constant temperature of 80° for the top and bottom, and 0° for the left- and right-hand sides. All of the initial steps in the solution procedure are the same as used previously except that the solution must be constructed using *cos* and *cosh*:

$$T = B \cos \lambda x \cosh \lambda y. \tag{7.86}$$

Since $T = 0$ for $x = L$, it is necessary that $\lambda L = \pi/2$, $3\pi/2$, $5\pi/2$, and so on. Therefore,

$$T = \sum_{n=0}^{\infty} B_n \cos \lambda_n x \cosh \lambda_n y, \tag{7.87}$$

where

$$\lambda_n = \frac{(1+2n)\pi}{2L}. \tag{7.88}$$

Of course, for $y = H$, $T = 80°$, so

$$80 = \sum_{n=0}^{\infty} B_n \cos \frac{(1+2n)\pi x}{2L} \cosh \frac{(1+2n)\pi H}{2L}. \tag{7.89}$$

The leading coefficients can now be determined using Fourier's definition, resulting in

$$B_n = \frac{320}{(1+2n)\pi \cosh \dfrac{(1+2n)\pi H}{2L}} \sin \frac{(1+2n)\pi}{2}, \tag{7.90}$$

and the distribution of T in the slab appears as shown in Figure 7.10.

We want to illustrate another slab problem with several rather interesting features. Suppose we have steady conduction in a slab that extends from $x = 0$ to $x = L$, and from $y = 0$ to $y = H$. Once again, we place the origin in the lower left-hand corner. The left edge and the top surface are both insulated, so at $x = 0$, $\partial T/\partial x = 0$, and at $y = H$, $\partial T/\partial y = 0$. The right edge ($x = L$) loses heat to the surroundings, so a Robin's-type boundary condition will be applied. The bottom of the slab has a temperature distribution: At $y = 0$, $T = f(x)$, and we will specify $f(x)$ later.

As in our previous examples,

$$T = (A \sin \lambda x + B \cos \lambda x)(C \sinh \lambda y + D \cosh \lambda y). \tag{7.91}$$

Because of the Neumann condition at the left edge of the slab, $A = 0$. Things appear a bit more difficult with respect to the y-surfaces (or edges), but there is an easy fix: We take

$$T = B \cos \lambda x \cosh \lambda (H - y). \tag{7.92}$$

Now we use the Robin's-type condition at the right-hand edge and you may want to verify that it results in the transcendental equation,

$$\lambda L \tan \lambda L = \frac{hL}{k}. \tag{7.93}$$

For our purposes, we will take the Biot modulus to be 1 and also let $H = L = 1$. The first 10 roots are shown here *along with* the coefficients, B_n, which will be determined as follows using eq. (7.94):

n	λ	B_n
1	0.8603	117.681
2	3.4256	−3.9488
3	6.4373	0.02904
4	9.5293	−1.1756 × 10^{-3}
5	12.6453	+1.4677 × 10^{-5}
6	15.7713	−7.3618 × 10^{-7}
7	18.9024	+1.1305 × 10^{-8}
8	22.0365	−5.6184 × 10^{-10}
9	25.1725	+1.0054 × 10^{-11}
10	28.3096	−4.033 × 10^{-13}

Once again, we see a case where we must use orthogonality—this is not a Fourier series problem! We now select $f(x) = 100 + 100x$; therefore, the temperature across the bottom of the slab varies from $100°$ to $200°$. Consequently, we can determine the B_ns from

$$B_n = \frac{\int_0^1 (100 + 100x)\cos \lambda_n x\, dx}{\cosh \lambda_n H \int_0^1 \cos^2 \lambda_n x\, dx}. \qquad (7.94)$$

An immediate question is raised: Is it possible that the first 10 roots of the transcendental equation, along with the accompanying 10 values for B_n, accurately portray the solution? If so, then we should be able to extract $f(x)$ for the bottom, where $y = 0$ using just B_1 through B_{10}. We will examine this behavior in Figure 7.11.

The transcendental eq. (7.93) occurs so regularly in applied mathematics that it is useful to have ready access to roots. The first 10 roots of $x\tan(x) = C$ have been computed for different values of C (ranging from 0.001 to 100) using Newton–Raphson and these results are presented in tabular form:

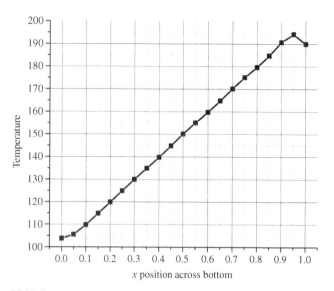

FIGURE 7.11. The temperature should vary across the bottom from $100°$ at $x = 0$ to $200°$ at $x = 1$. Therefore, at $x = 0.3$, $T = 130°$, and for $x = 0.75$, $T = 175°$, and so on. Clearly, the intermediate values are reasonably accurate, but the truncated series (with 10 terms) does not work so well at the ends. Additional terms will be necessary if we wish to improve our solution across the bottom.

$C = 0.8$	1	2	4	6	8	10
0.79103	0.86033	1.07687	1.26459	1.34955	1.39782	1.42887
3.37438	3.42562	3.64360	3.93516	4.11162	4.22636	4.30580
6.40740	6.43730	6.57833	6.81401	6.99236	7.12628	7.22811
9.50871	9.52933	9.62933	9.81188	9.96667	10.09492	10.20026
12.62963	12.64529	12.72230	12.86776	12.99881	13.11413	13.21419
15.75869	15.77129	15.83361	15.95363	16.06540	16.16746	16.25936
18.89188	18.90241	18.95468	19.05646	19.15314	19.24354	19.32703
22.02745	22.03650	22.08148	22.16965	22.25450	22.33509	22.41085
25.16452	25.17245	25.21190	25.28961	25.36502	25.43744	25.50638
28.30259	28.30964	28.34478	28.41419	28.48196	28.54756	28.61058

$C = 20$	40	60	80	100
1.49613	1.53250	1.54505	1.55141	1.55525
4.49148	4.59794	4.63529	4.65428	4.66577
7.49541	7.66466	7.72592	7.75732	7.77637
10.51167	10.73341	10.81720	10.86064	10.88713
13.54198	13.80484	13.90937	13.96435	13.99809
16.58640	16.87944	17.00262	17.06855	17.10931
19.64394	19.95755	20.09715	20.17334	20.22083
22.71311	23.03937	23.19308	23.27878	23.33272
25.79232	26.12497	26.29056	26.38496	26.44501
28.88002	29.21432	29.38965	29.49194	29.55774

$C = 0.001$	0.002	0.004	0.006	0.008	0.010	0.020
0.03162	0.04471	0.06320	0.07738	0.08932	0.09983	0.14095
3.14191	3.14223	3.14287	3.14350	3.14414	3.14477	3.14795
6.28334	6.28350	6.28382	6.28414	6.28446	6.28478	6.28637
9.42488	9.42499	9.42520	9.42541	9.42563	9.42584	9.42690
12.56645	12.56653	12.56669	12.56685	12.56701	12.56717	12.56796
15.70803	15.70809	15.70822	15.70835	15.70847	15.70860	15.70924
18.84961	18.84966	18.84977	18.84987	18.84998	18.85009	18.85062
21.99119	21.99124	21.99133	21.99142	21.99151	21.99160	21.99206
25.13278	25.13282	25.13290	25.13298	25.13306	25.13314	25.13354
28.27437	28.27440	28.27448	28.27455	28.27462	28.27469	28.27504

$C = 0.040$	0.060	0.080	0.1	0.2	0.4	0.6
0.19868	0.24253	0.27913	0.31105	0.43284	0.59324	0.70507
3.15427	3.16057	3.16685	3.17310	3.20393	3.26355	3.32037
6.28955	6.29272	6.29589	6.29906	6.31485	6.34613	6.37700
9.42902	9.43114	9.43326	9.43538	9.44595	9.46700	9.48793
12.56955	12.57114	12.57273	12.57432	12.58226	12.59811	12.61390
15.71051	15.71178	15.71305	15.71433	15.72069	15.73338	15.74605
18.85168	18.85274	18.85380	18.85486	18.86016	18.87075	18.88132
21.99297	21.99388	21.99479	21.99570	22.00024	22.00932	22.01839
25.13433	25.13513	25.13592	25.13672	25.14070	25.14865	25.15659
28.27575	28.27645	28.27716	28.27787	28.28141	28.28847	28.29554

We will conclude our discussion of *elliptic equations in rectangular coordinates* with a Poisson-type example; the solution procedure will lend itself to a variety of problems, including heat transfer with constant thermal energy production and pressure-driven viscous flow in a duct. Suppose we have slab of material that extends in the x-direction from $x = -A$ to $x = +A$ and in the y-direction from $y = -B$ to $y = +B$. We have steady conduction accompanied by thermal energy production throughout the interior. The production occurs at a *constant rate* such that

$$k\left[\frac{\partial^2 T}{\partial x^2} + \frac{\partial^2 T}{\partial y^2}\right] + P = 0. \qquad (7.95)$$

We will take the temperature at each edge of the slab to be zero. We referred earlier to the similarity to steady viscous flow in a duct which is governed by the equation

$$\mu\left[\frac{\partial^2 v_z}{\partial x^2} + \frac{\partial^2 v_z}{\partial y^2}\right] - \frac{dp}{dz} = 0. \qquad (7.96)$$

Indeed, with the no-slip (zero velocity) condition at the walls, this is *exactly* the same problem. In either case, if we can eliminate the inhomogeneity, we may be able to solve this problem just as we have in previous examples in this section. Let us now look at the conduction problem with *constant production* of thermal energy. Consider the effect of adding $-(P/2k)x^2$ to the usual product of functions of x and y; the result is that $-P/k$ appears on both sides of the equation, eliminating the problem caused by the production term. Consequently,

$$T = -\frac{P}{2k}x^2 + B\cos\lambda x\cosh\lambda y. \qquad (7.97)$$

Since $T = 0$ for $x = A$, we conclude that the solution must be written as

$$T = \frac{P}{2k}(A^2 - x^2) + \sum_{n=1}^{\infty} B_n \cos\lambda_n x \cosh\lambda_n y, \quad (7.98)$$

where the λ_ns come from $\cos(\lambda A) = 0$, of course. We select $P/k = 100$ and let $A = B = 1$; therefore,

$$\lambda = \frac{(2n+1)\pi}{2}, \quad \text{for } n = 0, 1, 2, 3, \dots. \qquad (7.99)$$

This is a Fourier series problem, and the coefficients, B_n, are determined from

$$B_n = \frac{-100}{\cosh\lambda_n} \int_0^1 (1 - x^2)\cos\lambda_n x\, dx. \qquad (7.100)$$

Of course, the maximum temperature will occur in the center of the slab as shown in Figure 7.12.

The same procedures we employed for the elliptic equations previously can be used in *cylindrical coordinates* as well. For example, consider a solid cylinder with diameter $2R$, which extends from $z = 0$ to $z = L$. The curved surface and the flat, circular end at $z = L$ are always maintained at

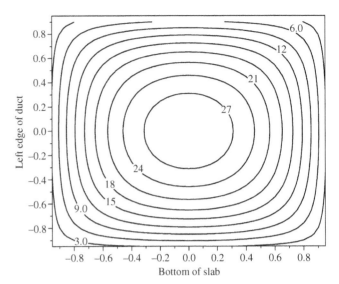

FIGURE 7.12. Temperature contours in a two-dimensional slab with constant production throughout the interior and the edges maintained at $0°$, as computed from eq. (7.98), eq. (7.99), and eq. (7.100).

$T = 0$. The end located at $z = 0$ is maintained at $T = T_0$ for all time. The governing equation for this case is

$$\frac{\partial^2 T}{\partial r^2} + \frac{1}{r}\frac{\partial T}{\partial r} + \frac{\partial^2 T}{\partial z^2} = 0. \qquad (7.101)$$

We take $T = f(r)g(z)$, which results in two ODEs:

$$f'' + \frac{1}{r}f' + \lambda^2 f = 0 \quad \text{and} \quad g'' - \lambda^2 g = 0. \quad (7.102)$$

The solutions for these two equations are

$$f = AJ_0(\lambda r) + BY_0(\lambda r) \quad \text{and} \quad g = C\sinh\lambda z + D\cosh\lambda z. \qquad (7.103)$$

Since the temperature must be finite at the center, $B = 0$. We can accommodate the boundary condition at the end of the cylinder at $z = L$ by taking

$$T = AJ_0(\lambda r)\sinh\lambda(L - z). \qquad (7.104)$$

Furthermore, since $T(r = R) = 0$, $J_0(\lambda R) = 0$, and

$$T = \sum_{n=1}^{\infty} A_n J_0(\lambda_n r)\sinh\lambda_n(L - z). \qquad (7.105)$$

We must get $T = T_0$ for $z = 0$, so utilizing orthogonality,

$$\frac{T_0}{\sinh\lambda_n L}\int_0^R rJ_0(\lambda_n r)dr = A_n\int_0^R rJ_0^2(\lambda_n r)dr. \quad (7.106)$$

The coefficients are therefore given by

$$A_n = \frac{2T_0}{\lambda_n R J_1(\lambda_n R)\sinh \lambda_n L}. \quad (7.107)$$

We will choose $R = 1$, $L = 2$, and $T_0 = 200°$, so we can examine the behavior of this solution; we fix $r = 1/2$ and use increasing values for z, beginning with $z = 1/16$:

z	1/16	1/8	1/4	1/2	1	3/2
$T°$	184.7	158.9	117.6	64.4	19.2	5.3

Application to Hyperbolic Equations

We pointed out at the beginning of this chapter that hyperbolic equations are usually associated with wave-type phenomena. Separation of variables can be applied to many of these problems, and we can illustrate this with the vibrating "string." It is a fascinating sidelight to note that vibrating string problems were solved by Euler and Bernoulli, among others, in the *middle of the eighteenth century*. Thus, we have 260 years' worth of experience with elementary hyperbolic problems to draw on.

Assume our string extends from $x = 0$ to $x = L$ and that both ends are fixed such that $\phi(x = 0) = 0$ and $\phi(x = L) = 0$ for all t. The string is given an initial displacement, $\phi(x, t = 0) = s(x)$, where the function $s(x)$ is specified. It may also have a distribution of initial velocity, $(\partial\phi/\partial t)(x, t = 0) = v(x)$. The equation of interest is

$$\frac{\partial^2\phi}{\partial x^2} = \frac{1}{c^2}\frac{\partial^2\phi}{\partial t^2}, \quad (7.108)$$

As usual, we take $\phi = f(x)g(t)$, and of course, this hyperbolic PDE meets all of the criteria for separation. Substitution and division by the product, fg, results in

$$\frac{f''g = \frac{1}{c^2}fg''}{fg}. \quad (7.109)$$

Consequently, we find two second-order ODEs:

$$f'' + \lambda^2 f = 0 \quad \text{and} \quad g'' + c^2\lambda^2 g = 0. \quad (7.110)$$

Of course, $f = A\sin\lambda x + B\cos\lambda x$, and since f must vanish for both $x = 0$ and $x = L$, we note $B = 0$ and $\lambda = n\pi/L$. We can also see by inspection that

$$g = C\sin c\lambda t + D\cos c\lambda t. \quad (7.111)$$

Therefore,

$$\phi = \sum_{n=1}^{\infty} \sin\lambda_n x[C_n \sin c\lambda_n t + D_n \cos c\lambda_n t]. \quad (7.112)$$

For $t = 0$, the string has some initial displacement, $\phi(x, t = 0) = f(x)$, and consequently,

$$f(x) = \sum_{n=1}^{\infty} D_n \sin\lambda_n x. \quad (7.113)$$

This is merely a Fourier sine series, so

$$D_n = \frac{2}{L}\int_0^L f(x)\sin\frac{n\pi x}{L}dx. \quad (7.114)$$

We can also accommodate a distribution of initial velocity by differentiating, $\partial\phi/\partial t$. Of course, *if the initial velocity is zero*, then $C_n = 0$, and the solution can be written as

$$\phi = \sum_{n=1}^{\infty}\frac{2}{L}\int_0^L f(x)\sin\frac{n\pi x}{L}dx \cdot \sin\frac{n\pi x}{L}\cos\frac{cn\pi t}{L}. \quad (7.115)$$

To illustrate what this solution will produce, we take $L = 10$, $c = 1$, and $f(x) = 10x - x^2$; some results are shown in Figure 7.13.

Now suppose we have a rectangular *membrane* that spans the gap between fixed supports located at $x = 0$ and $x = L$ and also at $y = 0$ and $y = H$. The appropriate equation is

$$\frac{\partial^2\phi}{\partial t^2} = s^2\left[\frac{\partial^2\phi}{\partial x^2} + \frac{\partial^2\phi}{\partial y^2}\right]. \quad (7.116)$$

We are interested in the response of the membrane to some initial displacement, which may be a function of both x and y, but the initial velocity of the membrane,

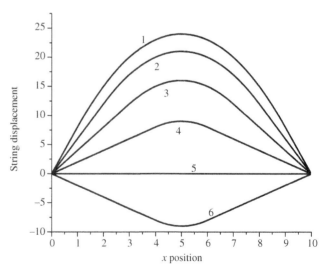

FIGURE 7.13. String displacement for ts of 1, 2, 3, 4, 5, and 6 from the solution of the hyperbolic PDE, eq. (7.95). The initial displacement was $f(x) = 10x - x^2$.

$\partial \phi / \partial t$, is zero. We begin by letting $\phi = f(x)g(y)h(t)$, and this results in

$$\frac{h''}{s^2 h} = \frac{f''}{f} + \frac{g''}{g} = -\lambda^2. \qquad (7.117)$$

Consequently, $h'' + s^2 \lambda^2 h = 0$ and $h = A \sin s\lambda t + B \cos s\lambda t$. The balance of the equation is rewritten so that

$$\frac{f''}{f} = -\frac{g''}{g} - \lambda^2, \qquad (7.118)$$

and we note that the left-hand side is a function of x only. Therefore, a second constant, η, is introduced such that

$$f'' + \eta^2 f = 0 \quad \text{and} \quad \frac{g''}{g} + \lambda^2 = \eta^2 \qquad (7.119)$$
$$\text{or} \quad g'' + (\lambda^2 - \eta^2)g = 0.$$

The solutions for the two additional equations are

$$f = C \sin \eta x + D \cos \eta x \qquad (7.120)$$

and

$$g = E \sin \sqrt{\lambda^2 - \eta^2}\, y + F \cos \sqrt{\lambda^2 - \eta^2}\, y. \qquad (7.121)$$

We now have

$$\begin{aligned} \phi = &(A \sin s\lambda t + B \cos s\lambda t)(C \sin \eta x + D \cos \eta x) \\ &\times (E \sin \sqrt{\lambda^2 - \eta^2}\, y + F \cos \sqrt{\lambda^2 - \eta^2}\, y), \end{aligned} \qquad (7.122)$$

but because the membrane is anchored at the edges, we must have $\phi = 0$ for both $x = 0$ and $y = 0$. This will require that $D = F = 0$. We also know that the initial velocity of the membrane is zero, that is, $\partial \phi / \partial t = 0$ for $t = 0$; therefore, $A = 0$ as well. What is left is a bit more tractable:

$$\phi = B \cos s\lambda t \sin \eta x \sin \sqrt{\lambda^2 - \eta^2}\, y. \qquad (7.123)$$

Now we have to ensure that the other two (supported) edges are fixed, so $\phi = 0$ for both $x = L$ and $y = H$. For the former, $\sin(\eta L) = 0$, so $\eta = m\pi/L$, and for the latter, $\sqrt{\lambda^2 - \eta^2} = n\pi/H$. Therefore,

$$\lambda^2 = \frac{n^2 \pi^2}{H^2} + \frac{m^2 \pi^2}{L^2},$$

and if we take $L = H = 1$, then

$$\phi = \sum_{m=1}^{\infty} \sum_{n=1}^{\infty} B_{mn} \cos(s\pi \sqrt{n^2 + m^2}\, t) \sin(m\pi x) \sin(n\pi y). \qquad (7.124)$$

The membrane has some initial displacement, so for $t = 0$, $\phi = \phi_0(x, y)$:

$$\phi_0(x, y) = \sum_{m=1}^{\infty} \sum_{n=1}^{\infty} B_{mn} \sin(m\pi x) \sin(n\pi y), \qquad (7.125)$$

which allows us to determine the needed coefficients:

$$B_{mn} = 4 \int_0^1 \int_0^1 \phi_0(x, y) \sin(m\pi x) \sin(n\pi y)\, dx dy. \qquad (7.126)$$

Depending on the form of the function, $\phi_0(x, y)$, it may be easier to evaluate the double integral, eq. (7.126), numerically. Algorithms for this purpose are discussed in Chapter 4.

The Schrödinger Equation   Before we leave our discussion of hyperbolic PDEs, we want to consider one of the most important developments of twentieth century physics. Erwin Schrödinger (1926) developed the equation that bears his name that, assuming it can be solved, yields the wave function for a system of interest. The Schrödinger equation revolutionized our thinking about particles and waves at small scales, and it made it very clear that classical Newtonian mechanics was wrong—only very slightly wrong at macroscopic scales to be sure—but very wrong at atomic scales. When Schrödinger was able to demonstrate that his model confirmed the discrete electron energy levels for hydrogen atoms that Niels Bohr had predicted more than a decade earlier, the proof was at hand. From that point on, quantum mechanics rapidly expanded our understanding of how atoms and particles behave.

Waves or particles or both? Thomas Young's dual-slit experiment with light 200 years ago revealed interference patterns, a sure sign that light was wavelike in its behavior (contrary to Newton's corpuscular theory, the idea that light consisted of small particles). By the late nineteenth and early twentieth centuries, evidence had begun to accumulate, suggesting that there were serious problems with classical mechanics. Atkins (1978) reviews how quantum mechanics resolved some of the phenomena known to be problematic, including the thermal properties of solids at low temperatures, the UV catastrophe, the photoelectric effect, and atomic and molecular spectra. The latter are particularly persuasive since the spectra reveal that a molecule can only absorb and emit light at specific, discrete frequencies. Then, in the 1920s, Louis de Broglie suggested that in addition to light (photons), other types of particles (like neutrons and electrons) would have a wavelength related to their momentum. And sure enough, it was discovered that interference patterns were generated by those particles as well; wavelike behavior was demonstrated in a variety of classic experiments, including the dual slit mentioned earlier and the diffraction of electrons from a crystal lattice (the Davisson–Germer experiment was a definitive moment in twentieth century physics).

If a particle such as an electron exhibits wavelike behavior, then it would seem that the simple wave (eq. 7.108) we considered previously might have an important connection to quantum mechanics. We will begin our exploration of this topic by rewriting eq. (7.108):

$$\frac{\partial^2 \phi}{\partial x^2} = \frac{1}{c^2}\frac{\partial^2 \phi}{\partial t^2}. \qquad (7.108)$$

and then setting $\phi = \psi(x)g(t)$ but with $g = \sin(2\pi ft)$. The result, after we divide by the product of $\psi(x)g(t)$, is

$$\frac{\psi''}{\psi} = -\frac{4\pi^2 f^2}{c^2}\psi \quad \text{or} \quad \frac{d^2\psi}{dx^2} + \frac{4\pi^2 f^2}{c^2}\psi = 0. \quad (7.127)$$

We will refer to ψ as the wave function. Now we assume that the total energy (E) of a particle is the sum of kinetic and potential (U) contributions: $E = (1/2)mv^2 + U$. We will let the momentum (p) of a particle be represented by $p = mv$, so that $v^2 = p^2/m^2$, and thus, $E = (1/2m)p^2 + U$. The wavelength, λ, is related to the momentum by Planck's constant: $\lambda = h/p$ (this is known as de Broglie's relationship). Since velocity can be written as the product of frequency and wavelength, $v = f\lambda$, we find

$$f^2 = 2m(E-U)\frac{v^2}{h^2}. \qquad (7.128)$$

The characteristic velocity in eq. (7.108), c, is replaced by v, so we obtain

$$\frac{d^2\psi}{dx^2} + \frac{8\pi^2 m}{h^2}(E-U)\psi = 0. \qquad (7.129)$$

This is the *time-independent Schrödinger equation* for one spatial dimension. We can generalize for three-dimensional problems:

$$\nabla^2\psi + \frac{8\pi^2 m}{h^2}(E-U)\psi = 0. \qquad (7.130)$$

It is common practice in quantum mechanics to replace h with $\hbar$; we note that $\hbar = h/2\pi$, such that $h^2 = 4\pi^2\hbar^2$. We will restrict our attention momentarily to the one-dimensional case for which E is a positive constant, but $U = 0$. Let us assume that this one-dimensional "box" extends from $x = 0$ to $x = L$, and at these ends, an infinite potential barrier is in place. Accordingly, we have

$$\frac{d^2\psi}{dx^2} + \beta\psi = 0, \text{ where } \beta = \frac{8\pi^2 mE}{h^2}. \qquad (7.131)$$

Therefore, $(D^2 + \beta)\psi = 0$ such that $(D + i\sqrt{\beta})(D - i\sqrt{\beta}) = 0$, and consequently,

$$\psi = A\sin\sqrt{\beta}x + B\cos\sqrt{\beta}x. \qquad (7.132)$$

We require that $\psi = 0$ at both $x = 0$ and $x = L$; from the former, $B = 0$, and from the latter,

$$\sin\sqrt{\beta}L = 0. \qquad (7.133)$$

Thus, $\sqrt{\beta}L = n\pi$ and $E_n = (n^2h^2)/(8mL^2)$. n is a *quantum number*, and the expression for E_n gives us the discrete, allowable energy states. This, in turn, means that the behavior of the wave function is described by $\sin(n\pi x/L)$; when $n = 1$, we get a half-wave over the interval $0 \leq x \leq L$, when $n = 2$, we get a complete cycle, and for $n = 3$, 1½ cycles, and so on.

Before we leave our example of the one-dimensional box, we should make note of an important consequence of quantum mechanics. Let us rewrite the solution for this case in a fully equivalent form:

$$\psi = C_1\exp\left(\frac{2\pi ix}{h}\sqrt{2m(E-U)}\right). \qquad (7.134)$$

We now imagine that our particle exits the one-dimensional box and enters the region where the potential barrier is very large (*but not infinite*). Under these conditions, $U > E$ and $2m(E - U)$ is negative. If we factor out the i resulting from the $\sqrt{(-1)}$, we find

$$\psi = C_1\exp\left(-\frac{2\pi x}{h}\sqrt{2m(U-E)}\right). \qquad (7.135)$$

This is a representation of the wave function *inside* the potential barrier, and it is important because Max Born recognized that the probability of finding the particle of interest at a particular position, x, was $|\psi(x)|^2$. This means that, based on eq. (7.135), there would be a small *but finite* probability that the particle in question could be found *outside* the box. This is called *tunneling*, and while it clearly does not apply to a marble sealed in a tin can, it does apply to electrons if they are confined by a finite potential barrier.

Now we turn our attention to the particle in a two-dimensional enclosure, but we let the potential well be circular such that $U = 0$ for $0 \leq r \leq R$, but $U = \infty$ for $r > R$. For this case,

$$\frac{1}{r}\frac{\partial}{\partial r}\left(r\frac{\partial\psi}{\partial r}\right) + \frac{1}{r^2}\frac{\partial^2\psi}{\partial\theta^2} + \frac{8\pi^2 m}{h^2}E\psi = 0. \quad (7.136)$$

Let us again try the product method, setting $\psi = f(r)g(\theta)$; the result is

$$-\frac{r^2}{f}\left(f'' + \frac{1}{r}f'\right) - \frac{8\pi^2 mE}{h^2}r^2 = \frac{g''}{g} = -\eta^2. \quad (7.137)$$

And in the familiar pattern, we obtain two ODEs:

$$g'' + \eta^2 g = 0 \quad (7.138)$$

with

$$f'' + \frac{1}{r}f' + \left(\frac{8\pi^2 mE}{h^2} - \frac{\eta^2}{r^2}\right)f = 0. \quad (7.139)$$

The solution for the first member of this pair is $g = A\sin\eta\theta + B\cos\eta\theta$. We simply choose to have $g = 0$ at $\theta = 0$ (think of the prime meridian at Greenwich) such that $B = 0$. And since $g(0°)$ must be the same as $g(360°)$, it is clear that we can have only integers for η (integer multiples of π). Now we rewrite eq. (7.139) in a more useful form (standard form for Bessel's differential equation):

$$r^2\frac{d^2 f}{dr^2} + r\frac{df}{dr} + \left(\frac{8\pi^2 mE}{h^2}r^2 - \eta^2\right)f = 0. \quad (7.140)$$

The solution for eq. (7.140) can be written as

$$f = C_1 J_\eta\left(\sqrt{\frac{8\pi^2 mE}{h^2}}r\right) + C_2 Y_\eta\left(\sqrt{\frac{8\pi^2 mE}{h^2}}r\right). \quad (7.141)$$

However, the wave function must be finite at the center (at $r = 0$) so $C_2 = 0$; furthermore, the wave function must be zero at $r = R$, and consequently,

$$J_\eta\left(\sqrt{\frac{8\pi^2 mE}{h^2}}R\right) = 0. \quad (7.142)$$

This constrains E to a series of distinct values (as expected), and we illustrate with the case for which $\eta = 1$. Since the zeros for J_1 occur at 3.83171, 7.01559, 10.17347, 13.32369, 16.47063, and so on, we find

$$\frac{8\pi^2 mE_1}{h^2} = \left(\frac{3.83171}{R}\right)^2 \quad (7.143)$$

or

$$E_1 = \left(\frac{3.83171}{R}\right)^2\frac{h^2}{8\pi^2 m}. \quad (7.144)$$

We have a complete description of the behavior of the wave function inside the circular potential well. An obvious extension of this problem is to make the box three-dimensional by allowing the cylinder to have some finite height in the z-direction. Wolfram™ has a very nice demonstration project that illustrates the behavior of the wave function for this case.

In the two previous examples, we employed infinitely deep potential wells for elementary one- and two-dimensional problems. We now move to a more realistic particle-in-a-box scenario in three dimensions; we want to develop a model for the hydrogen atom with one electron (or a *hydrogen-like* atom), so our "box" will actually be a spherical shell. But in this case, the potential energy of the electron will be $U = -(e^2/r)$, such that $U(r = 0) = -\infty$ and $U(r \to \infty) = 0$. The PDE for the wave function will be

$$\frac{1}{r^2}\frac{\partial}{\partial r}\left(r^2\frac{\partial\psi}{\partial r}\right) + \frac{1}{r^2\sin\theta}\frac{\partial}{\partial\theta}\left(\sin\theta\frac{\partial\psi}{\partial\theta}\right)$$
$$+ \frac{1}{r^2\sin^2\theta}\frac{\partial^2\psi}{\partial\phi^2} + \frac{8\pi^2 m}{h^2}\left(E + \frac{e^2}{r}\right)\psi = 0. \quad (7.145)$$

Note that this is a linear PDE and once again we have a candidate for separation; we will apply the product method and see if we can obtain radial and angular components. We will let $\psi(r, \theta, \phi) = \beta(r) \cdot \gamma(\theta, \phi)$ to begin. After a little work, we divide by the product of $\beta\gamma$ to obtain

$$\frac{r^2\frac{d^2\beta}{dr^2} + 2r\frac{d\beta}{dr}}{\beta} + \frac{1}{\gamma\sin\theta}\frac{\partial}{\partial\theta}\left(\sin\theta\frac{\partial\gamma}{\partial\theta}\right)$$
$$+ \frac{1}{\gamma\sin^2\theta}\frac{\partial^2\gamma}{\partial\phi^2} + \frac{8\pi^2 m}{h^2}(Er^2 + e^2 r) = 0. \quad (7.146)$$

The radial and angular portions can be taken to opposite sides and set equal to a constant of separation, say, $-\eta^2$. The $\beta(r)$ result is

$$r^2\frac{d^2\beta}{dr^2} + 2r\frac{d\beta}{dr} + \frac{8\pi m}{h^2}(Er^2 + e^2 r)\beta = \eta^2\beta. \quad (7.147)$$

It is standard procedure to replace the constant of separation (η^2) in eq. (7.140) with $\ell(\ell + 1)$; this is done for reasons that have been explained by Wieder (1973, p. 135) and the rationale will become apparent shortly. Now we can also separate the remaining equation for $\gamma(\theta, \phi)$ by setting $\gamma = C(\theta)D(\phi)$. Of course, another constant of separation arises and we let it be $-\alpha^2$. The polar (θ) part produces

$$\sin^2\theta\frac{d^2 C}{d\theta^2} + \sin\theta\cos\theta\frac{dC}{d\theta} + \left(\sin^2\theta \cdot \ell(\ell + 1) - \alpha^2\right)C = 0, \quad (7.148)$$

and the azimuth (ϕ) part yields

$$\frac{d^2 D}{d\phi^2} + \alpha^2 D = 0. \quad (7.149)$$

It is clear that eq. (7.149) is the easiest of the trio and we recognize

$$D = a_1 \exp(-i\alpha\phi) + a_2 \exp(+i\alpha\phi). \qquad (7.150)$$

Let us turn our attention to eq. (7.148); we let $z = \cos\theta$ such that $dC/d\theta = (dC/dx)(dx/d\theta)$, where $dx/d\theta = -\sin\theta$; we also replace $C(\theta)$ with $U(z)$. The result is

$$(1-z^2)\frac{d^2U}{dz^2} - 2z\frac{dU}{dz} + \left(\ell(\ell+1) - \frac{\alpha^2}{1-z^2}\right)U = 0, \quad (7.151)$$

which is an *associated Legendre differential equation*, with a solution that can be written in terms of Legendre polynomials. The product of the solutions for eq. (7.148) and eq. (7.149) is central to quantum mechanics, and it is referred to as a spherical harmonic (presented in normalized form):

$$\gamma_\ell^\alpha(\theta,\phi) = \varepsilon \left[\frac{(2\ell+1)}{4\pi}\frac{(\ell-|\alpha|)!}{(\ell+|\alpha|)!}\right]^{1/2} \exp(+i\alpha\phi)P_\ell^\alpha(\cos\theta),$$

$$(7.152)$$

where $\varepsilon = (-1)^\alpha$ for $\alpha \geq 0$. The P_ℓ^αs are associated Legendre functions obtained by differentiation of the Legendre polynomials; for example,

$$P_\ell^\alpha = (1-z^2)^{\alpha/2}\frac{d^\alpha P_\ell}{dz^\alpha}.$$

Legendre polynomials were introduced in Chapter 4 in our discussion of numerical quadrature and the first five Legendre polynomials are provided for the reader near the end of this chapter. For the constants of separation appearing in eq. (7.152), $\ell = 0, 1, 2,\ldots$ and $\alpha = 0, \pm 1, \pm 2,\ldots, \pm\ell$. Lastly, we consider the equation for the radial portion of the solution, eq. (7.147). If we divide by r^2 and restrict our attention to "large" values of r, then we get an equation that will be valid asymptotically (with the electron moved very far from the nucleus):

$$\frac{d^2\beta}{dr^2} + \frac{8\pi mE}{h^2}\beta \approx 0. \qquad (7.153)$$

Using differential operator notation again,

$$\left(D^2 + \frac{8\pi mE}{h^2}\right)\beta = (D+i\kappa)(D-i\kappa)\beta = 0$$

with $\kappa = \sqrt{8\pi mE/h^2}$. Therefore,

$$\beta = a_1 \exp\left(-i\sqrt{\frac{8\pi mE}{h^2}}r\right) + a_2 \exp\left(+i\sqrt{\frac{8\pi mE}{h^2}}r\right). \quad (7.154)$$

Obviously, we also need to know how the wave function behaves nearer to the nucleus. We can achieve a little simplification of eq. (7.147) by setting $\beta = \Omega/r$, which results in

$$\frac{d^2\Omega}{dr^2} + \frac{8\pi m}{h^2}\left(E + \frac{e^2}{r}\right)\Omega = \frac{\ell(\ell+1)}{r^2}\Omega. \qquad (7.155)$$

This equation can be solved for certain cases, for example, for the infinite spherical potential well, for which one obtains spherical Bessel and Neumann functions. However, in the instance of the hydrogen atom with its Coulomb potential, the equation is solved by expansion in a power series. This is made a little easier by the fact that we know something about the asymptotic radial behavior of the wave function. In principle at least, we can obtain an analytic solution for a particle in a spherical "box" which is the product of the three solutions, subject to any simplifying restrictions imposed, for example, eq. (7.153). You should make note of the fact that three quantum numbers have appeared quite naturally in the solution procedure; they are referred to as the *principal*, *azimuthal* (*often called orbital*), and *magnetic* quantum numbers.

APPLICATIONS OF THE LAPLACE TRANSFORM

You may recall from our previous discussion in Chapter 5 that the formal definition of the Laplace transform of a function of time, $f(t)$, is

$$L\{f(t)\} = F(s) = \int_0^\infty e^{-st}f(t)dt. \qquad (7.156)$$

The effect, of course, is that a continuous function of time is transformed to the s-plane. In our present context, the characteristic of the Laplace transform that is most important concerns time derivatives. Let $df/dt = f'(t)$:

$$L\{f'(t)\} = sL\{f(t)\} - f(t=0). \qquad (7.157)$$

In other words, we replace the derivative with *multiplication* by s, and subtract off the initial condition. It is obviously advantageous to formulate the problem in terms of "deviation" variables such that the initial value of f (for $t = 0$) is *zero*. Let us now see exactly what this will accomplish for us when applied to a parabolic PDE.

Assume we have a transient problem with molecular transport in one spatial direction in a semi-infinite medium such that

$$\frac{\partial\phi}{\partial t} = \kappa\frac{\partial^2\phi}{\partial x^2}. \qquad (7.158)$$

Our boundary conditions will have the form $\phi(x = 0, t) = 2$ and $\phi(x \rightarrow \infty, t) = 0$, with the initial condition $\phi(x, t = 0) = 0$. Naturally, this means that for positive t, ϕ (whatever that is) will flow—actually diffuse—into the medium from the left-hand boundary, where $x = 0$. Rearranging the equation and applying the Laplace transform yields the *subsidiary* equation:

$$\frac{d^2\phi(s)}{dx^2} - \frac{1}{\kappa} s\phi(s) = 0. \qquad (7.159)$$

If this ODE can be solved (with the corresponding boundary condtions), then we will obtain the *Laplace transform of the solution* of the PDE. If we can successfully invert that transform, we will find the solution we seek. An exponential solution is found for the ODE:

$$\phi(s) = C_1 \exp\left(-\sqrt{\frac{s}{\kappa}}x\right) + C_2 \exp\left(+\sqrt{\frac{s}{\kappa}}x\right). \quad (7.160)$$

Of course, this result cannot be unbounded in the x-direction, so $C_2 = 0$. We also transform the boundary condition:

For $x = 0$, $\phi(s) = 2/s$, and therefore,

$$\phi(s) = \frac{2}{s} \exp\left(-\sqrt{\frac{s}{\kappa}}x\right). \qquad (7.161)$$

In this case, we can turn immediately to a table of Laplace transforms (e.g., see the table in chapter 5 or section 29 in Abramowitz and Stegun, 1965), finding the pair

$$\frac{1}{s}\exp\left(-k\sqrt{s}\right) \qquad erfc\left(\frac{k}{2\sqrt{t}}\right). \quad (7.162)$$

Accordingly, the solution we seek is

$$\phi = 2 erfc\left(\frac{x}{\sqrt{4\kappa t}}\right). \qquad (7.163)$$

Let us illustrate this process again using the very same PDE (applied to the molecular transport of thermal energy) but with a more difficult boundary condition; our semi-infinite slab extends in the x-direction away from the interface located at $x = 0$. This time, the end of the slab at $x = 0$ is exposed to a fluid maintained at some elevated temperature, so we write

$$-k\frac{\partial T}{\partial x}\bigg|_{x=0} = h\left(T_\infty - T\big|_{x=0}\right). \qquad (7.164)$$

We divide by $-k$ and transform the boundary condition at $x = 0$:

$$\frac{\partial T(s)}{\partial x} - \frac{h}{k}T(s) = -\frac{hT_\infty}{k}\frac{1}{s}. \qquad (7.165)$$

The subsidiary equation is precisely the same as before:

$$\frac{d^2T(s)}{dx^2} - \frac{s}{\alpha}T(s) = 0, \qquad (7.166)$$

so

$$T(s) = C_1 \exp\left(-\sqrt{\frac{s}{\alpha}}x\right). \qquad (7.167)$$

Of course, $C_2 = 0$ since the transform must remain bounded. We differentiate $T(s)$ with respect to x and set $x = 0$; the transformed boundary condition is then used to find C_1:

$$C_1 = \frac{\dfrac{hT_\infty}{k}\dfrac{1}{s}}{\dfrac{h}{k} + \sqrt{\dfrac{s}{\alpha}}}. \qquad (7.168)$$

Therefore,

$$T(s) = \frac{\dfrac{hT_\infty}{k}\dfrac{1}{s}}{\dfrac{h}{k} + \sqrt{\dfrac{s}{\alpha}}} \exp\left(-\sqrt{\frac{s}{\alpha}}x\right) = \frac{HT_\infty}{s(H+Q)}\exp(-Qx),$$

$$(7.169)$$

where $H = h/k$ and $Q = \sqrt{s/\alpha}$. Once again, this is a form that we can find in a suitable table of Laplace transforms, and returning to the time domain, we can write

$$\frac{T}{T_\infty} = erfc\frac{x}{\sqrt{4\alpha t}} - \exp\left(\frac{hx}{k} + \frac{h^2\alpha t}{k^2}\right)erfc\left(\frac{x}{\sqrt{4\alpha t}} + \frac{h}{k}\sqrt{\alpha t}\right).$$

$$(7.170)$$

The utility of the Laplace transform for these transient conduction–diffusion problems in semi-infinite slabs is apparent. But can the technique also be applied to more difficult problems? The answer is a qualified yes, and we demonstrate this with a problem in which heat is transferred in the z-direction in a cylindrical rod with loss to a fluid surrounding the rod's surface. Heat flow is initiated by raising the temperature of the end of the rod at $z = 0$. The conductivity of the metal rod is large, so that the bulk of the *resistance* to heat flow is on the *fluid side* of the interface; therefore, we assume that the temperature of the solid rod does not vary (much) in the transverse or r-direction. An *approximate* model for this process is

$$\rho C_p \frac{\partial T}{\partial t} = k\frac{\partial^2 T}{\partial z^2} - \frac{2h}{R}(T - T_\infty), \qquad (7.171)$$

where $(2h/R)(T - T_\infty)$ accounts for loss at the surface of the rod. We define a new dependent variable, $\theta = T - T_\infty$, and divide by ρC_p, obtaining

$$\frac{\partial \theta}{\partial t} = \alpha \frac{\partial^2 \theta}{\partial z^2} - \frac{2h}{\rho C_p R}\theta, \qquad (7.172)$$

with the following boundary and initial conditions:

$$\theta(z = 0, t) = \theta_0, \quad \theta(z \to \infty, t) = 0, \quad \text{and} \quad \theta(z, t = 0) = 0.$$

Note that the second boundary condition given here is an *idealization*; for physically real situations, the medium that extends in the z-direction will certainly be of finite extent. However, if t is small, the medium may effectively appear to be very "deep" in the z-direction. The subsidiary equation is

$$\frac{d^2\theta(s)}{dz^2} - \frac{2h}{kR}\theta(s) - \frac{s}{\alpha}\theta(s) = 0. \qquad (7.173)$$

Once again, the solution for this equation can be found in a suitable table of Laplace transforms, allowing us to return directly to the time domain:

$$\theta = \frac{1}{2}\left[\exp\left(\sqrt{\frac{2h}{kR}}z\right)erfc\left(\frac{z}{\sqrt{4\alpha t}} + \sqrt{\frac{2h}{\rho C_p R}}t\right) + \exp\left(-\sqrt{\frac{2h}{kR}}z\right)erfc\left(\frac{z}{\sqrt{4\alpha t}} - \sqrt{\frac{2h}{\rho C_p R}}t\right)\right]. \qquad (7.174)$$

The reader may recognize that this is *exactly the same problem* as absorption into a quiescent liquid accompanied by a first-order, irreversible chemical reaction.

Now, suppose the solution of the subsidiary equation *cannot* be found in a table of transforms; is there any recourse? One possibility is through the application of the *inversion theorem*, which requires contour integration. The procedure is described by Carslaw and Jaeger (1959) in chapter 12, and they provide an illustrative example in section 12.6. In some cases, it is also possible that the solution of the subsidiary equation can be expanded into a series whose *individual* terms can be found in the table of transforms. For example, consider the quotient, $\cosh(bx)/\cosh(bL)$; we rewrite the hyperbolic functions so that

$$\cosh(bx)/\cosh(bL) = \frac{e^{bx} + e^{-bx}}{e^{bL} + e^{-bL}} = \frac{e^{bx} + e^{-bx}}{e^{bL}(1 + e^{-2bL})},$$

then use the binomial theorem to expand this into the series: $(e^{-b(L-x)} + e^{-b(L+x)})\sum_{n=0}^{\infty}(-1)^n e^{-2nbL}$. This technique can be useful *if* the resulting series converges rapidly enough.

APPROXIMATE SOLUTION TECHNIQUES

Occasionally, we will encounter a problem for which the techniques described earlier will not work, and an alternative numerical solution is either undesirable or simply not useful for the analyst's purpose. In such cases, we may be forced to seek an approximate analytic solution. We do have options in these circumstances and we will describe a couple of useful approaches here. For the reader unfamiliar with the approximate solution of PDEs, Villadsen and Michelsen (1978) is a good starting point.

Many of the methods that are available to us for this purpose have the same underlying theme: We choose a suitable polynomial that either automatically satisfies the boundary conditions or can easily be made to satisfy them. We then "adjust" the polynomial by determining values for the coefficients that—*in some sense*—give us the best possible performance. We can begin to think about this in the following way: Suppose we have a differential equation,

$$D(\phi) = 0, \quad \text{where } \phi = f(y) \quad \text{and} \quad a \le y \le b. \qquad (7.175)$$

We propose a trial function:

$$\phi_{trial} = \phi_0 + \sum_{i=1}^{n} c_i \phi_i(y). \qquad (7.176)$$

We define the *residual*, R, as

$$R = D(\phi_{trial}). \qquad (7.177)$$

If we could somehow force $R = 0$ for all y between a and b, we would have *the* solution! Of course, that really is not the objective; our aim is to find an analytic approximation that is reasonably accurate and cost-effective from the standpoint of time invested. Thus, we will settle for a compromise.

We can illustrate some of the principal ideas with a simple steady-state example from conduction. Imagine a slab of type 347 stainless steel for which one face is maintained at 0°F and the other at 1000°F. Over this temperature range, the thermal conductivity of 347 increases (almost linearly) by more than 60%. We let $k = a + bT$ and note that in rectangular coordinates,

$$\frac{d}{dy}\left[k(T)\frac{dT}{dy}\right] = 0. \qquad (7.178)$$

Therefore, the nonlinear differential equation of interest is

$$(a + bT)\frac{d^2T}{dy^2} + b\left(\frac{dT}{dy}\right)^2 = 0. \qquad (7.179)$$

Our boundary conditions for this problem are the following:

$$\text{At } y = 0, T = 0°\text{F}, \quad \text{and}$$

$$\text{at } y = h, T = 1000°\text{F}.$$

For convenience, we set $h = 1$ ft, and we arbitrarily propose

$$T = \sum C_n y^n, \quad \text{such that}$$

$$T = C_0 + C_1 y + C_2 y^2 + C_3 y^3 + \cdots. \tag{7.180}$$

If we set $C_0 = 0$, the boundary condition at $y = 0$ is automatically satisfied. We form the residual (R) by truncating eq. (7.180) and substituting the result into eq. (7.179):

$$[a + b(C_1 y + C_2 y^2 + C_3 y^3)](2C_2 + 6C_3 y) \\ + b(C_1 + 2C_2 y + 3C_3 y^2)^2 = R. \tag{7.181}$$

Our task now is to choose values for C_1, C_2, and C_3 that result in the smallest possible value for R. This minimization of R can take several different forms; for example, if we employ a weighting function, $W(y)$, and write

$$\int_0^h W(y)R\,dy = 0, \tag{7.182}$$

we have the method of weighted residuals (MWR). Finlayson (1980) points out that if we use the Dirac delta function for $W(y)$, then we are employing a simple collocation scheme where the residual will be zero at a few select points.

If we force the residual to be zero at the endpoints and also require eq. (7.180) to satisfy the boundary condition at $y = h$, then we have the three simultaneous algebraic equations:

$$2aC_2 + bC_1^2 = 0, \tag{7.183}$$

$$[a + b(C_1 + C_2 + C_3)](2C_2 + 6C_3) + b(C_1 + C_2 + C_3)^2 = 0, \tag{7.184}$$

and

$$1000 - C_1 - C_2 - C_3 = 0. \tag{7.185}$$

A solution is found by successive substitution:

$$C_1 = 1641.434, \quad C_2 = -920.838, \quad \text{and} \quad C_3 = 279.40.$$

We will also use a fourth-order Runge–Kutta scheme to solve eq. (7.179) numerically for comparison, and both solutions are shown in Figure 7.14.

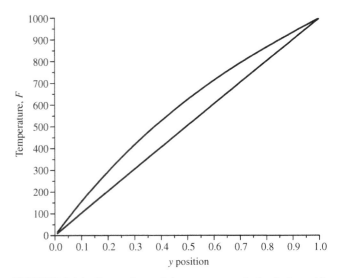

FIGURE 7.14. Comparison of the exact numerical solution with the collocation result (upper curve), which was obtained by requiring that $R = 0$ at both $y = 0$ and $y = h$. Although the approximate solution exhibits some similar behavior, it is *very* rough quantitatively.

It is essential that we note exactly what occurred here: We set the residual, R, to zero only at the *endpoints* of the interval, and this was done strictly for convenience. We cannot generally expect to obtain useful results this way.

Galerkin MWR Applied to a PDE

Let us look at an improved variant of the process described previously and apply it to a transient conduction (or diffusion) problem with a temperature- or concentration-dependent diffusivity. Such an equation might appear as

$$\frac{\partial \phi}{\partial t} = \frac{\partial}{\partial y}\left(\kappa(\phi)\frac{\partial \phi}{\partial y}\right). \tag{7.186}$$

We take $\kappa = 1 + \phi$ for simplicity, which results in the non-linear PDE,

$$\frac{\partial \phi}{\partial t} = (1 + \phi)\frac{\partial^2 \phi}{\partial y^2} + \left(\frac{\partial \phi}{\partial y}\right)^2. \tag{7.187}$$

Our boundary and initial conditions are $\phi(y = 0, t) = 1$, $\phi(y = 1, t) = 0$, and $\phi(y, t = 0) = 0$. In other words, we have a medium that initially has uniform (or zero) concentration or temperature; we elevate the concentration (or temperature) at the front face ($y = 0$), and diffusion into the medium commences. Finlayson (1980) points out that the MWR is well suited to this type of problem. We will use the Galerkin technique (named after the Russian mathematician Boris G. Galerkin) and begin by taking

$$\phi = 1 + b(t)y + c(t)y^2. \tag{7.188}$$

Of course, the boundary condition at the far face ($y = 1$) must be satisfied, so

$$-1 = b(t) + c(t) \quad \text{or} \quad c(t) = -1 - b(t). \quad (7.189)$$

We use this to eliminate $c(t)$ from eq. (7.188), resulting in the trial function,

$$\phi = 1 - y^2 + b(t)y(1 - y). \quad (7.190)$$

Now we take the original PDE and multiply by the weighting function (which in the Galerkin MWR is taken from the basis, or trial, functions) and integrate from $y = 0$ to $y = 1$. The left-hand side becomes

$$\int_0^1 y(1 - y)\frac{\partial \phi}{\partial t} dy, \quad (7.191)$$

where $\partial \phi / \partial t = b'y(1 - y)$. Therefore, this integral is simply

$$\int_0^1 \frac{db}{dt} y^2 (1 - y)^2 \, dy = \frac{1}{30}\frac{db}{dt}. \quad (7.192)$$

For the right-hand side of the equation, we differentiate the trial function as indicated in eq. (7.192), multiply by $y(1 - y)dy$, and obtain

$$\int_0^1 \left\{ (2 - y^2 + b(y - y^2))(-2 - 2b) + (b - 2y(1 + b))^2 \right\} y(1 - y)dy. \quad (7.193)$$

Equating the results of this integration with eq. (7.192) yields a first-order ODE for $b(t)$: $db/dt = -11 - 17b - b^2$. Although this equation could be integrated to produce an analytic solution, it is certainly easier to evaluate $b(t)$ numerically. Finlayson (1980) used the initial value (for b) of -2; however, advance knowledge of the numerical solution makes it possible to choose a "better" value in terms of the quality of the approximation at advanced times: We will employ $b(t = 0) = -3.325$. This results in $b(t = 0.10) = -1.3125$, and a comparison of the results from the Galerkin method with the actual numerical solution is shown in Figure 7.15.

This relatively simple approach to the solution of a nonlinear PDE has yielded acceptable results requiring determination of only one unknown function of time, $b(t)$. Naturally, the approximation could be improved by simply continuing the expansion (eq. 7.188), but at the risk of defeating the whole purpose; remember, our objective is to *quickly* find a suitable analytic approximation for $\phi(y)$.

The Rayleigh–Ritz Method

At the beginning of the twentieth century, Walther Ritz devised a method for approximating eigenfunctions based

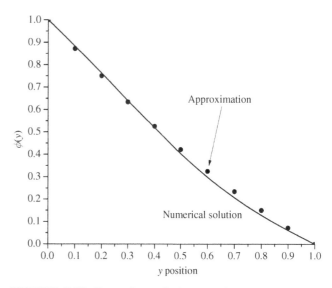

FIGURE 7.15. Comparison of the numerical solution (solid curve) of the nonlinear PDE for $t = 0.10$ with the approximation obtained from the Galerkin MWR.

on the minimization of certain integrals. This technique, commonly referred to as the Rayleigh–Ritz method, can be used to solve boundary-value problems governed by elliptic PDEs. The core of the procedure involves application of the *Dirichlet principle*, which concerns identification of a function that minimizes the integral:

$$I = \iiint (\phi_x^2 + \phi_y^2 + \phi_z^2)dxdydz. \quad (7.194)$$

It can be written for two dimensions (which will be of direct use to us) as

$$I = \iint |grad\phi|^2 \, dxdy. \quad (7.195)$$

Let us illustrate how this method works with an example adapted from chapter 12 in Weinberger (1965).

Consider the elliptic PDE,

$$\frac{\partial^2 \phi}{\partial x^2} + \frac{\partial^2 \phi}{\partial y^2} = 0, \quad (7.196)$$

defined over a triangular region for which $x > 0$, $y > 0$, and $x + 2y < 2$. For the bottom of the triangle,

$$\phi(x, y = 0) = x(2 - x), \quad (7.197)$$

which means that the maximum value at the bottom boundary occurs at $x = 1$: $\phi(x = 1, y = 0) = 1$. For the left-hand edge,

$$\phi(x=0, y)=0. \tag{7.198}$$

For the hypotenuse,

$$\phi(x=2-2y, y)=0. \tag{7.199}$$

Our plan is to select trial functions that satisfy the boundary conditions such that

$$\phi \cong \phi_0 + a_1\phi_1 + a_2\phi_2 + \cdots. \tag{7.200}$$

We hope to identify the constants, a_1, a_2, a_3, and so on, that give us the *best possible* approximation. This is to be achieved by finding values that give

$$\iint |grad(\phi_{\text{trial}} - \phi)|^2 \, dxdy, \tag{7.201}$$

the smallest obtainable value. We will now truncate our approximation, eq. (7.200), and demonstrate how this actually works. We use Dirichlet's principle to formulate an algebraic equation in a_1 (a quadratic in a_1) that can be differentiated and set equal to zero (please keep in mind that if we retained additional trial functions, we would obtain a set of equations for a_1, a_2, and so on, by setting the *partial derivatives* equal to zero):

$$\iint |grad(\phi_0 + a_1\phi_1)|^2 \, dxdy = \iint |grad\phi_0|^2 \, dxdy$$
$$+ 2a_1 \iint grad\phi_0 \cdot grad\phi_1 dxdy + a_1^2 \iint |grad\phi_1|^2 \, dxdy. \tag{7.202}$$

Consequently, the optimal value for the coefficient, a_1, is

$$a_1 = \frac{\iint grad\phi_0 \cdot grad\phi_1 dxdy}{\iint |grad\phi_1|^2 \, dxdy}. \tag{7.203}$$

The trial functions must satisfy the boundary conditions, and we will first try $\phi_0 = x(2 - x - 2y)$ and $\phi_1 = xy(2 - x - 2y)$. Therefore, we will differentiate with respect to x, then y, for the numerator:

$$\frac{\partial \phi_0}{\partial x} = 2(1 - x - y) \quad \text{and} \quad \frac{\partial \phi_0}{\partial y} = -2x, \tag{7.204}$$

and

$$\frac{\partial \phi_1}{\partial x} = 2y(1 - x - y) \quad \text{and} \quad \frac{\partial \phi_1}{\partial y} = x(2 - x - 4y). \tag{7.205}$$

The double integral in the numerator is then

$$\int_0^1 \int_0^{2-2y} 4y(1 - x - y)^2 - 2x^2(2 - x - 4y) dxdy, \tag{7.206}$$

and the form for the denominator is

$$\int_0^1 \int_0^{2-2y} 4y^2(1 - x - y)^2 + x^2(2 - x - 4y)^2 \, dxdy. \tag{7.207}$$

The resulting quotient is $-3/5$, resulting in

$$\phi \cong x(2 - x - 2y) - \frac{3}{5}xy(2 - x - 2y) = x\left(1 - \frac{3}{5}y\right)(2 - x - 2y). \tag{7.208}$$

The elliptic PDE given by eq. (7.196) was also solved numerically so that the quality of the Rayleigh–Ritz approximation could be better assessed, and these numerical results are shown in Figure 7.16.

Now we will compute several values from the approximate solution (eq. 7.208) for comparison:

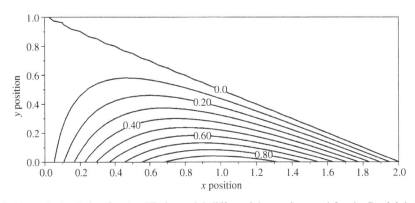

FIGURE 7.16. Numerical solution for the elliptic partial differential equation used for the Rayleigh–Ritz example.

x	y	$\phi(x, y)$
0.1	0.8	0.0156
0.4	0.6	0.1024
0.6	0.3	0.3936
0.8	0.2	0.5632
1.0	0.2	0.5280
1.2	0.2	0.4224
1.4	0.1	0.5264
1.7	0.06	0.2950

You will immediately note that the Rayleigh–Ritz approximation has produced reasonable results; for most of the points provided above, the error is less than 10%, and in many cases, it is only about 2%. In particular, if we look at $(x, y) = (0.6, 0.3)$, the numerical solution yields about 0.4, and for $(x, y) = (1.0, 0.2)$, it produces about 0.53 (the approximate solution has an error that is less than 0.4%).

Collocation

You may have noticed that in the introductory example for this section—the solution of eq. (7.179)—a number of quite arbitrary choices were made; these include the polynomial itself and the location of the points where we forced the residual to be zero. A critical question concerns the placement of the collocation points—an equidistant or haphazard siting is likely to be less than optimal. Therefore, we should contemplate changes to the procedure that may improve the outcome. Suppose we begin by selecting a polynomial that *automatically satisfies the boundary conditions*. In addition, if we use orthogonal polynomials, and place the collocation points at the roots of one or more of the terms, we will significantly decrease the burden placed on the analyst. We are now describing what Villadsen and Stewart (1967) called *interior collocation*.

We can illustrate our first improvement with a nonlinear ODE example from fluid mechanics. Suppose we have a non-Newtonian fluid in a rectangular duct, subjected to a constant pressure gradient. If the fluid exhibits power-law behavior, then one of the possibilities is

$$\frac{d^2v_x}{dy^2} = -C_0\sqrt{\frac{dv_x}{dy}}. \qquad (7.209)$$

The boundary conditions are the following:

$$\text{At } y = 0, v_x = 0 \quad \text{and}$$

$$\text{at } y = 1, v_x = 0.$$

We can avoid any difficulties caused by the sign change on the velocity gradient by noting that at $y = 1/2, dv_x/dy = 0$. For this example, we choose the polynomial

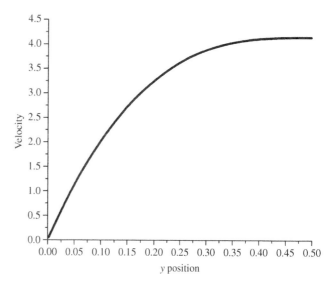

FIGURE 7.17. Exact numerical solution for non-Newtonian flow through a rectangular duct with $C_0 = -20$.

$$v_x = c_1(y - y^2) + c_2(y - y^2)^2 + c_3(y - y^2)^3 + \cdots. \qquad (7.210)$$

The conditions at $y = 0$ and $y = 1/2$ are automatically satisfied. We will select $C_0 = -20$ and find the exact numerical solution (provided in Figure 7.17) so we have a basis for comparison.

The reader should complete this example and compare his or her result with the previously computed profile. Note that it is necessary for $c_1 = 24.91347$ (the reader should confirm this); this value results in an excellent approximation. How many terms must one retain in the assumed polynomial to get accurate results? If we terminate the polynomial with the c_2-term, and require the residual to be zero only at $y = 1/4$, we actually find that

$$c_1 = 46.52397 \quad \text{and} \quad c_2 = -21.68451.$$

Although the resulting shape is correct, this solution is unacceptable because the centerline velocity is roughly twice the correct value. It is clear that we should contemplate further improvements for this technique.

Polynomials are said to be orthogonal on the interval (a, b) with respect to the weighting function, $W(x)$, if

$$\int_a^b W(x)P_n(x)P_m(x)dx = 0, \quad \text{where } n \neq m. \qquad (7.211)$$

Let us consider the first few Legendre polynomials on the interval $(-1, 1)$ for problems that *lack symmetry*. We want to explore how orthogonality may work to our advantage.

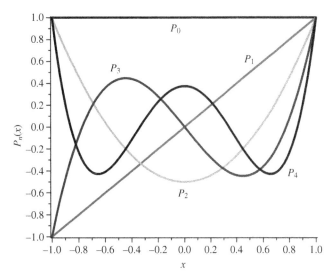

FIGURE 7.18. Legendre polynomials, P_0 through P_4, on the interval -1 to 1.

$$P_0 = 1 \quad P_1 = x \quad P_2 = \frac{1}{2}(3x^2 - 1) \quad P_3 = \frac{1}{2}(5x^3 - 3x)$$

$$P_4 = \frac{1}{8}(35x^4 - 30x^2 + 3).$$

You may want to confirm, for example, that

$$\int\limits_{-1}^{+1} P_1(x)P_2(x)dx = \frac{1}{2}\left[\frac{3}{4}x^4 - \frac{1}{2}x^2\right]_{-1}^{+1} = 0. \quad (7.212)$$

The first five Legendre polynomials are shown in Figure 7.18.

Note that if we were to locate collocation points at $x = \pm 1/\sqrt{3}$, then $P_2 = 0$. Similarly, for $x = \pm\sqrt{(3/5)}$, $P_3 = 0$ A further improvement can be obtained by making the dependent variables the function values *at the collocation points* rather than the coefficients appearing in the polynomial representation. This modified procedure was described by Villadsen and Stewart (1967) and also explained very clearly by Finlayson (1980, pp. 73–74).

Let us now suppose that we have a boundary-value problem *with symmetry* about the centerline where

$$\frac{d^2\phi}{dx^2} + f(x, \phi) = 0. \quad (7.213)$$

The independent variable, x, extends from -1 to 1, and the field variable, ϕ, has a set value (say, 1) at the endpoints. Naturally, at the centerline, $d\phi/dx = 0$. Accordingly, we propose

$$\phi = \phi(\pm 1) + (1 - x^2)\sum C_n P_n(x^2), \quad (7.214)$$

where the P_ns are Jacobi polynomials for a slab:

$n = 0$	1	
$n = 1$	$(7.1 - 5x^2)$	± 0.447214
$n = 2$	$(7.1 - 14x^2 + 21x^4)$	$\pm 0.2852315,$
		± 0.7650555
$n = 3$	$(7.1 - 27x^2 + 99x^4 - 85.8x^6)$	$\pm 0.209299,$
		$\pm 0.5917,$
		± 0.87174

At this point, eq. (7.214) is substituted into eq. (7.213) to form the residual. We can solve this set of equations for the coefficients (the C_ns) or we can develop an alternative set of equations written in terms of the function values (ϕ_ns) at the collocation points. The reader is encouraged to try both approaches for this example.

Orthogonal Collocation for Partial Differential Equations

Orthogonal collocation has also been used to solve elliptic PDEs of the form

$$\frac{\partial^2\phi}{\partial x^2} + \frac{\partial^2\phi}{\partial y^2} = f(x, y), \quad (7.215)$$

on the unit square, $x(0, 1)$ and $y(0, 1)$. Examples of the method's application are provided by Houstis (1978), Prenter and Russell (1976), and Villadsen and Stewart (1967). Please note that an elliptic equation for any rectangular region $x(a, b)$ and $y(c, d)$, can be mapped into the unit square by employing the transformation,

$$x \to \frac{x - a}{b - a} \quad \text{and} \quad y \to \frac{y - c}{d - c}.$$

This broadens the applicability of the technique considerably. Now, let us suppose for illustration that eq. (7.215) has a solution given by

$$\phi = 3e^x e^y (x - x^2)(y - y^2), \quad (7.216)$$

which can be plotted to yield the results shown in Figure 7.19:

Prenter and Russell (1976) solved this problem using bicubic Hermite polynomials, and their results indicate very favorable performance relative to the Ritz–Galerkin method. Furthermore, in some cases, the use of collocation with Hermite polynomials has outperformed solution of elliptic equations by the finite-difference method. Section 22 in Abramowitz and Stegun is a good starting point for the reader interested in the use of Hermite polynomials.

In an example provided by Villadsen and Stewart (1967), the Poisson equation,

$$\frac{\partial^2\phi}{\partial x^2} + \frac{\partial^2\phi}{\partial y^2} = -1, \quad (7.217)$$

(for Poiseuille flow through a duct) was solved on the square $(-1 < x < +1)$, $(-1 < y < +1)$ by taking

$$\phi = (1 - x^2)(1 - y^2)\sum\sum A_{ij}P_i(x^2)P_j(y^2). \quad (7.218)$$

If the expansion is limited to the Jacobi polynomial, $P_1 = (1 - 5x^2)$, and the collocation point is placed at $(x_1, y_1) = (0.447214, 0.447214)$, then

$$\phi \cong \tfrac{5}{16}(1 - x^2)(1 - y^2). \quad (7.219)$$

This solution is plotted in Figure 7.20 along with the correct numerical solution for easy comparison. Note that the truncated approximation is surprisingly good.

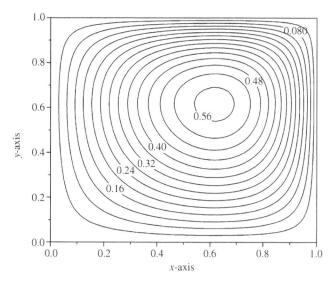

FIGURE 7.19. Solution for the elliptic partial differential equation, $(\partial^2\phi/\partial x^2) + (\partial^2\phi/\partial y^2) = 6xye^xe^y(xy + x + y - 3)$.

Villadsen and Stewart refined this rough solution by including $P_2 = (1 - 14x^2 + 21x^4)$ in the expansion with the three collocations points located at $(x, y) \rightarrow (0.2852315, 0.2852315)$, $(0.7650555, 0.2852315)$, and $(0.7650555, 0.7650555)$. The improved result was

$$\phi \cong (1 - x^2)(1 - y^2)\big[0.31625 - 0.013125(1 - 5x^2 + 1 - 5y^2) \\ + 0.00492(1 - 5x^2)(1 - 5y^2)\big].$$

$$(7.220)$$

Several collocation schemes for elliptic PDEs are available through a FORTRAN-based system called ELLPACK. The development of this software was initiated in 1976 and the effort was coordinated by John Rice of Purdue. Support for the project came from the National Science Foundation, the Department of Energy, and the Office of Naval Research; collocation modules include COLLOCATION, HERMITE COLLOCATION, and INTERIOR COLLOCATION. See the ELLPACK home page for recent developments of this software. ELLPACK allows a user with a minimal knowledge of FORTRAN to solve elliptic PDEs rapidly; even more importantly, the analyst can compare different solution techniques for accuracy and computational speed. Rice and Boisvert (1985) is an excellent starting point for the analyst interested in ELLPACK.

THE CAUCHY–RIEMANN EQUATIONS, CONFORMAL MAPPING, AND SOLUTIONS FOR THE LAPLACE EQUATION

Earlier in this chapter, we discussed the solution of elliptic PDEs using separation of variables. We now want to illustrate a very different approach that can be applied to a

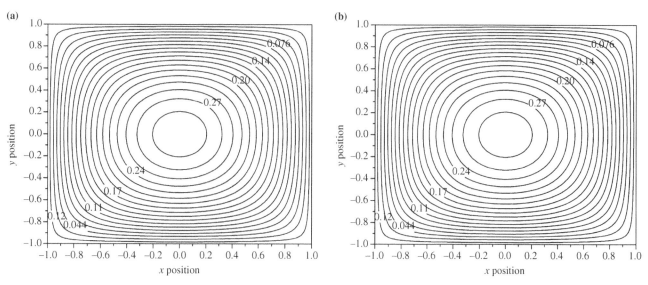

FIGURE 7.20. Comparison of the approximate solution (left) with the correct numerical solution (right).

limited class of elliptic PDEs; we will restrict our attention to the two-dimensional Laplace equation. Functions that satisfy this PDE are said to be harmonic, and it is important for us to remember that only *conservative* fields can be represented by the two-dimensional Laplace equation.

Our initial focus in this section is the function

$$w = f(z) = \phi(x, y) + i\psi(x, y). \qquad (7.221)$$

If w is *analytic* in a region denoted by R, and if ϕ and ψ are related by the *Cauchy–Riemann equations* (see eq. 7.224), then the real and imaginary parts are solutions for the two-dimensional Laplace equation. Indeed, if these conditions are met, then *any* analytic function $f(z)$ is the solution for *some problem* governed by the Laplace equation. Under these circumstances, we refer to the $f(z)$s as solutions for *potential* problems, and we will explore this technique using a topic familiar to students of hydrodynamics, ideal fluid flow.

Although we are using *ideal potential flow* as the framework for our discussion, the technique we present here will be applicable to other types of problems as well including electrostatics, steady two-dimensional diffusion and heat conduction, as well as two-dimensional scattering of electromagnetic waves. Let us begin by clarifying exactly what we mean by an ideal flow: We stipulate that the *fluid* is inviscid and incompressible, and that the *flow* is irrotational; a useful mnemonic device in this context is to think of the *three Is* (inviscid, incompressible, irrotational). For a two-dimensional ideal flow, the velocity vector components can be obtained by differentiation of the *velocity potential*, ϕ, in the corresponding directions:

$$v_x = \frac{\partial \phi}{\partial x} \quad \text{and} \quad v_y = \frac{\partial \phi}{\partial y}. \qquad (7.222)$$

Next we define the *stream function*, ψ, such that

$$v_x = -\frac{\partial \psi}{\partial y} \quad \text{and} \quad v_y = \frac{\partial \psi}{\partial x}. \qquad (7.223)$$

We think of a streamline (a curve of constant ψ) as the path followed by a fluid particle. If streamlines are converging locally, then the flow is accelerating in that region; if the streamlines are diverging, the fluid velocity is decreasing. Evidently, the velocity potential and the stream function are related:

$$\frac{\partial \phi}{\partial x} = -\frac{\partial \psi}{\partial y} \quad \text{and} \quad \frac{\partial \phi}{\partial y} = \frac{\partial \psi}{\partial x}. \qquad (7.224)$$

These are the *Cauchy–Riemann equations*, and they *guarantee* that *any analytic function* of the complex variable, z, where $z = x + iy$, is the solution for *some* potential flow problem; that is, given a function of the complex variable, z, which we will write as $w = f(z)$, we have a mapping between the x-y plane and the $\phi - \psi$ plane; we need only to equate the real and imaginary parts:

$$\phi + i\psi = f(z). \qquad (7.225)$$

This branch of mathematics is known as conformal mapping due to the fact that angles are preserved; in the $\phi - \psi$ plane, velocity potential lines and streamlines intersect at right angles, just as lines of constant x and y do in the x-y plane. We specified an incompressible fluid for which $\nabla \cdot v = 0$, so if we differentiate eq. (7.222) appropriately, then

$$\frac{\partial^2 \phi}{\partial x^2} + \frac{\partial^2 \phi}{\partial y^2} = 0. \qquad (7.226)$$

We also indicated that the flow was to be irrotational, which means that $\nabla \times v = 0$, and if we differentiate eq. (7.223) accordingly, then

$$\frac{\partial^2 \psi}{\partial x^2} + \frac{\partial^2 \psi}{\partial y^2} = 0. \qquad (7.227)$$

Thus, *both* the velocity potential and the stream function are governed by the Laplace equation and—most importantly—if $w(z)$ is a single-valued function of z over the region, R, and if it is differentiable at every point in that region, then every analytic $w(z)$ gives us a solution for the Laplace equation. But it is critical for us to emphasize that had we included fluid friction (a *dissipative* process) in this study of moving fluids, then the Laplace equation would no longer be applicable since we would not have a conservative field.

Let us illustrate the process we have in mind with an elementary example. We set

$$w = f(z) = z^2 + \frac{1}{z^2} = (x + iy)^2 + \frac{1}{(x + iy)^2}$$
$$= \frac{x^2 - y^2}{x^4 + 2x^2y^2 + y^4} - \frac{2xyi}{x^4 + 2x^2y^2 + y^4}. \qquad (7.228)$$

The function is clearly not analytic at $z = 0$, so we exclude that point from the region of interest. Therefore,

$$\phi = (x^2 - y^2)\left[1 + \frac{1}{x^4 + 2x^2y^2 + y^4}\right]$$

and

$$\psi = 2xy\left[1 - \frac{1}{x^4 + 2x^2y^2 + y^4}\right]. \qquad (7.229)$$

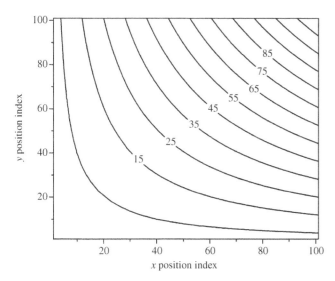

FIGURE 7.21. Plot of the stream function for the complex potential given by eq. (7.228).

So we have identified the velocity potential and the stream function for some potential flow, but we are faced with the immediate question: Exactly what problem governed by the Laplace equation is this a solution for? We will find out by plotting values of ψ for (x, y) pairs in the first quadrant, then we construct appropriate contours. The result is shown in Figure 7.21.

Notice that according to the definition of the stream function given by eq. (7.223), the flow in this case is right to left (the fluid enters the figure at the right-hand boundary and leaves through the upper surface).

We have found a solution for the Laplace equation by a backward process that is easy enough to execute but might not be very useful under more general circumstances. We can, of course, simply write down more functions of z: $w = f(z)$, and identify the results by plotting $\psi(x, y)$. The reader is encouraged to explore this approach and an interesting case (flow over a circular obstruction, or log) is given by

$$w = f(z) = \pi R U \coth\left(\frac{\pi R}{z}\right), \qquad (7.230)$$

where R is the radius of the obstruction and U is the intensity of the approaching flow. Should you wish to try this, start by writing $\coth(x) = (e^x + e^{-x})/(e^x - e^{-x})$, and note that the polar form for a complex number is $x + iy = r(\cos\theta + i\sin\theta)$. Of course, $r = \sqrt{x^2 + y^2}$.

Because the Laplace equation is linear, we can also use superposition to combine individual solutions, building complex potentials (or stream functions) for more complicated problems. For example, we could take a horizontal potential flow around a cylinder, for which

$$\psi = V_\infty \sin\theta\left(r - \frac{R^2}{r}\right),$$

and add to that the vortex, $\psi = K \ln r$. This combination will produce flow about a right circular cylinder with rotation and will result in a vertical lift being generated (the *Magnus effect*). If we combine a source with a uniform flow (both in polar form for convenience) we get

$$\psi = -\frac{K}{2\pi}\theta + Vr\sin\theta,$$

which is flow about a half-body. As you can see, one can obtain the velocity potential and the stream function for many situations of interest by merely combining elementary solutions.

The methodology employed in the earlier example and additionally recommended for a student exercise will certainly work, but it is not very useful when we are seeking a solution for the Laplace equation for a *particular* problem. Moreover, in many cases, this backward or indirect process would just be a needless duplication of effort. Many conformal mappings are known and compilations exist that we can consult directly. One example is the book *Dictionary of Conformal Representations* by Kober (1952). We can find, for example, an extensive collection of functions of the type $w = f(z) = z^a$, where a is real, in part two of Kober's book. Part three is devoted to exponential (and related) functions, for example, $w = e^z$.

Let us explore the use of such a dictionary with an example: We will consider ideal flow into a channel, or alternatively, the potential field (actually equipotential lines) accompanying two charged plates of finite size separated by a distance, $2b$. This problem appears in Kober (1952, section 11.5, pp. 116 and 117), from which we find

$$z = w + \exp(w) \quad \text{or} \quad x + iy = \phi + i\psi + \exp(\phi + i\psi). \qquad (7.231)$$

Since $e^{i\psi} = \cos\psi + i\sin\psi$, it is easy to show that

$$x = \ln\left(\frac{y - \psi}{\sin\psi}\right) + \left(\frac{y - \psi}{\tan\psi}\right). \qquad (7.232)$$

The form of this equation suggests that we select a constant value for ψ, allow y to range through a plausible sequence of values, and compute the corresponding x-positions. This process will allow us to prepare an appropriate plot, which is provided here as Figure 7.22.

We conclude that conformal mapping is an easy approach to the solution of a limited class of problems described by the Laplace equation. But as we pointed out

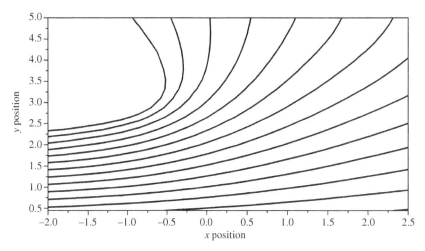

FIGURE 7.22. A partial construction of streamlines for a potential flow entering a channel or a canal. This view shows just the upper right-hand corner of the field.

previously, the indirect approach of writing down an analytic complex potential and then determining what problem is solved by it is not very practical. There are a couple of techniques that can be applied more broadly. For example, let us consider the case in which we have a constant potential along some parametrically defined curve. Specifically, suppose we know that the potential is constant over an ellipse for which $x = A\cos t$ and $y = B\sin t$; therefore,

$$z = x + iy = A\cos w + iB\sin w, \qquad (7.233)$$

where $w = \phi + i\psi$. With a bit of work, we can show that

$$A^2 = \frac{x^2}{\left[\cosh\psi + \dfrac{B}{A}\sinh\psi\right]^2} + \frac{y^2}{\left[-\sinh\psi + \dfrac{B}{A}\cosh\psi\right]^2}. \qquad (7.234)$$

Thus, the "streamlines" or equipotential lines are confocal ellipses, as expected. Another approach that can be quite useful when the potential is known on a *polygonal boundary* is the Schwarz–Christoffel formula; Smith (1953) provides several examples of its application and Bieberbach's (1953) discussion is helpful as well. The SC formula is

$$w = \int\limits_{0}^{z} \frac{dt}{\Pi(t - a_k)^{\alpha_k}}, \quad \text{with } \Sigma\alpha_k = 2. \qquad (7.235)$$

Thus, if the half-plane were to be mapped onto the interior of a triangle with *exterior* angles $\alpha_1\pi$, $\alpha_2\pi$, and $\alpha_3\pi$, then

$$w = \int\limits_{0}^{z} \frac{dt}{(t - a_1)^{\alpha_1}(t - a_2)^{\alpha_2}(t - a_3)^{\alpha_3}}, \qquad (7.236)$$

where all of the αs are greater than zero and their sum, $a_1 + a_2 + a_3 = 2$. a_1, a_2, and so on, are the vertices mapped onto the real axis of the z-plane. We can look at an elementary case given by Lamb (1945) for which two finite points are chosen on the real axis at ± 1:

$$w = A\int \frac{dt}{\sqrt{t^2 - 1}} = A\cosh^{-1}(t) + B. \qquad (7.237)$$

Lamb shows how this method can be used to model a Borda entrance (in two dimensions) and he provides several other interesting results obtained with the Schwarz–Christoffel formula in chapter 4 of *Hydrodynamics*.

CONCLUSION

Many important problems involving molecular (or diffusive) transport arising in engineering and the applied sciences can be solved using the product method or separation of variables. Moreover, extensive collections of these solutions exist (e.g., Crank and Carslaw and Jaeger); frequently, an analyst can consult such resources and directly adapt an existing solution to their needs. This does not mean that every problem that can be solved using the product method has already been solved. There will always be variations that present a new challenge. But, once a student understands the technique, he/she will be much better able to assess what is possible and what is prohibitively difficult. In the more general case of nonlinear PDEs, one must either accept the limitations of an approximate analytic solution or proceed to a numerical simulation. The importance of the latter has grown rapidly—and pretty much in step with the expanding availability of computing power. The numerical solution of PDEs is the subject of the next chapter.

PROBLEMS

7.1. Solve the boundary-value problem

$$\frac{\partial T}{\partial t} = \alpha \frac{\partial^2 T}{\partial y^2}, \quad \text{for } 0 \le y \le 10,$$

given $\alpha = 1$ and $T(y = 0, t) = 0$, $T(y = 10, t) = 45$, and $T(y, t = 0) = 20$.

7.2. Solve the boundary-value problem

$$\frac{\partial T}{\partial t} = \alpha \frac{\partial^2 T}{\partial y^2}, \quad \text{for } 0 \le y \le 10,$$

given $\alpha = 2$ and $T(y = 0, t) = 20$, $T(y = 10, t) = 10$, and $T(y, t = 0) = 5 + 5y$.

7.3. Solve the boundary-value problem

$$\frac{\partial T}{\partial t} = \alpha \frac{\partial^2 T}{\partial y^2}, \quad \text{for } 0 \le y \le 10,$$

given $\alpha = 1/4$ and $T(y = 0, t) = 10$, and for $y = 10$:

$$-k \frac{\partial T}{\partial y}\bigg|_{y=10} = h\left(T|_{y=10} - T_\infty\right).$$

The initial condition is $T(y, t = 0) = 30$ and the Biot number, $hL/k = 1/4$.

7.4. Find the distribution of S over the annular region $R_1 \le r \le R_2$, where S is governed by the potential equation:

$$\frac{\partial^2 S}{\partial r^2} + \frac{1}{r}\frac{\partial S}{\partial r} + \frac{1}{r^2}\frac{\partial^2 S}{\partial \theta^2} = 0.$$

The constant values ot the edges are $S(r = R_1) = 100$ and $S(r = R_2) = 10$.

7.5. Repeat Problem 7.4 but with $R_1 = 1$, $R_2 = 3$, $S(r = R_1) = 100$, and

$$-\kappa \frac{\partial S}{\partial r}\bigg|_{r=R_2} = \beta(S_{r=R_2} - 10).$$

We know that $\beta/\kappa = 1/4$.

7.6. We are investigating a problem governed by the Poisson equation:

$$\frac{\partial^2 \phi}{\partial x^2} + \frac{\partial^2 \phi}{\partial y^2} = -1.$$

Suppose $\phi(x, y) = a_0 + a_1 x + a_2 y + a_3 x^2 + a_4 xy + a_5 y^2$. Find the relationships between coefficients in this polynomial.

7.7. The Laplace equation is applied to a rectangular region that extends from $x = 0$ to $x = L$ and from $y = 0$ to $y = H$:

$$\frac{\partial^2 U}{\partial x^2} + \frac{\partial^2 U}{\partial y^2} = 0.$$

For the left and right sides ($x = 0$ and $x = L$), $U = 1$. For the top and bottom edges ($y = 0$ and $y = H$), the flux is zero, $\partial U/\partial y = 0$ at the bottom, and $\partial U/\partial y = 1$ at the top. Find the distribution, $U(x, y)$.

7.8. Find the distribution of temperature in a slab of material that extends in the x-direction: $0 \le x \le L$. The governing equation is

$$\frac{\partial T}{\partial t} = \alpha \frac{\partial^2 T}{\partial x^2},$$

given that $T(x = 0, t) = T_0$, $(\partial T/\partial x)(x = L, t) = 0$, and $T(x, t = 0) = T_i$.

7.9. Find the solution for

$$\frac{\partial \phi}{\partial t} = \kappa \left[\frac{\partial^2 \phi}{\partial r^2} + \frac{1}{r}\frac{\partial \phi}{\partial r}\right]$$

for the annular region, $R_1 \le r \le R_2$, where $R_1 = 1$ and $R_2 = 4$. We know that $\phi(r = R_1, t) = 200$, $\phi(r, t = 0) = 10$, and

$$-\alpha \frac{\partial \phi}{\partial r} = \beta \phi(r = R_2).$$

7.10. Steady viscous flow in a duct is driven by a sliding upper surface, moving with constant velocity in the z-direction. By Newton's law of friction, the velocity at the other three walls will be zero, of course. The governing equation is

$$0 = \frac{\partial^2 V_z}{\partial x^2} + \frac{\partial^2 V_z}{\partial y^2}, \quad \text{for } 0 \le x \le L \text{ and } 0 \le y \le H.$$

For $x = 0$, $V = 0$, for $y = 0$, $V = 0$, for $x = L$, $V = 0$, and for $y = H$, $V = 10$. Let $H = L = 1$, and find $V(x, y)$.

7.11. Find $T(x, y)$ in a two-dimensional slab with the origin placed in the lower left-hand corner. The top, the right edge, and the left edge are all maintained at $50°$. For the bottom ($y = 0$),

$$T = f(x) = 100.$$

For the two-dimensional slab, $H = 2$ and $L = 1$.

7.12. A long cylinder has a uniform initial temperature, $T_i = 75°$. At $t = 0$, the surface (at $r = R$) is rapidly cooled to $0°$. Find $T(r, t)$:

$$\frac{\partial T}{\partial t} = \alpha \left[\frac{\partial^2 T}{\partial r^2} + \frac{1}{r} \frac{\partial T}{\partial r} \right].$$

The thermal diffusivity, α, is 2, and the cylinder radius is 2.

7.13. A long hollow cylinder has an initial temperature distribution of $T = f(r)$ for $R_1 \leq r \leq R_2$. For all positive ts, the surfaces at R_1 and R_2 are maintained at $0°$. Find $T(r, t)$ for two different cases:

$$f(r) = T_0, \quad \text{and then} \quad f(r) = 10 + 10 \frac{(r - R_1)}{(R_2 - R_1)}.$$

7.14. Consider a diffusion tube that extends from $z = -L$ to $z = +L$. The concentration of the species of interest is governed by

$$\frac{\partial C}{\partial t} = D \frac{\partial^2 C}{\partial z^2}.$$

Of course, the ends are impermeable, so for $z = \pm L$, $\partial C/\partial z = 0$. For the initial condition, $-L \leq z \leq 0$, $C = 1$ and $0 \leq z \leq +L$, $C = 0$. Find $C(z, t)$.

7.15. Given

$$k \left[\frac{\partial^2 T}{\partial x^2} + \frac{\partial^2 T}{\partial y^2} \right] + P = 0,$$

find $T(x, y)$ given $P/k = 200$. The origin is placed at the center, and all four edges are maintained at $50°$. The slab is square with $L = H = 2$.

7.16. Find the distribution of the variable, ϕ, in a circular disk given

$$\phi(r = R, 0 < \theta < \pi) = 1 \quad \text{and} \quad \phi(r = R, -\pi < \theta < 0) = 0.$$

The governing equation is

$$\frac{1}{r} \frac{\partial}{\partial r} \left(r \frac{\partial \phi}{\partial r} \right) + \frac{1}{r^2} \frac{\partial^2 \phi}{\partial \theta^2} = 0.$$

7.17. A porous slab, with surfaces located at $x = \pm b$, is initially saturated with solvent. We wish to model the diffusion process within the slab for a drying problem where the loss of solvent at the surface(s) is described by

$$-D \frac{\partial C}{\partial x} \bigg|_{x=b} = K(C_{x=b} - C_\infty).$$

The governing equation is

$$\frac{\partial C}{\partial t} = D \frac{\partial^2 C}{\partial x^2}.$$

Find $C(x, t)$, and then find an expression for the total amount of solvent lost from the slab over a time, t. Remember: The slab loses solvent from *both* surfaces.

7.18. A taut string is secured (fixed and stationary) at $y = 0$ and $y = 9$. At a point corresponding to $y = 3$, the string is displaced ½ unit in the transverse direction and then released. Find $u(y, t)$ using the product method given:

$$\frac{\partial^2 u}{\partial y^2} = \frac{1}{c^2} \frac{\partial^2 u}{\partial t^2}.$$

7.19. A drum head is a membrane (or skin) stretched over a circular mounting rim of radius, R. The displacement of the membrane in response to an initial forcing function is governed by

$$\frac{\partial^2 \phi}{\partial t^2} = c^2 \left[\frac{\partial^2 \phi}{\partial r^2} + \frac{1}{r} \frac{\partial \phi}{\partial r} + \frac{1}{r^2} \frac{\partial^2 \phi}{\partial \theta^2} \right].$$

Use the product method to find a solution for this problem given that

$$\phi(r = R, \theta, t) = 0, \frac{\partial \phi}{\partial t}(r, \theta, t = 0) = 0, \text{ and}$$

$$\phi(r, \theta, t = 0) = F(r).$$

7.20. A solid sphere of radius, $R = 3$ cm, has an initial temperature of $45°C$. At $t = 0$, the sphere is plunged into a large, cooled bath maintained at $0°C$. If the heat transfer coefficient between the fluid phase and the surface of the sphere is 0.02 cal/(cm^2 s $°C$), and if the thermal diffusivity, α, is 0.9 cm^2/s, find $T(r, t)$ and plot the temperature distributions at $t = 1/2$, 1, 2, and 4 seconds. The density and heat capacity of the material are 1.74 and 0.24, respectively (cgs units). The governing equation for this problem is

$$\rho C_p \frac{\partial T}{\partial t} = k \left[\frac{1}{r^2} \frac{\partial}{\partial r} \left(r^2 \frac{\partial T}{\partial r} \right) \right].$$

7.21. A circular disk of radius, R, is well insulated on the flat, circular faces (top and bottom sides). At $t = 0$, the edge of the disk is rapidly cooled to $0°$ and heat flows in the r-direction toward the rim:

$$\frac{\partial T}{\partial t} = \alpha \left[\frac{\partial^2 T}{\partial r^2} + \frac{1}{r} \frac{\partial T}{\partial r} \right].$$

Find $T(r, t)$ given $T(r, t = 0) = T_0(1 - r^2)$ and $R = 1$.

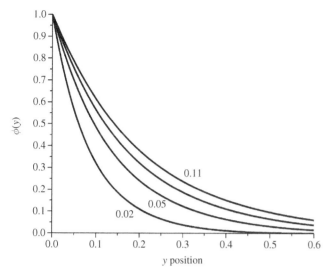

FIGURE 7.23. Computed numerical solutions for ts of 0.02. 0.05, 0.08, and 0.11. Note that the horizontal axis has been truncated to better show the time evolution.

7.22. Use the Galerkin MWR technique to find an approximate solution for

$$\frac{\partial \phi}{\partial t} = \frac{\partial}{\partial y}\left[\kappa \frac{\partial \phi}{\partial y}\right], \quad \text{for } 0 \le y \le 1,$$

given $\kappa = (\phi_0 + 4\phi)^{-1}$. The boundary and initial conditions are $\phi(y = 0, t) = 1$, $\phi(y = 1, t) = 0$, and $\phi(y, t = 0) = 0$. Figure 7.23 contains some numerical results for this problem and it is provided as follows to assist students with their work.

7.23. We have a long, solid cylinder in which thermal energy is produced at a uniform rate (per unit volume) corresponding to S. The surface of the cylindrical solid will be maintained at $T = 0$ for all time, t. The governing equation has the form

$$\frac{\partial T}{\partial t} = \alpha \left[\frac{1}{r}\frac{\partial}{\partial r}\left(r \frac{\partial T}{\partial r}\right)\right] + \frac{S}{\rho C_p}.$$

We want to find the analytic solution for this problem. Begin by verifying that the steady-state solution for this situation is

$$T_{ss} = \frac{S}{4k}(R^2 - r^2).$$

Then, let the dependent variable, T, be written as the sum of steady-state and transient parts: $T = T_{ss} + T_1$. Use this sum to eliminate the inhomogeneity and demonstrate that

$$\frac{\partial T_1}{\partial t} = \alpha \left[\frac{\partial^2 T_1}{\partial r^2} + \frac{1}{r}\frac{\partial T_1}{\partial r}\right].$$

Now complete your solution and check your result with section 7.9 in Carslaw and Jaeger (1959).

7.24. Consider a solid sphere with a radius of 1, in which thermal energy production occurs at a constant rate:

$$\rho C_p \frac{\partial T}{\partial t} = k\left[\frac{\partial^2 T}{\partial r^2} + \frac{2}{r}\frac{\partial T}{\partial r}\right] + S_0.$$

The source term, S_0, will have units of calorie per cubic centimeter second. We want to solve this problem analytically (and later numerically) and then plot our solutions. Two substitutions are required to find a solution by separation of variables. The first, which is already familiar to us, is $T = \theta/r$. However, this step leaves the inhomogeneity, which must be dealt with by setting

$$\theta = \phi - \frac{S_0 r^3}{6k}.$$

The parametric values we will employ are

$$k = 0.025, \quad \alpha = 0.075, \quad R = 1, \quad \text{and} \quad S_0 = 2.675.$$

We are particularly interested in values of the parameter, $\alpha t/R^2 = 0.05$, 0.1, and 0.25. Find the temperature profiles for each and plot them on the same figure. Use the dimensionless group for the dependent variable, $\frac{6k}{S_0}\frac{T}{R^2}$. Assume that that the initial temperature of the sphere is 0 and that the surface is maintained at zero for all time.

7.25. Suppose we have a finite cylinder (of length, L) that is at some initial temperature, T_i. At $t = 0$, the end of the cylinder (at $z = 0$) is instantaneously heated to T_0. The governing equation is

$$\rho C_p \frac{\partial T}{\partial t} = k\left[\frac{1}{r}\frac{\partial}{\partial r}\left(r \frac{\partial T}{\partial r}\right) + \frac{\partial^2 T}{\partial z^2}\right].$$

The surface of the cylinder, at $r = R$, loses heat to the surroundings such that

$$-k\left(\frac{\partial T}{\partial r}\right)_{r=R} = h(T_R - T_\infty),$$

and we will assume that $T_\infty = 0$. The far end of the cylinder, at $z = L$, is maintained at $T = 0$ for all time. Begin by finding the *steady-state temperature distribution* in the cylinder and check your result with section 8.3 in Carslaw and Jaeger (1959). Then, explore the transient problem and find an analytic solution if you can.

7.26. A viscous fluid, which is initially at rest, lies on a planar surface corresponding to the x-axis. The fluid extends very far in the vertical (or $y-$) direction. At $t = 0$, the planar wall begins to oscillate such that $V_x(y = 0) = V_0 \cos(\omega t)$. Find the velocity distribution in the fluid assuming that the flow is governed by

$$\frac{\partial V_x}{\partial t} = \nu \frac{\partial^2 V_x}{\partial y^2}.$$

This scenario is referred to as *Stokes' second problem* and you may want to consider using the Laplace transform.

7.27. We have a potential field in spherical coordinates with ϕ symmetry such that

$$\frac{1}{r^2} \frac{\partial}{\partial r}\left(r^2 \frac{\partial \psi}{\partial r}\right) + \frac{1}{r^2 \sin\theta} \frac{\partial}{\partial \theta}\left(\sin\theta \frac{\partial \psi}{\partial \theta}\right) = 0.$$

We want to find an analytic solution for this problem. Begin by proposing $\psi = f(r)g(\theta)$ and carry out the separation. The equation for f will be of the Cauchy type and the equation for g can be transformed into Legendre's differential equation such that the solution can be written in terms of Legendre polynomials. This process is not trivial, and you may want to consult chapter 12 in Spiegel (1971) for assistance. Express the solution as the product of f and g, but do not worry about boundary conditions.

REFERENCES

Abramowitz, M. and I. A. Stegun. *Handbook of Mathematical Functions*, Dover Publications, New York (1965).

Atkins, P. W. *Physical Chemistry*, W. H. Freeman and Company, San Francisco (1978).

Bieberbach, L. *Conformal Mapping*, Chelsea Publishing Company, New York (1953).

Carslaw, H. S. *An Introduction to the Theory of Fourier's Series and Integrals*, 3rd revised edition, Dover Publications, New York (1950).

Carslaw, H. S. and J. C. Jaeger. *Conduction of Heat in Solids*, 2nd edition, Oxford Clarendon Press, Oxford (1959).

Crank, J. *The Mathematics of Diffusion*, 2nd edition, Oxford Clarendon Press, Oxford (1975).

Finlayson, B. A. *Nonlinear Analysis in Chemical Engineering*, McGraw-Hill, New York (1980).

Glasgow, L. A. *Transport Phenomena: An Introduction to Advanced Topics*, John Wiley & Sons, New York (2010).

Houstis, E. N. Collocation Methods for Linear Elliptic Problems. *BIT*, 16:301 (1978).

Kober, H. *Dictionary of Conformal Representations*, Dover Publications, New York (1952).

Korner, T. W. *Fourier Analysis*, Cambridge University Press, Cambridge (1989).

Lamb, H. *Hydrodynamics*, 6th edition, Dover Publications, New York (1945).

Powers, D. L. *Boundary Value Problems*, 2nd edition, Academic Press, New York (1979).

Prenter, P. M. and R. D. Russell. Orthogonal Collocation for Elliptic Partial Differential Equations. *SIAM Journal of Numerical Analysis*, 13:923 (1976).

Rice, J. R. and R. F. Boisvert. *Sovling Elliptic Problems Using ELLPACK*, Springer-Verlag, New York (1985).

Schrödinger, E. Quantisierung als Eigenwertproblem. *Annalen der Physik*, 79:489 (1926).

Smith, L. P. *Mathematical Methods for Scientists and Engineers*, Dover Publications, New York (1953).

Spiegel, M. R. *Advanced Mathematics for Engineers and Scientists*, Schaum's Outline Series, McGraw-Hill, New York (1971).

Spiegel, M. R. *Theory and Problems of Fourier Analysis*, Schaum's Outline Series, McGraw-Hill, New York (1974).

Villadsen, J. and M. L. Michelsen. *Solution of Differential Equation Models by Polynomial Approximation*, Prentice-Hall, Englewood Cliffs, NJ (1978).

Villadsen, J. and W. E. Stewart. Solution of Boundary-Value Problems by Orthogonal Collocation. *Chemical Engineering Science*, 22:1483 (1967).

Weinberger, H. F. *A First Course in Partial Differential Equations*, John Wiley & Sons, New York (1965).

Wieder, S. *The Foundations of Quantum Theory*, Academic Press, New York (1973).

8

NUMERICAL SOLUTION OF PARTIAL DIFFERENTIAL EQUATIONS

INTRODUCTION

Our usual approach will involve discretization of partial differential equations (PDEs), followed by solution of the resulting algebraic equations. Discretization is key to both finite-difference methods (FDMs) and finite-element methods (FEMs). The two approaches require the same level of numerical effort, but the latter is particularly useful for problems involving irregular shapes and boundaries (an introduction to FEM will be provided in Chapter 11). On the other hand, FDMs are much less software-dependent, and for simple problems, FDM solutions can be obtained with a broad spectrum of hardware–software combinations, even through the use of commonplace tools like spreadsheet programs. Thus, the analyst can solve many important practical problems without commercial modeling software, without high-level language proficiency, without compiler experience, and without mesh generation and refinement.

We should anticipate that when we solve a PDE numerically, we may not obtain a completely accurate solution. Of course, we expect discrepancies arising from both roundoff and truncation, and a common view is that we are solving the given PDE with some acceptable level of error. There is a second viewpoint that is useful in the context of certain computations, and it reveals a more insidious problem that we need to recognize: *When we discretize a PDE, we are actually creating a PDE that may have additional terms; that is, we end up with an equation that is not the original model for the phenomenon of interest. Clearly, we need to understand how those additional terms affect the solution.*

We will give a very brief introduction to this topic here, but the interested reader should consult Chapter 6 in Anderson (1995) for detail. Consider the fragmentary equation for a transient problem with convective transport:

$$\frac{\partial \phi}{\partial t} + V \frac{\partial \phi}{\partial x} = \cdots. \tag{8.1}$$

One possible discretization can be written as

$$\frac{\phi_{i,j+1} - \phi_{i,j}}{\Delta t} + V \frac{\phi_{i,j} - \phi_{i-1,j}}{\Delta x} \cong \cdots, \tag{8.2}$$

where the index $j + 1$ refers to the new time step, $t + \Delta t$. Please note that the x-direction gradient of ϕ is written in the *upwind* (backward) form; the need for this particular difference will be explained later. If we now expand $\phi_{i,j+1}$ and $\phi_{i-1,j}$ in Taylor series,

$$
\begin{aligned}
\phi_{i,j+1} = \phi_{i,j} &+ \left(\frac{\partial \phi}{\partial t}\right)_{i,j} \Delta t + \left(\frac{\partial^2 \phi}{\partial t^2}\right)_{i,j} \frac{(\Delta t)^2}{2} \\
&+ \left(\frac{\partial^3 \phi}{\partial t^3}\right)_{i,j} \frac{(\Delta t)^3}{6} + \cdots
\end{aligned} \tag{8.3}
$$

$$
\begin{aligned}
\phi_{i-1,j} = \phi_{i,j} &- \left(\frac{\partial \phi}{\partial x}\right)_{i,j} \Delta x + \left(\frac{\partial^2 \phi}{\partial x^2}\right)_{i,j} \frac{(\Delta x)^2}{2} \\
&- \left(\frac{\partial^3 \phi}{\partial x^3}\right)_{i,j} \frac{(\Delta x)^3}{6} + \cdots,
\end{aligned} \tag{8.4}
$$

Applied Mathematics for Science and Engineering, First Edition. Larry A. Glasgow.
© 2014 John Wiley & Sons, Inc. Published 2014 by John Wiley & Sons, Inc.

and then substitute the results into the original eq. (8.1), we recover the original terms *but with the addition of new ones.* Anderson shows through a process of differentiation and subtraction that the time derivatives that appear in "new" terms in the equation can be replaced by derivatives with respect to *x*, ultimately resulting in

$$\frac{\partial \phi}{\partial t} + V \frac{\partial \phi}{\partial x} = A \frac{\partial^2 \phi}{\partial x^2} + B \frac{\partial^3 \phi}{\partial x^3} + \cdots. \qquad (8.5)$$

The original equation is recovered on the left-hand side, but new derivatives appear on the right. The even derivatives are *dissipative*, and in computational fluid dynamics (CFD), they are referred to as *artificial viscosity*; they exert a stabilizing influence on the computation. The odd derivatives are *dispersive*, and they can create distortions and in some cases destabilize a computation. Let us emphasize the essential point of this discussion: The discretization process we employ can produce additional terms in the PDE. We may, in fact, be solving a PDE that differs from the actual model (1) of the phenomenon of interest. Though this sounds ominous, it may be beneficial in particular circumstances; artificial viscosity, for example, can be used intentionally to make an unstable computational scheme stable. But to be absolutely clear, if we pursue this course (rendering an unstable computation stable by adding artificial viscosity), we are adopting the viewpoint that finding *some* kind of numerical solution is better than not finding one at all.

Finite-Difference Approximations for Derivatives

Finite-difference approximations allow us to develop algebraic representations for differential equations. Consider the following Taylor series expansions:

$$y(x+h) = y(x) + hy'(x) + \frac{h^2}{2} y''(x) + \frac{h^3}{6} y'''(x) + \cdots \qquad (8.6)$$

and

$$y(x-h) = y(x) - hy'(x) + \frac{h^2}{2} y''(x) - \frac{h^3}{6} y'''(x) + \cdots. \qquad (8.7)$$

When we add the two equations together, we obtain

$$y(x+h) + y(x-h) = 2y(x) + h^2 y''(x) + f(h^4) + \cdots.$$

If we discard all of the terms involving h^4 (and up), we get

$$y''(x) \cong \frac{y(x+h) - 2y(x) + y(x-h)}{h^2}. \qquad (8.8)$$

This second-order, central difference approximation for the second derivative has a leading error on the order of h^2. If h is small, this approximation should be good. For example, let $y = x\sin x$ thus, $dy/dx = \sin x + x\cos x$ and $(d^2y)/(dx^2) = 2\cos x - x\sin x$.

Now let $x = 0.3$: $y = 0.088656$, $dy/dx = 0.582121$, and $(d^2y)/(dx^2) = 1.822017$; then choose $h = 0.01$:

$$\frac{d^2 y}{dx^2} \cong \frac{0.094568 - 2(0.088656) + 0.082926}{(0.01)^2} = 1.820.$$

This is about 0.11% less than the analytic value for the second derivative. By simply combining Taylor series expansions, we can build any number of approximations and for derivatives of any order. Furthermore, these approximations can be forward, backward, centered, or skewed. Some of the more useful forms are compiled for you as follows. Note that **F** $\Rightarrow$ forward, **C** $\Rightarrow$ central, **B** $\Rightarrow$ backward, and *h* is convenient shorthand for Δx:

First order:

$$\mathbf{F} \qquad y_i' = \frac{1}{h}(y_{i+1} - y_i) \qquad (8.9)$$

$$\mathbf{B} \qquad y_i' = \frac{1}{h}(y_i - y_{i-1}) \qquad (8.10)$$

Second order:

$$\mathbf{F} \qquad y_i' = \frac{1}{2h}(-3y_i + 4y_{i+1} - y_{i+2}) \qquad (8.11)$$

$$y_i'' = \frac{1}{h^2}(y_i - 2y_{i+1} + y_{i+2}) \qquad (8.12)$$

$$\mathbf{C} \qquad y_i' = \frac{1}{2h}(y_{i+1} - y_{i-1}) \qquad (8.13)$$

$$y_i'' = \frac{1}{h^2}(y_{i+1} - 2y_i + y_{i-1}) \qquad (8.14)$$

$$\mathbf{B} \qquad y_i' = \frac{1}{2h}(3y_i - 4y_{i-1} + y_{i-2}) \qquad (8.15)$$

$$y_i'' = \frac{1}{h^2}(y_i - 2y_{i-1} + y_{i-2}) \qquad (8.16)$$

Third order:

$$\mathbf{F} \qquad y_i' = \frac{1}{6h}(2y_{i+3} - 9y_{i+2} + 18y_{i+1} - 11y_i) \qquad (8.17)$$

$$y_i'' = \frac{1}{h^2}(-y_{i+3} + 4y_{i+2} - 5y_{i+1} + 2y_i) \qquad (8.18)$$

$$y_i''' = \frac{1}{h^3}(y_{i+3} - 3y_{i+2} + 3y_{i+1} - y_i) \qquad (8.19)$$

B $\quad y_i' = \frac{1}{6h}(11y_i - 18y_{i-1} + 9y_{i-2} - 2y_{i-3}) \qquad (8.20)$

$$y_i'' = \frac{1}{h^2}(2y_i - 5y_{i-1} + 4y_{i-2} - y_{i-3}) \qquad (8.21)$$

$$y_i''' = \frac{1}{h^3}(y_i - 3y_{i-1} + 3y_{i-2} - y_{i-3}) \qquad (8.22)$$

Fourth order:

F $\quad y_i' = \frac{1}{12h}(-3y_{i+4} + 16y_{i+3} - 36y_{i+2} + 48y_{i+1} - 25y_i)$

$$\qquad (8.23)$$

$$y_i'' = \frac{1}{12h^2}(11y_{i+4} - 56y_{i+3} + 114y_{i+2} - 104y_{i+1} + 35y_i)$$

$$\qquad (8.24)$$

$$y_i''' = \frac{1}{2h^3}(-3y_{i+4} + 14y_{i+3} - 24y_{i+2} + 18y_{i+1} - 5y_i)$$

$$\qquad (8.25)$$

$$y_i'''' = \frac{1}{h^4}(y_{i+4} - 4y_{i+3} + 6y_{i+2} - 4y_{i+1} + y_i) \qquad (8.26)$$

C $\quad y_i' = \frac{1}{12h}(-y_{i+2} + 8y_{i+1} - 8y_{i-1} + y_{i-2}) \qquad (8.27)$

$$y_i'' = \frac{1}{12h^2}(-y_{i+2} + 16y_{i+1} - 30y_i + 16y_{i-1} - y_{i-2})$$

$$\qquad (8.28)$$

$$y_i''' = \frac{1}{2h^3}(y_{i+2} - 2y_{i+1} + 2y_{i-1} - y_{i-2}) \qquad (8.29)$$

$$y_i'''' = \frac{1}{h^4}(y_{i+2} - 4y_{i+1} + 6y_i - 4y_{i-1} + y_{i-2}) \qquad (8.30)$$

B $\quad y_i' = \frac{1}{12h}(25y_i - 48y_{i-1} + 36y_{i-2} - 16y_{i-3} + 3y_{i-4})$

$$\qquad (8.31)$$

$$y_i'' = \frac{1}{12h^2}(35y_i - 104y_{i-1} + 114y_{i-2} - 56y_{i-3} + 11y_{i-4})$$

$$\qquad (8.32)$$

$$y_i''' = \frac{1}{2h^3}(5y_i - 18y_{i-1} + 24y_{i-2} - 14y_{i-3} + 3y_{i-4})$$

$$\qquad (8.33)$$

$$y_i'''' = \frac{1}{h^4}(y_i - 4y_{i-1} + 6y_{i-2} - 4y_{i-3} + y_{i-4}) \qquad (8.34)$$

Boundaries with Specified Flux

Consider a conduction problem in a slab for which the right-hand boundary is insulated; thus, $q_x = 0$; this is an example of a Neumann boundary condition ($\partial T/\partial x = 0$). Let the nodal point on the boundary be represented by the index, n, and let the temperatures for $n - 2$ and $n - 1$ be 50° and 45°, respectively. We can determine the temperature at the boundary by setting the derivative equal to zero. However, if we use a first-order, backward difference in this situation:

$n - 2$	$n - 1$	n
50°	45°	?°

then $T_n = 45°$, a result that is clearly unphysical because the temperature "profile" on this row has a discontinuity in slope. One alternative is to employ eq. (8.15):

$$T_n = \frac{1}{3}(-50 + 4(45)) = 43.333°. \qquad (8.35)$$

Of course, a third- or fourth-order backward difference could be used as well; if we go with third-order and set the temperature at $n - 3$ to 56°, we find $T_n = 42.909°$.

We should also examine the use of a Robin's-type boundary condition (a boundary condition of the third kind) for a solid–fluid interface:

$$-k_s \frac{\partial T}{\partial x} = h_f(T_n - T_\infty). \qquad (8.36)$$

Let the Biot modulus, Bi $= \Delta x h_f/k_s$; then, *one possible expression for* T_n *is*

$$T_n = \frac{2BiT_\infty + 4T_{n-1} - T_{n-2}}{3 + 2Bi}. \qquad (8.37)$$

If we select Bi $= 1$ and $T_\infty = 20°$ and use the temperatures given earlier for the $n - 1$ and $n - 2$ positions, then

$$T_n = \frac{2(20) + 4(45) - 50}{5} = 34°. \qquad (8.38)$$

Note how eq. (8.37) is affected when Bi is very low—the result is exactly the same as eq. (8.35)!

ELLIPTIC PARTIAL DIFFERENTIAL EQUATIONS

Our main focus in this section is on Laplace- and Poisson-type elliptic PDEs that apply to equilibrium phenomena. Examples include steady-state conduction in a slab:

$$\frac{\partial^2 T}{\partial x^2} + \frac{\partial^2 T}{\partial y^2} = 0, \qquad (8.39)$$

steady viscous flow in a two-dimensional duct,

$$\frac{\partial^2 V_z}{\partial x^2} + \frac{\partial^2 V_z}{\partial y^2} = \frac{1}{\mu}\frac{dp}{dz}, \qquad (8.40)$$

and the Laplacian of the stream function for two-dimensional potential flow:

$$\frac{\partial^2 \psi}{\partial x^2} + \frac{\partial^2 \psi}{\partial y^2} = 0. \qquad (8.41)$$

We should begin this part of our discussion by looking at a solution procedure for eq. (8.39). Suppose we have a square slab of material with prescribed temperatures on all four edges ($400°$ across the top and $100°$ for both sides and the bottom); we wish to find the interior temperature distribution, $T(x, y)$. We discretize the slab using $\Delta x = \Delta y$ (a square mesh) with five nodes in each direction. Since the boundary temperatures are known, we have nine interior nodes where the temperature must determined:

400	400	400	400	400
100				100
100				100
100				100
100	100	100	100	100

If we approximate eq. (8.39) with second-order central differences, we find

$$T_{i+1,j} + T_{i-1,j} + T_{i,j+1} + T_{i,j-1} - 4T_{i,j} = 0. \qquad (8.42)$$

Thus, we have an elementary problem in which we must find the solution for nine simultaneous, linear algebraic equations. This can be accomplished in many different ways, and we choose to employ Crout's (also known as Cholesky's) method. The reader may wish to verify that the solution for the given problem is the following:

400	400	400	400	400
100	228.6	258	228.6	100
100	156.3	175	156.3	100
100	121.4	129.5	121.4	100
100	100	100	100	100

This raises an interesting question: How accurate is this solution? For example, is the temperature at the center of the slab really $175°$? Since this is a problem for which the analytic solution is known (see eq. 8.82 in Chapter 7), we can test the given result using the infinite series very easily. To four decimal places, the center temperature from the analytic solution is $174.9995°$. This shows that our approximate solution for the center ($175°$) is unusually accurate; we can also find the actual temperatures immediately above and

below this point to get a clearer picture of the overall quality of the solution. One node above the center the analytic solution produces $262.158°$ (as opposed to $258°$) and one node below, we find $128.624°$ (as opposed to $129.5°$). Our numerical solution is surprisingly close considering the coarse discretization that was employed.

Many elliptic PDEs can be solved in this manner, and since the coefficient matrix is usually sparse, such problems can be solved very efficiently. Although our slab example used only nine interior nodes, much larger problems can be solved in the same way. Some care must be exercised in such cases, however, because roundoff error can accumulate and corrupt the solution. If very large sets of simultaneous equations are to be solved using an elimination method, it may be necessary to either use greater precision in the calculations or, alternatively, to incorporate *error equations* into the procedure. We will give a brief sketch of this process here, but the interested reader may want to consult chapter 3 in James et al. (1977).

Suppose we have a set of simultaneous equations:

$$a_{11}X_1 + a_{12}X_2 + a_{13}X_3 + \cdots = C_1$$

$$a_{21}X_1 + a_{22}X_2 + \cdots, \quad \text{etc.}$$

When we solve these equations, we obtain a set of *approximate* values for the X_ns that we will represent like this: Y_1, Y_2, and so on. We take these approximate values back to the set of equations and compute the new constants; that is,

$$a_{11}Y_1 + a_{12}Y_2 + a_{13}Y_3 + \cdots = D_1. \qquad (8.43)$$

If roundoff errors have been generated, then $D_1 \neq C_1$, $D_2 \neq C_2$, and so on. Now we presume that the desired values for the X_ns can be obtained by *adding a correction to the approximate solution*: $X_1 = Y_1 + \Delta X_1$, $X_2 = Y_2 + \Delta X_2$, and so on. The correction expressions are substituted into the original algebraic equations, replacing the unknown X_ns:

$$a_{11}(Y_1 + \Delta X_1) + a_{12}(Y_2 + \Delta X_2) + a_{13}(Y_3 + \Delta X_3) + \ldots = C_1. \qquad (8.44)$$

Next, we subtract the set of equations obtained with the initial estimates, resulting in

$$a_{11}\Delta X_1 + a_{12}\Delta X_2 + a_{13}\Delta X_3 + \cdots = C_1 - D_1$$

$$a_{21}\Delta X_1 + a_{22}\Delta X_2 + a_{23}\Delta X_3 + \cdots = C_2 - D_2, \quad \text{etc.} \qquad (8.45)$$

The solution of this set of equations produces the corrections that are added to the original estimates: $X_1 = Y_1 + \Delta X_1$, $X_2 = Y_2 + \Delta X_2, \ldots$ This process can be repeated any number of times should that prove necessary.

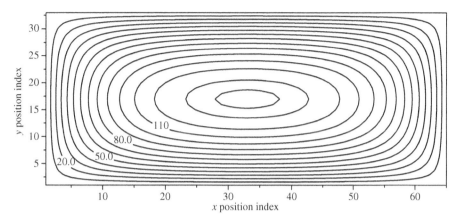

FIGURE 8.1. Velocity distribution in a rectangular duct computed with the Gauss–Seidel iterative method. The duct measures 8×4 cm with $dp/dz = -3$ dyne/cm^2/cm.

An Iterative Numerical Procedure: Gauss–Seidel

There are alternative solution techniques that can be applied to elliptic PDEs and we will now examine a straightforward iterative scheme; let us consider laminar flow in a rectangular duct for this example. By using the second-order central difference approximations for the second derivatives (where the i and j indices represent the x- and y-directions, respectively), eq. (8.40) can be written as

$$\frac{1}{\mu}\frac{dp}{dz} \cong \frac{V_{i+1,j} - 2V_{i,j} + V_{i-1,j}}{(\Delta x)^2} + \frac{V_{i,j+1} - 2V_{i,j} + V_{i,j-1}}{(\Delta y)^2}. \quad (8.46)$$

If the discretization employs a square mesh ($\Delta x = \Delta y$), then we can isolate the term with the largest numerical coefficient, with the convenient result:

$$V_{i,j} \approx \frac{1}{4}\left[V_{i+1,j} + V_{i-1,j} + V_{i,j+1} + V_{i,j-1} - \frac{(\Delta x)^2}{\mu}\frac{dp}{dz}\right]. \quad (8.47)$$

Please note that the z-direction subscript has been dropped from velocity to minimize clutter. This approximation is the basis for a simple Gauss–Seidel iterative computational scheme for the solution of such problems. In this case, of course, the velocity is zero on the boundaries, so we merely apply the algorithm to all of the interior points, row by row. The newly computed values are employed as soon as they become available (which distinguishes Gauss–Seidel from the Jacobi iterative method). As an example, consider the case of laminar flow in a rectangular duct 8 cm wide and 4 cm high; the pressure gradient is -3 dyne/cm^2/cm and the viscosity is 0.04 g/(cm s). All of the nodal velocities will be initialized to zero to start the computation.

For the specified pressure gradient, the centerline (maximum) velocity will be about 139 cm/s. The computed velocity distribution is shown in Figure 8.1 as a contour plot.

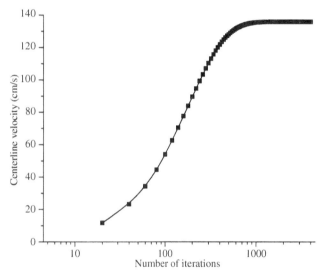

FIGURE 8.2. Centerline velocity as a function of the number of iterations for the solution of the Poisson equation for laminar flow in a rectangular duct as approximated by eq. (8.47).

In a computation of this type, a key issue is the number of iterations required to attain convergence. For the example shown here, we can monitor the evolution of centerline velocity during the calculations; this is illustrated in Figure 8.2. Keep in mind that we initialized all of the interior nodes at zero velocity. We could certainly enhance the progress toward convergence by starting the computation with a better initial estimate, that is, providing a more suitable distribution for $V_{i,j}$.

Note that a reasonably accurate value is obtained with about 1000 iterations, and after 3000 iterations, the third decimal place is essentially fixed.

Improving the Rate of Convergence with Successive Over-Relaxation (SOR)

The rate of convergence of iterative solutions can be accelerated significantly through use of the *extrapolated Liebmann method* (also known as successive over-relaxation, or SOR). In this technique, the change that *would have been produced* by a single Gauss–Seidel iteration is increased through the use of an accelerating factor, which is usually denoted by ω. SOR can be implemented easily in the previous example by a slight modification of eq. (8.47):

$$V_{i,j}{}^{(\text{new})} \approx V_{i,j} + \tfrac{1}{4}\omega \bigg[V_{i+1,j} + V_{i-1,j} + V_{i,j+1}$$
$$+ V_{i,j-1} - 4V_{i,j} - \frac{(\Delta x)^2}{\mu} \frac{dp}{dz} \bigg]. \qquad (8.48)$$

The $V_{i,j}$s appearing on the right-hand side of eq. (8.48) are from the latest available calculations, of course. You can see immediately that if $\omega = 1$, this is identically the Gauss–Seidel algorithm. For *over-relaxation*, ω will have a value between 1 and 2; the rate of convergence is very sensitive to the value of the acceleration parameter. Please see Smith (1965) for additional discussion. Frankel (1950) has shown that, for large rectangular domains such as that used in our example,

$$\omega_{\text{opt}} \approx 2 - \sqrt{2}\pi \left(\frac{1}{p^2} + \frac{1}{q^2} \right)^{1/2}, \qquad (8.49)$$

where p and q are the number of nodal points used in the x- and y-directions, respectively. For our case, $p = 65$ and $q = 33$, so $\omega_{\text{opt}} \approx 1.85$. The consequences of a poor choice are shown clearly in Figure 8.3, where the number of iterations required to achieve a desired degree of convergence is reported. While it is apparent that SOR can significantly reduce the computational effort required to solve elliptic PDEs, the acceleration parameter, ω, must be chosen carefully to obtain the greatest possible benefit. We should make one other observation regarding ω: In the iterative solution of *nonlinear* PDEs, stability can sometimes be maintained by using under-relaxation, that is, by setting $\omega < 1$.

We will now illustrate the application of SOR with a very detailed example. Assume we have a mild steel slab; the left-hand side is maintained at 1000°, the bottom at 500°, and the top is insulated. The right-hand side loses heat to the surroundings according to

$$-k \left.\frac{\partial T}{\partial x}\right|_{x=L} = h(T|_{x=L} - T_\infty). \qquad (8.50)$$

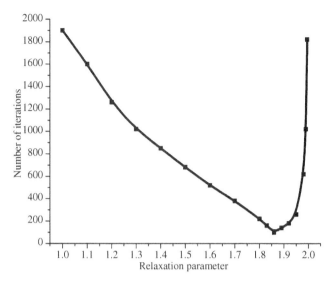

FIGURE 8.3. Number of iterations required to achieve $\varepsilon = 2 \times 10^{-7}$ as a function of ω. A Poisson-type equation for laminar flow in a rectangular duct is being solved and the minimum is located at about $\omega \approx 1.86$.

This is a steady-state problem so the temperature in the interior of the slab is governed by

$$\frac{\partial^2 T}{\partial x^2} + \frac{\partial^2 T}{\partial y^2} = 0. \qquad (8.51)$$

Results obtained from a typical program structure are shown in the contour plot provided in Figure 8.4.

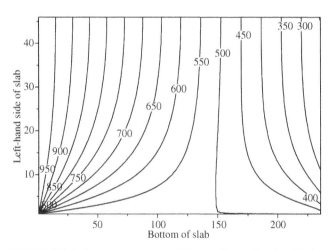

FIGURE 8.4. Isotherms computed for a mild steel slab with the left-hand side maintained at 1000° and the bottom at 500°. The top surface is insulated and the right-hand side loses thermal energy to the surroundings according to Newton's law of cooling.

```
        #COMPILE EXE
#DIM ALL
        REM *** Illustration of SOR computation for heat transfer in a steel slab
        REM *** Left-hand boundary maintained at 1000, bottom at 500, top is insulated, and right side loses heat
            GLOBAL L,HT,h,k,dx,dy,iter,w,Tsur,i,j AS SINGLE
FUNCTION PBMAIN
        DIM T(251,201) AS SINGLE
        dx=0.1:dy=0.1:k=0.108:h=0.0203:Tsur=100:w=1.98
        REM *** initialize temperature
            FOR j=1 TO 201
                T(1,j)=1000
                NEXT j
            FOR i=2 TO 251
                T(i,1)=500
                NEXT i
            iter=0
50 REM *** continue
        FOR j=2 TO 200
        FOR i=2 TO 250
            T(i,j)=T(i,j)+w/4*(T(i+1,j)+T(i-1,j)+T(i,j+1)+T(i,j-1)-4*T(i,j))
                NEXT i:NEXT j
        REM *** top boundary (insulated)
            FOR i=2 TO 250
                T(i,201)=(4*T(i,200)-T(i,199))/3
                NEXT i
        REM *** right-hand boundary–Robin's type BC
            FOR j=2 TO 200
                T(251,j)=(-h*Tsur+k/(2*dx)*(-4*T(250,j)+T(249,j)))/(-3*k/(2*dx)-h)
                NEXT j
                    iter=iter+1
                        PRINT iter,T(150,100)
                        IF iter>6000 THEN 200 ELSE 50
200 REM *** continue
        OPEN "c:STslab1.dat" FOR OUTPUT AS #1
            FOR j=1 TO 201
                FOR i=1 TO 251
                    WRITE#1,i,j,T(i,j)
                        NEXT i:NEXT j
        CLOSE:END
```

As we indicated previously, the progress of such a computation (i.e., the rate at which one obtains a satisfactory solution) is very sensitive to the value selected for the relaxation parameter, ω. We will illustrate this by changing ω and monitoring the value of $T(150,100)$ at exactly 500 iterations:

ω	$T(150, 100)$
1.0	0.0122
1.1	0.0672
1.2	0.2837
1.3	0.9799

ω	$T(150, 100)$
1.4	2.912
1.5	7.774
1.6	19.39
1.7	46.86
1.8	112.55
1.85	174.57
1.90	270.81
1.95	418.41
1.98	513.84
1.99	523.79

The initial value for all of the interior nodes was 0°. Note that the correct temperature at this location, $T(150, 100)$, is 523.8°.

Although we have purposefully tried to minimize connecting our discussions to specific computational software, the student needs to be aware that there are *many* commercial packages that have capabilities for elementary PDEs. We will illustrate one such option here, using Mathcad™. Suppose we have a two-dimensional (square) slab of material with the edges (T, B, L, R) maintained at the following temperatures: 200°, 10°, 50°, and 200°. The temperature in the slab will be governed by the Laplace equation, that is, eq. (8.48):

$$\frac{\partial^2 T}{\partial x^2} + \frac{\partial^2 T}{\partial y^2} = 0.$$

Our discretization for this PDE is exactly the same as eq. (8.42), of course:

$$T_{i+1,j} + T_{i-1,j} + T_{i,j+1} + T_{i,j-1} - 4T_{i,j} = 0.$$

The Mathcad function we will employ is relax($a, b, c, d, e, f, u, rjac$). Note that a is the matrix of coefficients on $(i + 1, j)$, and all of them are 1s. b is the matrix of coefficients for $(i - 1, j)$, and again, these values are all 1s. c is the matrix of coefficients on $(i, j + 1)$, all 1s, and d is the matrix of coefficients for $(i, j - 1)$, also 1s. The matrix of coefficients for the central temperature (i, j) in the pattern is e, and of course, those values are all -4. f would correspond to the source term, if one were present. In our case, all of the fs are zero. The matrix u contains the constant boundary temperatures, and estimates for the interior nodes. $rjac$ is a constant between 0 and 1 that affects the rate of convergence of the relaxation algorithm. For the simple problem we are considering here, the Mathcad procedure is not much affected by the choice of $rjac$. Using 25 interior nodes (so the coefficient matrices are all 7×7), we obtain the following result:

relax($a, b, c, d, e, f, u, 0.5$) =

$$\begin{pmatrix} 200 & 200 & 200 & 200 & 200 & 200 & 200 \\ 50 & 123.747 & 153.013 & 168.152 & 179.073 & 189.353 & 200 \\ 50 & 91.977 & 120.151 & 140.523 & 158.788 & 178.341 & 200 \\ 50 & 74.009 & 95.093 & 115 & 137.214 & 165.221 & 200 \\ 50 & 58.967 & 71.212 & 87.169 & 109.848 & 145.33 & 200 \\ 50 & 40.646 & 43.619 & 52.617 & 69.679 & 106.252 & 200 \\ 50 & 10 & 10 & 10 & 10 & 10 & 200 \end{pmatrix}.$$

Naturally, the first question one should ask concerns the reliability of this computation: How accurate is it? To address this, we will refine the mesh and compute the interior temperatures ourselves using one of the algorithms we have already discussed. Using 13 nodes in each direction (and

single precision), and reporting only every other node for ease of comparison with the preceding results, we obtain

200	200	200	200	200	200	200
50	123.77609	153.79272	168.72809	179.47177	189.59695	200
50	91.29428	120.20753	140.88583	159.26443	178.76828	200
50	73.57849	94.815605	114.99994	137.57727	165.87845	200
50	58.76750	70.73545	86.72105	109.79236	146.24150	200
50	40.40302	42.99232	51.81481	68.67136	106.22386	200
50	10	10	10	10	10	200

These data indicate that the Mathcad solution obtained with *relax* is reasonably accurate, even with the coarse discretization we employed. The largest discrepancies between the two sets of results are on the order of 1.5%, and those errors appear on the bottom interior row. Of course, we can use exactly the same discretization (employing 13×13 matrices) in Mathcad, and when we do so, we get the following results on the diagonal (starting in the lower left-hand corner and proceeding toward the upper right): 50, 40.403, 70.737, 115.002, 159.266, 189.597, and 200. These numbers are virtually identical with our do-it-yourself computation—the largest discrepancy is smaller than 0.002%.

PARABOLIC PARTIAL DIFFERENTIAL EQUATIONS

An Elementary, Explicit Numerical Procedure

Suppose we have viscous fluid that extends far in the y-direction, initially at rest near a plane wall that is set in motion with velocity, V_0, at time, $t = 0$; thus, $V_x(y = 0, t) = V_0$. Letting $V = V_x/V_0$,

$$\frac{\partial V}{\partial t} = \nu \frac{\partial^2 V}{\partial y^2}. \tag{8.52}$$

This scenario is known as Stokes' first problem, and the analytic solution is just

$$\frac{V_x}{V_0} = 1 - erf\left(\frac{y}{\sqrt{4\nu t}}\right),$$

where the error function $erf(\eta) = \left(2/\sqrt{\pi}\right)\int_0^\eta \exp(-\eta^2)d\eta$:

η	0	0.1	0.2	0.4	0.8	1.6	3.2
$erf(\eta)$	0.00	0.1125	0.2227	0.4284	0.7421	0.9764	1.0000

An *explicit* algorithm is easily developed for eq. (8.52); using a first-order forward difference for the time derivative followed by isolation of the V value on the new time step results in

TABLE 8.1. Explicit Computation with Unstable Parametric Choice(s)

t	$i = 1$	$i = 2$	$i = 3$	$i = 4$	$i = 5$	$i = 6$	$i = 7$
0	1	0	0	0	0	0	0
Δt	1	0.6	0	0	0	0	0
$2\Delta t$	1	0.48	0.36	0	0	0	0
$3\Delta t$	1	0.72	0.216	0.216	0	0	0
$3\Delta t$	1	0.5856	0.5184	0.0864	0.1296	0	0
$4\Delta t$	1	0.7939	0.2995	0.3715	0.0259	0.0777	0
$5\Delta t$	1	0.6209	0.6394	0.1210	0.2644	0	0.0467
$6\Delta t$	1	0.8594	0.3173	0.5181	0.0197	0.1866	−0.0093
$7\Delta t$	1	0.6185	0.7630	0.0986	0.4189	−0.0311	0.1306

TABLE 8.2. Explicit Computation with Stable Parametric Choice(s)

t	$i = 1$	$i = 2$	$i = 3$	$i = 4$	$i = 5$	$i = 6$	$i = 7$
0	1	0	0	0	0	0	0
Δt	1	0.4	0	0	0	0	0
$2\Delta t$	1	0.48	0.16	0	0	0	0
$3\Delta t$	1	0.56	0.224	0.064	0	0	0
$4\Delta t$	1	0.6016	0.2944	0.1024	0.0256	0	0
$5\Delta t$	1	0.6381	0.3405	0.1485	0.0461	0.0102	0
$6\Delta t$	1	0.6638	0.3872	0.1843	0.0727	0.0205	0.0041
$7\Delta t$	1	0.6859	0.4158	0.2190	0.0965	0.0348	0.0090

$$V_{i,j+1} = \frac{\Delta t \nu}{(\Delta y)^2}\left[V_{i+1,j} - 2V_{i,j} + V_{i-1,j}\right] + V_{i,j}. \qquad (8.53)$$

Equation (8.53) is attractive because of its simplicity; it is easy to understand and easy to execute, but it poses a potential problem. To ensure stability, it is necessary that

$$\frac{\Delta t \nu}{(\Delta y)^2} \le \frac{1}{2}. \qquad (8.54)$$

We will illustrate this using eq. (8.53) by choosing $\nu = 0.05$ cm^2/s, $\Delta y = 0.1$ cm, and $\Delta t = 0.12$ second; of course, this guarantees that we are over the limit of ½ (actually 0.6). We can put the calculation into Table 8.1 and monitor the evolution of the nodal velocities, which will reveal the consequence of our choices. Since the analytic solution for this problem is known, we have a convenient comparison available.

The problem revealed by Table 8.1 is easy to resolve. We change our parametric choices to yield $\Delta t \nu/(\Delta y)^2 = 0.4$ and repeat the calculation.

This is an important lesson. If we need good spatial resolution, Δy will be small and Δt will need to be *very small*, perhaps prohibitively small. Fortunately, we do have options that will work well for this type of problem. Before we consider them, however, we will look specifically at the entry in Table 8.2 for $i = 4$ and $t = 7\Delta t$ (which is 0.2190);

the analytic solution for this particular point is $1 - erf(0.8964) = 0.205$, so the discrepancy produced by the explicit computation amounts to a little less than 7%. Though larger than we would like, this would still be satisfactory for many applications.

The Crank–Nicolson Method

Consider a transient diffusion problem in two spatial dimensions:

$$\frac{\partial C}{\partial t} = D\left[\frac{\partial^2 C}{\partial x^2} + \frac{\partial^2 C}{\partial y^2}\right], \qquad (8.55)$$

where C is the molar concentration of the species of interest and D is the diffusivity.

If we were to solve this problem using the explicit approach described in the previous section, we would have to choose Δt such that

$$\Delta t D\left(\frac{1}{(\Delta x)^2} + \frac{1}{(\Delta y)^2}\right) \le \frac{1}{2}. \qquad (8.56)$$

If the problem required enhanced spatial resolution, then the time-step size, Δt, would need to be very small, and the required computational effort might be excessive (particularly in view of the large time required for the concentration field in many diffusion problems to develop, for example, in liquids, $D \approx 10^{-5}$ cm^2/s). To illustrate, suppose that $\Delta x = \Delta y = 1/20$ and $D = 1/10$; then, $80\,\Delta t \le 1/2$ and Δt must be less than 0.00625. Fortunately, there are alternatives and the Crank–Nicolson method is one option.

In the Crank–Nicolson approach, a first-order forward difference is used for the time derivative, and the second derivatives (the molecular transport terms) are written twice, once on the present time-step row, t, and once for $t + \Delta t$. The arithmetic average of the two values is used in the computation. Let the i, j, and k indices correspond to x, y, and t, respectively. The scheme can be written out as

$$\frac{C_{i,j,k+1} - C_{i,j,k}}{\Delta t} \cong \frac{D}{2}\left[\frac{C_{i+1,j,k} - 2C_{i,j,k} + C_{i-1,j,k}}{(\Delta x)^2} \right.$$
$$+ \frac{C_{i,j+1,k} - 2C_{i,j,k} + C_{i,j-1,k}}{(\Delta y)^2}$$
$$+ \frac{C_{i+1,j,k+1} - 2C_{i,j,k+1} + C_{i-1,j,k+1}}{(\Delta x)^2}$$
$$\left. + \frac{C_{i,j+1,k+1} - 2C_{i,j,k+1} + C_{i,j-1,k+1}}{(\Delta y)^2} \right].$$
$$(8.57)$$

Of course, this algorithm is implicit, which means that a set of simultaneous algebraic equations must be solved to advance to the new time-step row, $k + 1$, that is, $t + \Delta t$. Note that the computational pattern involves five points: the central node, i, j, then left and right, and up and down. The Crank–Nicolson method is stable for *any* value of Δt. We employ a square mesh so that $\Delta x = \Delta y$ and isolate the $k + 1$ values on the left-hand side of the equation:

$$C_{i,j,k+1}\left[\frac{1}{\Delta t} + \frac{2D}{(\Delta x)^2} \right]$$
$$- \frac{D}{2(\Delta x)^2}(C_{i+1,j,k+1} + C_{i-1,j,k+1} + C_{i,j+1,k+1} + C_{i,j-1,k+1})$$
$$= \frac{D}{2(\Delta x)^2}[C_{i+1,j,k} + C_{i-1,j,k} + C_{i,j+1,k} + C_{i,j-1,k} - 4C_{i,j,k}]$$
$$+ \frac{C_{i,j,k}}{\Delta t}.$$
$$(8.58)$$

An attractive feature of this approach is that the coefficients for the computational pattern on the new time-step row are simply

i, j+1

$$\frac{-D}{2(\Delta x)^2}(\text{T, B, L, R})$$

i −1, j   **i, j**   **i +1, j**

$$\frac{1}{\Delta t} + \frac{2D}{(\Delta x)^2}(\text{Center})$$

i, j −1

Let us illustrate the advantages offered by Crank–Nicolson with an example. Suppose we have a slab of material with a thermal diffusivity (α) of 0.03 cm^2/s, which is roughly characteristic of minerals like fluorite and quartz. The slab measures 6×6 cm and it has an initial temperature

of 0°. At $t = 0$, the temperature of the left-hand side is instantaneously raised to 1000° and the top edge to 600°. The other two edges are maintained at 0° for all time. In this case, of course, the governing equation is

$$\frac{\partial T}{\partial t} = \alpha\left[\frac{\partial^2 T}{\partial x^2} + \frac{\partial^2 T}{\partial y^2} \right],$$

which is completely analogous to eq. (8.55). We are interested in the temperature distribution in the slab at $t = 50$ seconds. We first compute the result explicitly, using $\Delta t = 0.01$ second, which corresponds to 5000 time steps. The results are shown in the following array:

600	600	600	600	600	600	600
1000	644.62	461.64	374.39	324.39	245.17	0
1000	607.64	348.74	217.09	153.31	96.47	0
1000	570.37	280.35	133.69	70.72	36.25	0
1000	525.29	231.49	91.31	36.06	14.12	0
1000	404.95	154.94	53.92	17.69	5.51	0
1000	0	0	0	0	0	0

Now we carry out the computation a second time, but we use Crank–Nicolson with $\Delta t = 50$ seconds; that is, we employ just *one time step*! We should not expect the two sets of results to compare favorably:

600	600	600	600	600	600	600
1000	831.73	520.05	430.11	389.99	314.98	0
1000	715.85	311.76	183.87	134.90	89.89	0
1000	674.36	242.93	103.88	55.70	29.52	0
1000	637.82	205.64	71.53	28.78	11.86	0
1000	521.68	144.47	43.21	14.42	4.92	0
1000	0	0	0	0	0	0

By no means is this acceptable. But remember that we have *replaced 5000 time steps* (explicit) *with just one* (Crank–Nicolson). If we reduce the total time by a factor of 10, that is, we carry out the calculations to $t = 5$ seconds using both the explicit technique with $\Delta t = 0.01$ second and a single 5 s step with Crank–Nicolson, the typical discrepancy is just a few percent. And, if we drop down to 2 seconds to compare 200 time steps (explicit) with just one (Crank–Nicolson), we find that the typical difference for values in the first interior column (at $t = 2$ seconds) is less than 0.5%; this is illustrated as follows:

Explicit	Crank–Nicolson
88.39	88.21
57.79	57.52
56.87	56.67
56.82	56.61
55.25	55.13

Because Crank–Nicolson is so easy to use in one spatial dimension, the reader is encouraged to try applying the method to the following slab example. The initial temperature of the semi-infinite slab is zero; at $t = 0$, the temperature of the front face is elevated to 500°. Given a thermal diffusivity of 0.12 cm²/s, we compute the temperature distribution in the slab using both the analytic solution,

$$\theta = erfc\left(\frac{y}{\sqrt{4\alpha t}}\right),$$

and Crank–Nicolson with a single time step. After 16 seconds, the temperature profiles appear as shown here:

y (cm)	0	1	2	3	4	5	6	7
T, analytic	500	303.5	154.2	65.9	23.9	7.5	2.1	0.5
T, CN	500	375	140.6	52.7	19.8	7.4	2.8	1.0

Again, the reader should note that the Crank–Nicolson calculation employed *just one* 16-second time step; he/she might also consider repeating the calculation but with $\Delta t = 1$ s.

Alternating-Direction Implicit (ADI) Method

The Peaceman and Rachford (1955) or ADI method can be particularly useful for the types of parabolic PDEs we have been discussing, and it is more efficient than Crank–Nicolson. Let the indices i, j, and k represent x, y, and t, respectively. We will use transient conduction in two spatial dimensions for our example:

$$\frac{\partial T}{\partial t} = \alpha\left[\frac{\partial^2 T}{\partial x^2} + \frac{\partial^2 T}{\partial y^2}\right].\qquad(8.59)$$

The first half of the ADI algorithm is used to advance to the $k + 1$ time step:

$$\frac{T_{i,j,k+1} - T_{i,j,k}}{\alpha \Delta t} = \frac{T_{i+1,j,k+1} - 2T_{i,j,k+1} + T_{i-1,j,k+1}}{(\Delta x)^2}$$
$$+ \frac{T_{i,j+1,k} - 2T_{i,j,k} + T_{i,j-1,k}}{(\Delta y)^2},\qquad(8.60a)$$

and the second half takes us to $k + 2$:

$$\frac{T_{i,j,k+2} - T_{i,j,k+1}}{\alpha \Delta t} = \frac{T_{i+1,j,k+1} - 2T_{i,j,k+1} + T_{i-1,j,k+1}}{(\Delta x)^2}$$
$$+ \frac{T_{i,j+1,k+2} - 2T_{i,j,k+2} + T_{i,j-1,k+2}}{(\Delta y)^2}.\qquad(8.60b)$$

Note that neither step can be repeated unilaterally. Let us examine a simple application. A two-dimensional slab of material is at a uniform initial temperature of 100°. At $t = 0$, one face (the bottom) is instantaneously heated to 400°. Let $\Delta x = \Delta y = 1$, as well as $\alpha = 1$ and $\Delta t = 1/8$. We rewrite eq. (8.60a) isolating the $k + 1$ terms on the right-hand side:

$$-T_{i,j+1,k} + \left(2 - \frac{(\Delta x)^2}{\alpha \Delta t}\right)T_{i,j,k} - T_{i,j-1,k}$$
$$= T_{i+1,j,k+1} - \left(2 + \frac{(\Delta x)^2}{\alpha \Delta t}\right)T_{i,j,k+1} + T_{i-1,j,k+1}.\qquad(8.61)$$

Now we will illustrate the process with a simple square slab; the top, left, and right sides are all maintained at 100°. The bottom will be set to 400°. The nine interior nodes are initialized at 100°.

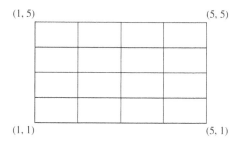

We apply eq. (8.61) at the interior points, row by row; the first horizontal sweep results in

100	100	100
100	100	100
133.67	136.73	133.67

for the nine interior points. Now we recast eq. (8.60b) for application to the columns to advance to the $k + 2$ time step:

$$-T_{i+1,j,k+1} + \left(2 - \frac{(\Delta x)^2}{\alpha \Delta t}\right)T_{i,j,k+1} - T_{i-1,j,k+1}$$
$$= T_{i,j+1,k+2} - \left(2 + \frac{(\Delta x)^2}{\alpha \Delta t}\right)T_{i,j,k+2} + T_{i,j-1,k+2}.\qquad(8.62)$$

We solve the simultaneous equations that result from applying this algorithm to the columns and obtain

100.55	100.6	100.55
105.5	106	105.5
154.42	159.37	154.42

If the total number of equations is modest, then a direct elimination scheme can be used for solution. The coefficient matrix follows the tridiagonal pattern (with 1, −10, 1 for the selected parameters), so the process is easy to automate. Smith (1965) states that for rectangular regions the ADI method requires about *25 times less work* than an explicit computation. Carrying out the procedure to $t = 1.75$ yields

114.91	120.25	114.91
146.35	161.01	146.35
221.06	247.42	221.06

for the interior nodes. Chung (2002) notes that this scheme is unconditionally stable, which makes it very attractive for problems in which the time evolution is slow; that is, we can employ a *very* large Δt relative to the elementary explicit technique and still obtain acceptable accuracy.

Three Spatial Dimensions

Naturally, the solution techniques for parabolic PDEs that we have discussed in this section can be extended to three dimensions as we shall now demonstrate. Consider a cube of solid material, measuring 10 cm on each side, initially at some uniform temperature, T_i. At $t = 0$, the temperatures of the four vertical faces are instantaneously changed to elevated values. In particular, the front face will be 400°, the right-hand face 200°, the left-hand face 1000°, and the back 600°. The bottom of the cube is insulated and the top horizontal surface will lose thermal energy to the surroundings by Newton's law of cooling. A sketch of the arrangement appears in Figure 8.5.

The governing equation for this case is just

$$\frac{\partial T}{\partial t} = \alpha \left[\frac{\partial^2 T}{\partial x^2} + \frac{\partial^2 T}{\partial y^2} + \frac{\partial^2 T}{\partial z^2} \right]. \qquad (8.63)$$

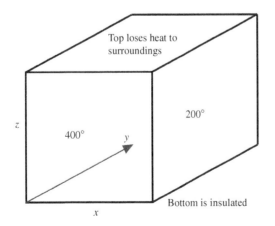

FIGURE 8.5. Cube of material with four vertical sides maintained at different temperatures for all $t > 0$.

We will take the thermal conductivity of the medium to be 0.075 cal/(cm s °C) and discretize the equation letting Δx, Δy, and Δz all be 0.16667 cm. Accordingly, the number of interior mesh points will be 205,379. We will use a first-order forward difference for the time derivative and second-order central differences for the conduction terms:

$$\frac{T_{i,j,k,2} - T_{i,j,k,1}}{\Delta t} \cong \frac{\alpha}{(\Delta x)^2} \big[T_{i+1,j,k,1} + T_{i-1,j,k,1} + T_{i,j+1,k,1} + T_{i,j-1,k,1} \\ + T_{i,j,k+1,1} + T_{i,j,k-1,1} - 6T_{i,j,k,1} \big] \qquad (8.64)$$

Our intent is to solve the equation explicitly by forward-marching in time. We will employ just two values for the time index, 1 and 2, corresponding to the old and new time steps (this is done to minimize storage requirements). We will take the thermal diffusivity, α, to be 0.088 cm²/s and the time step, Δt, to be 0.01 second, resulting in

$$\Delta t \alpha \left[\frac{1}{(\Delta x)^2} + \frac{1}{(\Delta y)^2} + \frac{1}{(\Delta z)^2} \right] = 0.095, \qquad (8.65)$$

which is much less than the limit for stability (recall that the limit is 1/2). We can get a sense of how $T(x, y, z, t)$ develops by looking at the *top surface* of the cube at $t = 7.5, 15, 30,$ and 60 seconds; these results are shown in Figure 8.6.

The sequence of contour plots shown in Figure 8.6 reveals the speed with which thermal energy is conveyed throughout the cube. Although the explicit method was used to solve this problem, the execution time was not prohibitively long, despite the fact that each time step required approximately 212,000 calculations. Since Δt was 0.01 second, about 1.27×10^9 calculations were required to reach $t = 60$ seconds. The ease with which this problem was solved suggests that many heat and mass transfer problems involving three spatial dimensions can be handled exactly this way.

HYPERBOLIC PARTIAL DIFFERENTIAL EQUATIONS

Perhaps the best-known example of a hyperbolic PDE is the "wave" equation; for one spatial dimension, it can be written as

$$\frac{\partial^2 u}{\partial t^2} = c^2 \frac{\partial^2 u}{\partial x^2}. \qquad (8.66)$$

Of course, our immediate thought with respect to a physical interpretation might center on a vibrating string. But

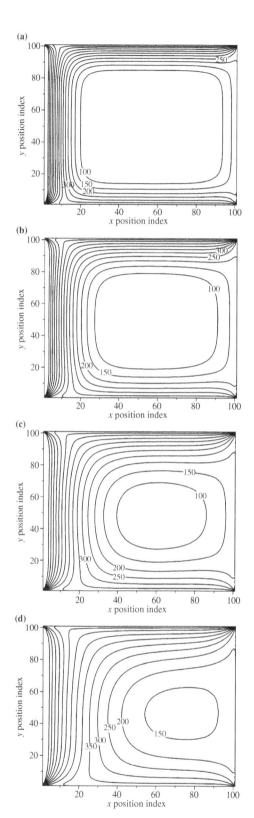

FIGURE 8.6. Evolution of the temperature distribution *on the top surface* of the cube; the four contour plots correspond to 7.5, 15, 30, and 60 s (top to bottom). For the two-dimensional top view shown here, the right-hand edge is maintained at 200°, the left-hand side at 1000°, the bottom at 400°, and the top at 600°.

wavelike behavior can be found for many different phenomena, including electrical and magnetic fields and even nerve impulses (for the latter, the interested reader should explore the *FitzHugh–Nagumo model*).

Because the wave equation has been around since the middle of the eighteenth century, much is known about its solutions. In fact, the reader is encouraged to apply the variable transformation,

$$r = x + ct \quad \text{and} \quad s = x - ct, \tag{8.67}$$

to eq. (8.66) to reproduce d'Alembert's solution process from 1747. This approach is of particular interest to us because it represents a special case of the technique we wish to discuss, the *method of characteristics*. The name of this technique arises from the fact that, at every point in the *x-t* plane, two characteristic directions can be identified for which *ordinary* differential equations can be used to "solve" eq. (8.66) in a stepwise process.

Before we begin that discussion, we will illustrate several important points using an extremely simple first-order "constant coefficient advection" (first-order wave equation) model:

$$\frac{\partial \phi}{\partial t} + 3 \frac{\partial \phi}{\partial x} = 0. \tag{8.68}$$

We let $\phi(x, t = 0) = 0$ and introduce the disturbance (at $x = 0$) that will propagate in the *x*-direction. We can use this particular model to underscore some of the problems that one may encounter with hyperbolic PDEs. First, we will introduce a finite-duration impulse (with finite amplitude) and solve eq. (8.68) numerically using an explicit approach. Since the "velocity" in the *x*-direction is "3," we will solve the equation for specific times of 1/3, 2/3,..., 5/3; thus, the advected disturbance should be centered at 1, 2,..., 5.

The result depicted in Figure 8.7 is probably not what you expected. We can make the nature of the problem even clearer by inputting a unit step change at $x = 0$, letting the "sharp-edged" step be carried along in the positive *x*-direction; this is illustrated in Figure 8.8.

It may be apparent to you that this result also fails to meet expectations; for a homogeneous wave equation in one dimension, the shape of the traveling wave should not change! We will now demonstrate what *should have* transpired. We first do this using a familiar technique, the Laplace transform, which will eliminate the time derivative. Applying the transform and solving the subsidiary equation results in

$$\phi(s) = C_1 \exp\left(-\frac{sx}{3}\right). \tag{8.69}$$

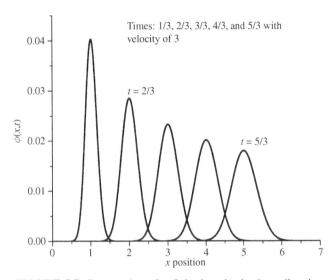

FIGURE 8.7. Propagation of a finite impulse in the x-direction due to a constant velocity of 3.

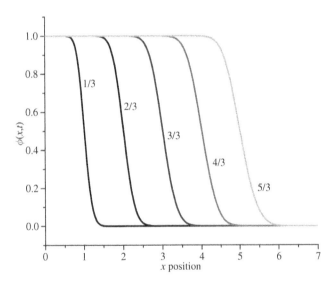

FIGURE 8.8. Advection of a unit step in the x-direction due to the constant velocity, 3.

The unit step is put in at the left-hand boundary; that is, at $x = 0$, $\phi(s) = 1/s$. Therefore,

$$\phi(s) = \frac{1}{s}\exp\left(-\frac{sx}{3}\right). \quad (8.70)$$

We can invert directly by consulting a table of transforms. Letting $k = x/3$, we find

$$\phi = 0, \text{ for } 0 < t < k \quad \text{and} \quad \phi = 1, \text{ if } t > k.$$

In other words, the unit step disturbance propagates downstream *unchanged* according to the analytic solution— exactly like the behavior of an idealized plug-flow tubular reactor! It is apparent that we need a more accurate solution technique than the explicit procedure and one that has been much used in the solution of hyperbolic PDEs is the method of characteristics. The reader may find additional detail helpful; consultation with Smith (1965) or Sarra (2003) is recommended.

The Method of Characteristics

We direct our attention toward a particular curve given by $(x(s), t(s))$. We let the derivatives dt/ds and dx/ds be equated to the coefficients on the $\partial\phi/\partial t$ and $\partial\phi/\partial x$ terms in eq. (8.68). Please note what the consequence of this action is! Therefore,

$$\frac{dt}{ds} = 1 \quad \text{and} \quad \frac{dx}{ds} = 3. \quad (8.71)$$

For the latter, we find $x = 3s + C_1$; for $s = 0$, we have $x = x_0$, and thus, $x_0 = x - 3s$. From the former of this pair, we find $t = s + C_2$, where $C_2 = 0$; that is, for this type of problem, there is only *one* characteristic equation to solve. Therefore, $x_0 = x - 3t$ and $\phi = f(x - 3t)$; we have identified the transformation from (x, t) to (x_0, s).

We will now look at an example that permits us to more fully gauge the usefulness of the method of characteristics. Consider the behavior of an ideal string suspended between supports located at $x = 0$ and $x = L$, where $L = 2$. The velocity of propagation will be taken as 1 (i.e., $c = 1$); therefore,

$$\frac{\partial^2 u}{\partial t^2} = \frac{\partial^2 u}{\partial x^2}. \quad (8.72)$$

The initial shape of the string is specified, $f(x, 0) = \sin(\pi x)$, and the initial velocity is zero. The analytic solution for this case is known and we will want to make use of it:

$$u(x, t) = \sum_{n=1}^{\infty}\left[\int_0^2 \sin \pi x \sin \frac{n\pi x}{2}\,dx\right]\sin \frac{n\pi x}{2}\cos \frac{n\pi t}{2}. \quad (8.73)$$

Some results from this equation are presented graphically in Figure 8.9 for specific x positions corresponding to $x = 1/16, 1/8, 1/4, 1/2$.

We begin with values taken from a "curve" along which the $u(x, t)$s are known.

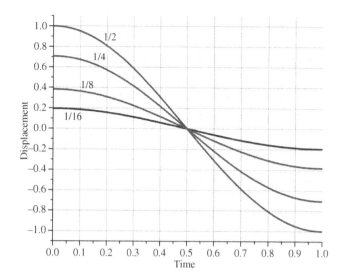

FIGURE 8.9. Analytic solution for the wave (eq. 8.72) for specific x positions, 1/16, 1/8, 1/4, and 1/2.

The slopes of the "characteristic" directions are obtained from the roots of the quadratic equation:

$$\left(\frac{dt}{dx}\right)^2 - 1 = 0, \qquad (8.74)$$

and of course, these values are $+1$ and -1. We use these slopes to extrapolate from known positions, P and Q, to a new position, R. The initial points for this example are selected from Figure 8.9: Let $(x_P, t_P) = (1/8, 0.2)$ and $(x_Q, t_Q) = (1/4, 0.2)$. The new point is identified from the linear approximations,

$$t_R - t_P = +1(x_R - x_P) \quad \text{and} \quad t_R - t_Q = -1(x_R - x_Q). \qquad (8.75)$$

The solutions for the simultaneous eq. (8.75) are $x_R = 3/16$ and $t_R = 0.2625$. We now use the differential relationships along the characteristics to obtain new estimates for $p = \partial u/\partial x$ and $q = \partial u/\partial t$. These slopes (for positions P and Q), in turn, allow us to estimate the change in the dependent variable, u; we use the average of the initial and projected slopes to compute this change (and hence, the new value for the displacement, u). It is easy to show that

$$q_R = \frac{1}{2}[p_Q - p_P + q_P + q_Q].$$

For the points selected from Figure 8.9, $p_P = 2.12$, $p_Q = 0.9$, $q_P = -0.714$, and $q_Q = -1.438$. Therefore, $q_R = -1.686$ and $p_R = 1.148$. Since the change in the dependent variable, u, is just $du = p\,dx + q\,dt$, we find that

$$u_R - u_P = \frac{1}{2}(p_P + p_R)(x_R - x_P) + \frac{1}{2}(q_P + q_R)(t_R - t_P),$$

which yields $u_R = 0.3371$. The reader should turn immediately to the results presented in Figure 8.9 and estimate the value of u at this new point, R—the value is approximately 0.34. In this case, the linear extrapolation combined with the use of the average slopes has produced a very good estimate.

Smith (1965) shows how this estimate for the dependent variable u can be subsequently refined by using the average slopes to improve the coordinates of the projected position—which are, in turn, used to get improved slopes. In this manner, very accurate solutions for hyperbolic PDEs can be obtained through iteration; however, the technique is not the easiest to automate and for that reason is probably not used as commonly as it once was. There is an FDM for solution of some hyperbolic PDEs that is extremely easy to implement and it is described in the next section.

The Leapfrog Method

Let us now return to the familiar wave equation with one spatial dimension,

$$\frac{\partial^2 u}{\partial t^2} = c^2 \frac{\partial^2 u}{\partial x^2}, \qquad (8.76)$$

and formulate one possible finite-difference approximation for it:

$$\frac{u_{i,j+1} - 2u_{i,j} + u_{i,j-1}}{(\Delta t)^2} \cong c^2 \frac{u_{i+1,j} - 2u_{i,j} + u_{i-1,j}}{(\Delta x)^2}. \qquad (8.77)$$

We now isolate the value on the new time-step row:

$$u_{i,j+1} \cong \frac{c^2(\Delta t)^2}{(\Delta x)^2}[u_{i+1,j} - 2u_{i,j} + u_{i-1,j}] + 2u_{i,j} - u_{i,j-1}. \qquad (8.78)$$

Since c has dimensions of velocity, l/t, it is clear that the quotient,

$$\frac{c^2(\Delta t)^2}{(\Delta x)^2},$$

is dimensionless. In fact, it is the Courant number, Co, squared of course. Notice what happens if we select $Co = 1$; the finite-difference approximation is now simply

$$u_{i,j+1} \cong u_{i+1,j} + u_{i-1,j} - u_{i,j-1}. \qquad (8.79)$$

This very compact expression forms the basis for what is called the "leapfrog" method and it will allow us to solve certain wave-equation problems. You may notice, however, that the algorithm requires values for *two previous* time steps; that is, it is not self-starting.

Now let us suppose that we have initial values for both position and velocity such that

$$u(x, 0) = f(x) \quad \text{and} \quad \frac{\partial u}{\partial t}(x, 0) = g(x). \quad (8.80)$$

We let $u_{i,1} = f_i$ and since

$$\frac{\partial u}{\partial t} \cong \frac{u_{i,j+1} - u_{i,j-1}}{2\Delta t},$$

if we set $j = 1$, then this derivative is just

$$\frac{u_{i,2} - u_{i,0}}{2\Delta t} = g_i. \quad (8.81)$$

We take this result back to the leapfrog algorithm and (letting $j = 1$) isolate $u_{i,2}$. Therefore,

$$u_{i,2} = \frac{1}{2}[f_{i+1} + f_{i-1} + 2\Delta t g_i]. \quad (8.82)$$

This allows us to get the computation started.

We will now illustrate how this works with a typical example. Suppose we have a "string" stretched between supports located at $x = 0$ and $x = L$. The "string" is perfectly elastic and is under great tension such that the gravitational force is unimportant. The displacement (deflection) of the string is described by the wave equation:

$$\frac{\partial^2 u}{\partial t^2} = c^2 \frac{\partial^2 u}{\partial x^2}. \quad (8.83)$$

The string has an initial displacement and an initial velocity given by $u(x, 0) = f(x)$ and $(\partial u/\partial t)|_{t=0} = g(x)$, respectively. We will take $L = 10$, an initial velocity of zero, but an initial deflection described by $u = x$ for $0 < x < 1$ and $u = 1 - (x - 1)/9$ for $1 < x < 10$. We will use the leapfrog method to compute the string's displacement as a function of time. Note that the initial deflection propagates to the right (and down) as illustrated in Figure 8.10.

ELEMENTARY PROBLEMS WITH CONVECTIVE TRANSPORT

Our focus in this section concerns problems in which convection is important; that is, problems where momentum, heat, mass, and so on, are transported by virtue of a nonzero velocity vector component. Examples of the terms of interest are

$$\rho\left(v_x \frac{\partial v_x}{\partial x} + v_y \frac{\partial v_x}{\partial y} + v_z \frac{\partial v_x}{\partial z}\right) = \dots \text{ for } x\text{-momentum},$$

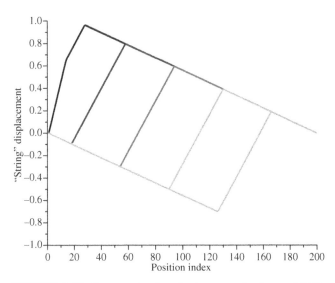

FIGURE 8.10. Computed string displacement for times of 0.06, 0.36, 0.72, 1.08, and 1.44; we can render these ts dimensionless using ct/L and the corresponding values are 0.03, 0.18, 0.36, 0.54, and 0.72. The position index of 200 corresponds to $L = 10$.

$$\rho C_p\left(v_x \frac{\partial T}{\partial x} + v_y \frac{\partial T}{\partial y} + v_z \frac{\partial T}{\partial z}\right) = \dots \text{ for thermal energy},$$

and

$$\left(v_x \frac{\partial C_A}{\partial x} + v_y \frac{\partial C_A}{\partial y} + v_z \frac{\partial C_A}{\partial z}\right) = \quad (8.84a,b,c)$$

... for mass (concentration of species "A").

Inclusion of such terms in a model represents a serious escalation in difficulty as we shall see. However, a limited class of such problems can be solved readily using techniques that are already familiar to us.

Imagine a situation in which a chemical species with concentration C is carried in the z-direction by fluid motion. Furthermore, assume that any mixing that occurs—possibly as a result of turbulence—can be represented as though it were diffusive in character. One model for this phenomenon can be written as

$$\frac{\partial C}{\partial t} + V \frac{\partial C}{\partial z} = D \frac{\partial^2 C}{\partial z^2}. \quad (8.85)$$

This is an *axial dispersion model* in which the velocity in the z-direction is taken to be constant. Let the index i represent z position and j represent time; one possible discretization for this equation is

$$\frac{C_{i,j+1} - C_{i,j}}{\Delta t} \cong D \frac{C_{i+1,j} - 2C_{i,j} + C_{i-1,j}}{(\Delta z)^2} - V \frac{C_{i,j} - C_{i-1,j}}{\Delta z}. \quad (8.86)$$

Please note that an *upwind* difference has been used in the convective transport term. The significance of this choice will be discussed in the next section—for now, we will simply operate on the assumption that this is appropriate. If we multiply by Δt and add $C_{i,j}$ to both sides of the equation, we obtain an elementary explicit algorithm:

$$C_{i,j+1} \cong \Delta t \left[D \frac{C_{i+1,j} - 2C_{i,j} + C_{i-1,j}}{(\Delta z)^2} - V \frac{C_{i,j} - C_{i-1,j}}{\Delta z} \right] + C_{i,j}.$$

(8.87)

The reader might wonder why we have not chosen a "better" approximation for the first derivative, $\partial C/\partial z$. For example, we could select a second-order central difference where

$$\frac{\partial C}{\partial z} \cong \frac{C_{i+1,j} - C_{i-1,j}}{2\Delta z}.$$

(8.88)

Please consider trying this change yourself; it is worth taking a little time to discover that it will not work! The result of the computation is unphysical.

Now assume that some finite pulse is put into the flow at the entrance where $z = 0$; our intent is to model the behavior of this pulse as it is carried downstream by the flow. However, there is an obvious limitation to our modeling approach: If the length of the test section is L, then at a time corresponding to L/V (actually a little before), the tracer pulse will reach the outflow boundary. Naturally, an outflow boundary condition of $C(z = L, t) = 0$ would be violated. This a common, recurring problem in the computational solution of transport problems—an outflow condition is necessary, but specification of the wrong one will constrain the solution and produce an incorrect result. An obvious "remedy" is to stop the computation before the tracer pulse arrives at the far end of the computational domain. Let us explore how this very simple computational model performs; we set $V = 2$, $D = 1/4$, $\Delta z = 0.2$, and $\Delta t = 0.005$. The inlet concentration is set to 2 for $0 < t < 0.4$, and then it reverts to zero. The evolution of the input pulse is illustrated in Figure 8.11.

Since the fluid velocity is 2, the successive peaks are centered at axial positions corresponding to $2t$, of course. The dispersion is causing the attenuation of the initial pulse height and the broadening of the distributions; note that by $t = 50$, the tracer is covering axial positions from 80 to 120, which is double the width seen at $t = 10$. The significance of the dispersion model illustrated in this example is that eq. (8.85) allows us to characterize the performance of a flow reactor. By monitoring an inert tracer injected into the reactor, we can evaluate V and D (and thus the Peclet number). These results can be used in turn through the inclusion of chemical kinetics to predict performance (i.e.,

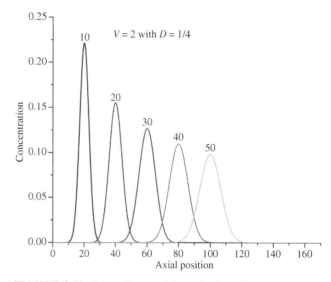

FIGURE 8.11. Dispersion model results for a flow reactor with Pe $= VL/D = 6400$ at times corresponding to 10, 20, 30, 40, and 50. Note that Pe is a kind of Peclet number for mass transfer.

expected conversion). The model itself, eq. (8.85), can be augmented to account for radial dispersion by incorporating two dispersion coefficients:

$$\frac{\partial C}{\partial t} + V \frac{\partial C}{\partial z} = \frac{D_R}{r} \frac{\partial}{\partial r} \left(r \frac{\partial C}{\partial r} \right) + D_L \frac{\partial^2 C}{\partial z^2}.$$

(8.89)

We turn now to an example of convective heat transport. Consider laminar flow in a duct (formed by parallel walls); the lower wall is located at $y = 0$ and the upper wall at $y = B$. The flow is fully developed and the fluid enters at a uniform (low) temperature. The heated walls are maintained at a higher temperature and heat transfer to the fluid occurs as the fluid moves in the x-direction. This is a steady-state problem with the velocity distribution given by

$$v_x = \frac{1}{2\mu} \frac{dp}{dx}(y^2 - By).$$

(8.90)

The appropriate energy equation for this case is

$$\rho C_p v_x \frac{\partial T}{\partial x} = k \left[\frac{\partial^2 T}{\partial x^2} + \frac{\partial^2 T}{\partial y^2} \right].$$

(8.91)

Note that conduction terms in both the transverse and flow directions have been included. The latter can be neglected if the product of the Reynolds and Prandtl numbers (RePr) is greater than about 100, and for many liquids, this is likely to be the case. Therefore,

$$\frac{\partial T}{\partial x} = \frac{\alpha}{A(y^2 - By)} \frac{\partial^2 T}{\partial y^2},$$

(8.92)

where $A = (1/2\mu)(dp/dx)$. Such problems provide us with an opportunity for simple solution by forward-marching in the x-direction; we discretize the equation so that

$$T_{i+1,j} \cong \frac{\Delta x \alpha}{A(y^2 - By)}\left[\frac{T_{i,j+1} - 2T_{i,j} + T_{i,j-1}}{(\Delta y)^2}\right] + T_{i,j}. \quad (8.93)$$

We now have an algorithm that allows us to march downstream, computing new temperatures as we go, given the initial uniform inlet temperature distribution. Let us illustrate how this is going to work with an example; we select an inlet temperature of 10°C and maintain the parallel walls at 50°C. We take the fluid to be water with $\alpha = 0.00147$ cm²/s and Pr = 6.8. The pressure gradient, dp/dx, is set to -0.1778 dyne/cm²/cm. We expect the temperature distribution to evolve slowly—after all, we are relying on molecular conduction to carry the thermal energy into the interior of the fluid. In fact, Figure 8.12 shows that the bulk fluid temperature is only 21.2°C at $x = 160$ cm; even by $x = 400$ cm, the bulk fluid temperature is only 29.94°C.

In the last example of this section, we would like to treat the case of mass transfer in the annular space between concentric cylinders. The fluid contained within is initially at rest, but at $t = 0$, the outer cylinder (located at $r = R_2$) begins to rotate with constant angular velocity, ω. Of course, this means that the tangential velocity vector component at the wall is $v_\theta(r = R_2) = \omega R_2$. We place a small spot of inert tracer adjacent to the inner wall; we are interested in how this

material is transported as the velocity distribution develops. The two governing equations are

$$\frac{\partial v_\theta}{\partial t} = \nu\left[\frac{\partial^2 v_\theta}{\partial r^2} + \frac{1}{r}\frac{\partial v_\theta}{\partial r} - \frac{v_\theta}{r^2}\right] \quad (8.94)$$

and

$$\frac{\partial C}{\partial t} + \frac{v_\theta}{r}\frac{\partial C}{\partial \theta} = D\left[\frac{\partial^2 C}{\partial r^2} + \frac{1}{r}\frac{\partial C}{\partial r}\right]. \quad (8.95)$$

Notice that we have omitted molecular (diffusional) transport in the θ-direction as it should be small relative to convective transport once the velocity distribution begins to develop. Our solution strategy is as follows: We will solve the discretized version of eq. (8.94) explicitly by forward-marching in time. Each time we compute a new velocity distribution, we will use $v_\theta(r, t)$ in the discretized version of eq. (8.95) to calculate C over the entire array of r and θ positions: $R_1 \leq r \leq R_2$ and $0 \leq \theta \leq 2\pi$. Of course, we must use upwind differencing for the convective transport term

$$\left(\frac{v_\theta}{r}\frac{\partial C}{\partial \theta}\right)$$

since we are using the explicit technique. A complete pass through *both* equations corresponds to a time step. We merely repeat the process until the desired final time is attained. An elementary code for the computational process is included here:

```
#COMPILE EXE
#DIM ALL
    REM *** Convective transport example
        GLOBAL i,j,dt,dth,dr,R1,R2,kvis,D,w,d2vdr2,dvdr,r
            pos AS SINGLE
        GLOBAL tt,zz,d2cdr2,dcdr,dcdth AS SINGLE
FUNCTION PBMAIN
    DIM v(91,2) AS SINGLE
    DIM C(91,91,2) AS SINGLE
        dr=0.03333:dth=0.069813:kvis=0.02:D=0.001:
            w=3:R1=5:R2=8
        dt=0.001
    REM *** initialize velocity and concentration fields
        FOR i=1 TO 90
            v(i,1)=0
        NEXT i
            v(91,1)=w*R2
        FOR i=1 TO 90
            FOR j=1 TO 90
            c(i,j,1)=0
            NEXT j:NEXT i
            FOR i=1 TO 5
                FOR j=1 TO 5
```

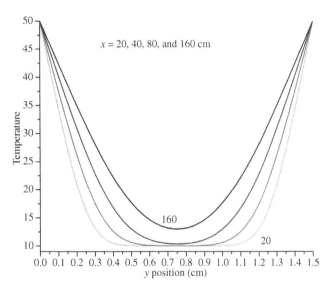

FIGURE 8.12. Heat transfer to fully developed laminar flow between parallel walls, 1.5 cm apart. Temperature distributions are shown for x positions of 20, 40, 80, and 160 cm. The walls are maintained at 50°C and the fluid (water) enters the heated section with a uniform temperature of 10°C. The fluid properties are taken as constant, although that is a bit of a stretch since the viscosity of water at 50°C is only about 0.55 cp.

```
                c(i,j,1)=1
            NEXT j:NEXT i
        tt=0
100 REM *** continue
REM *** compute revised velocities
    FOR i=2 TO 90
      rpos=R1+(i-1)*dr
      d2vdr2=(v(i+1,1)-2*v(i,1)+v(i-1,1))/dr∧2
      dvdr=(v(i+1,1)-v(i-1,1))/(2*dr)
      v(i,2)=dt*kvis*(d2vdr2+1/rpos*dvdr-v(i,1)/
        rpos∧2)+v(i,1)
      NEXT i
    FOR i=2 TO 90
      v(i,1)=v(i,2)
      NEXT i
REM *** begin computation for concentration field
    FOR j=2 TO 90
      FOR i=2 TO 90
        d2cdr2=(c(i+1,j,1)-2*c(i,j,1)+c(i-1,j,1))/dr∧2
        dcdr=(c(i+1,j,1)-c(i-1,j,1))/(2*dr)
        dcdth=(c(i,j,1)-c(i,j-1,1))/dth
        rpos=R1+(i-1)*dr
    c(i,j,2)=dt*D*(d2cdr2+1/rpos*dcdr)-v(i,1)/
      rpos*dcdth*dt+c(i,j,1)
            NEXT i:NEXT j
          FOR j=2 TO 90
            FOR i=2 TO 90
            c(i,j,1)=c(i,j,2)
            NEXT i:NEXT j
          FOR i=2 TO 90
            c(i,91,1)=c(i,90,1)
            c(i,1,1)=c(i,91,1)
              NEXT i
    tt=tt+dt
        PRINT c(6,6,1),c(7,7,1),c(8,8,1),c(9,9,1)
          IF tt<55 THEN 100 ELSE 300
300 REM *** continue
      OPEN "c:MTcylind.dat" FOR OUTPUT AS #1
      FOR j=1 TO 91
        FOR i=1 TO 91
          WRITE#1,i,j,c(i,j,1)
          NEXT i:NEXT j
      CLOSE:END
```

Figure 8.13 illustrates the progress of the tracer (contaminant) plume as the velocity distribution develops. Remember, the outer cylinder is put in motion at $t = 0$ with $v_\theta(r = R_2) = \omega R_2$. The inner cylinder is fixed (stationary) so at very small t, mass transfer occurs mainly by molecular diffusion.

In the three examples of convective transport provided earlier, we were able to find solutions with familiar, elementary numerical procedures. Of course, in all cases, we had only one nonzero component of the velocity vector. The situation for transport involving two-dimensional flows is more difficult, but a very useful procedure for such problems is described in the next section.

A NUMERICAL PROCEDURE FOR TWO-DIMENSIONAL VISCOUS FLOW PROBLEMS

We now will describe a very powerful technique that can be used to solve a host of important flow and transport problems in *two dimensions*. We begin by writing down the governing equations for the motion of an incompressible fluid:

$$\frac{\partial v_x}{\partial t} + v_x \frac{\partial v_x}{\partial x} + v_y \frac{\partial v_x}{\partial y} = -\frac{1}{\rho}\frac{\partial p}{\partial x} + \nu\left[\frac{\partial^2 v_x}{\partial x^2} + \frac{\partial^2 v_x}{\partial y}\right], \quad (8.96)$$

$$\frac{\partial v_y}{\partial t} + v_x \frac{\partial v_y}{\partial x} + v_y \frac{\partial v_y}{\partial y} = -\frac{1}{\rho}\frac{\partial p}{\partial y} + \nu\left[\frac{\partial^2 v_y}{\partial x^2} + \frac{\partial^2 v_y}{\partial y^2}\right], \quad (8.97)$$

and

$$\frac{\partial v_x}{\partial x} + \frac{\partial v_y}{\partial y} = 0. \quad (8.98)$$

These three equations must be solved simultaneously for the general problem we are contemplating. The main difficulty in such cases—and one that plagues CFD—is the determination of $p(x, y, t)$. As the flow field evolves, $p(x, y, t)$ must change to ensure that continuity is satisfied (i.e., $\nabla \cdot v = 0$). In the approach we are about to describe, the problem of $p(x, y, t)$ is circumvented.

We now cross differentiate eq. (8.96) and eq. (8.97), resulting in

$$\frac{\partial^2 v_x}{\partial t \partial y} + \frac{\partial v_x}{\partial y}\frac{\partial v_x}{\partial x} + v_x\frac{\partial^2 v_x}{\partial x \partial y} + \frac{\partial v_y}{\partial y}\frac{\partial v_x}{\partial y} + v_y\frac{\partial^2 v_x}{\partial y^2}$$
$$= -\frac{1}{\rho}\frac{\partial^2 p}{\partial x \partial y} + \nu\left[\frac{\partial^3 v_x}{\partial x^2 \partial y} + \frac{\partial^3 v_x}{\partial y^3}\right] \quad (8.99)$$

and

$$\frac{\partial^2 v_y}{\partial t \partial x} + \frac{\partial v_x}{\partial x}\frac{\partial v_y}{\partial x} + v_x\frac{\partial^2 v_y}{\partial x^2} + \frac{\partial v_y}{\partial x}\frac{\partial v_y}{\partial y} + v_y\frac{\partial^2 v_y}{\partial y \partial x}$$
$$= -\frac{1}{\rho}\frac{\partial^2 p}{\partial y \partial x} + \nu\left[\frac{\partial^3 v_y}{\partial x^3} + \frac{\partial^3 v_y}{\partial y^2 \partial x}\right]. \quad (8.100)$$

The *vorticity vector component*, which for two-dimensional flow in the *x-y* plane is a measure of rotation about the *z*-axis, is defined by

$$\omega_z = \left(\frac{\partial v_y}{\partial x} - \frac{\partial v_x}{\partial y}\right). \quad (8.101)$$

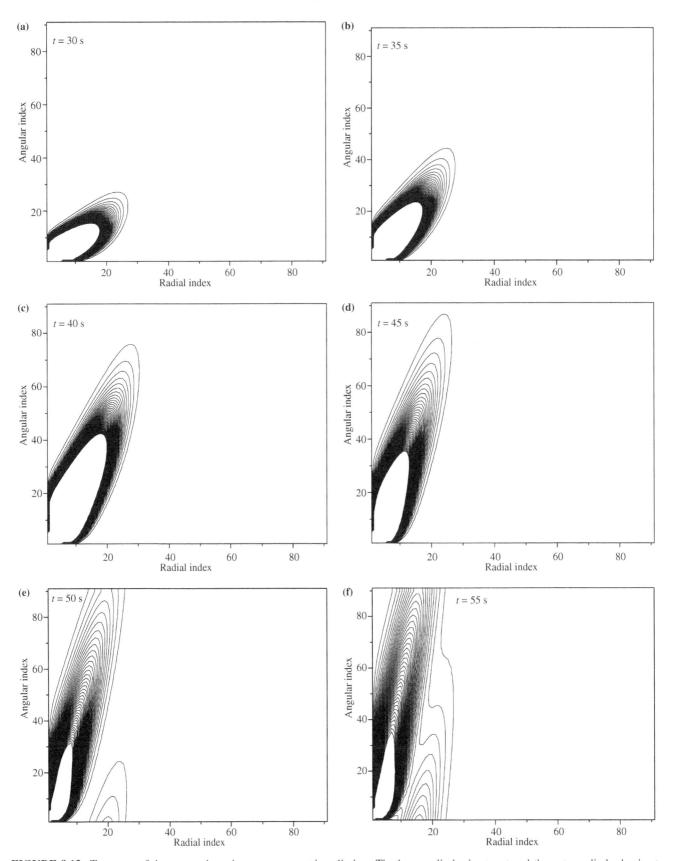

FIGURE 8.13. Transport of the tracer plume between concentric cylinders. The inner cylinder is at rest and the outer cylinder begins to rotate with constant angular velocity at $t = 0$.

Note what happens when we subtract eq. (8.99) from eq. (8.100), employing the vorticity definition:

$$\frac{\partial \omega}{\partial t} + v_x \frac{\partial \omega}{\partial x} + v_y \frac{\partial \omega}{\partial y} = \nu \left[\frac{\partial^2 \omega}{\partial x^2} + \frac{\partial^2 \omega}{\partial y^2} \right]. \quad (8.102)$$

This is the *vorticity transport equation*. We have eliminated pressure, and if we now introduce the stream function,

$$v_x = \frac{\partial \psi}{\partial y} \quad \text{and} \quad v_y = -\frac{\partial \psi}{\partial x}, \quad (8.103)$$

we can guarantee that continuity will automatically be satisfied and we can use the vorticity definition, eq. (8.101), to obtain the relationship between ω and ψ:

$$-\omega = \frac{\partial^2 \psi}{\partial x^2} + \frac{\partial^2 \psi}{\partial y^2}. \quad (8.104)$$

One of the reasons that the vorticity approach is so powerful is that vorticity is only created at the boundaries of the flow—not in the interior! A solution procedure suggests itself: We use the vorticity distribution to get the stream function, use the stream function to get the velocity vector components, and then use the new velocities to get an updated vorticity distribution, and repeat. We now illustrate the use of this method for the computation of the transient flow of a confined viscous fluid off of a step. In this situation, the available flow area doubles and the motion of the fluid (which is initially at rest) is driven by the lateral or sliding motion of the upper surface. The ultimate Reynolds number at the entrance is 75, although the computation is not allowed to proceed that long since the vortex generated on the face of the step continues to grow, eventually reaching the outflow boundary. Typical results from a sequence of calculations as described earlier are shown in Figure 8.14.

An attractive feature of the vorticity approach is that only elementary solution procedures are required; we solve eq. (8.102) explicitly and we solve eq. (8.104) iteratively using SOR. Furthermore, this technique can be easily extended to include heat or mass transfer. For example, suppose we wanted to model nonisothermal conditions combined with a two-dimensional flow:

$$\frac{\partial T}{\partial t} + v_x \frac{\partial T}{\partial x} + v_y \frac{\partial T}{\partial y} = \alpha \left[\frac{\partial^2 T}{\partial x^2} + \frac{\partial^2 T}{\partial y^2} \right]. \quad (8.105)$$

Note the similarity between this eq. (8.105) and the vorticity transport equation. That suggests that the solution procedure used for eq. (8.102) could be used here as well. In fact, both equations can be solved using the elementary explicit technique if we are appropriately careful with respect to numerical stability issues that arise due to the

presence of the convective transport terms. Consider the case of the x-direction transport of thermal energy; for explicit solution, the convective transport term, $v_x(\partial T/\partial x)$, is taken to the right-hand side and the equation is discretized. We multiply by Δt and observe that a dimensionless quotient has now appeared: $\Delta t v_x/\Delta x$. This is the Courant number, Co (that we saw in the previous section), and its value *must be* between 0 and 1 if the elementary explicit scheme is to work successfully. Furthermore, the temperature gradient or derivative, $\partial T/\partial x$, must be written in the "upwind" form so that a disturbance is only *advected in the direction of the fluid motion* (by the term "advection," we mean transport by fluid motion in a particular direction). Let us illustrate by assuming that the velocity in the x-direction can be taken to be a positive constant, V, then the first two terms in eq. (8.105) are written as

$$\frac{\partial T}{\partial t} = -V \frac{\partial T}{\partial x} + \cdots, \quad \text{which become} \quad \frac{T_{i,j+1} - T_{i,j}}{\Delta t}$$

$$\cong -V \frac{T_{i,j} - T_{i-1,j}}{\Delta x} + \cdots, \quad (8.106)$$

and therefore,

$$T_{i,j+1} \cong (1 - Co)T_{i,j} + CoT_{i-1,j} \cdots. \quad (8.107)$$

We will now explore the application of this technique to flow resulting from natural convection in enclosures (a Rayleigh–Bénard problem). A viscous fluid, initially at rest and at uniform temperature, is contained within a rectangular enclosure. At $t = 0$, the temperature of the bottom surface is elevated and buoyancy-driven fluid motion ensues. The warmer fluid will rise and the cooler fluid will fall, setting up a pattern of recirculation. We assume that the rectangular vessel is two-dimensional and that the Boussinesq approximation is adequate for a description of the buoyancy effect (this means that the fluid density is actually treated as constant in the equations of motion, but a buoyancy force term is appended). Buoyancy affects the vertical (or y-component) of the Navier–Stokes equation, which must be modified by the addition of $\rho g \beta \Delta T$, where β is the coefficient of volumetric expansion (e.g., for an ideal gas, this is merely $1/T$, so at 300 K, $\beta = 0.0033$). For these example calculations, the Prandtl and Grashof numbers are Pr = 6.75 and Gr = 1000, respectively. The width-to-height ratio for the rectangular duct is 2.47, and the size of this ratio determines the number of convection rolls that will ultimately appear in the enclosure.

Of course, it is entirely appropriate to question the accuracy of this computation: Do the results realistically portray buoyancy-driven flows in two-dimensional enclosures? We can obtain comparisons from the literature for confirmation; see the interferogram (figure 139, p. 82) in Van Dyke (1982), taken from the work of Oertel and Kirchartz (1979). The

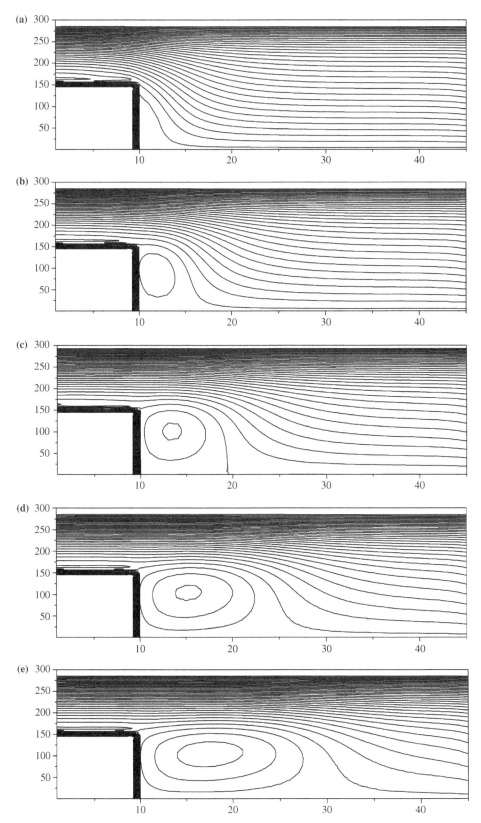

FIGURE 8.14. Transient viscous flow off of a step with an ultimate Reynolds number of 75 (based on the mean inflow velocity). The fluid is confined and is initially at rest, and the motion is being driven by a sliding upper surface. The velocity of the upper surface increases linearly with time until it reaches the predetermined value.

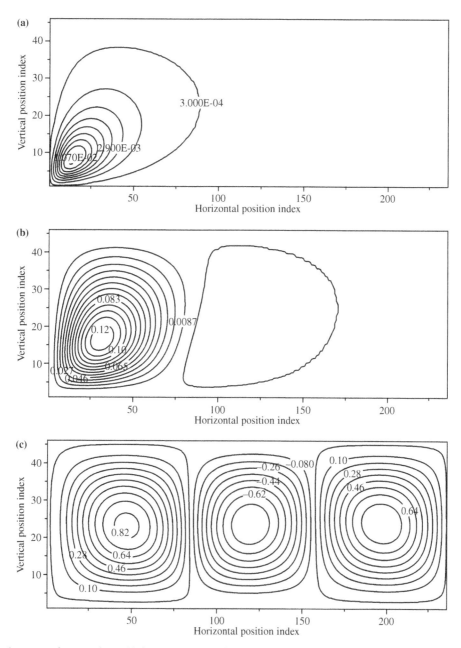

FIGURE 8.15. Development of convection rolls in an enclosure with heating at the bottom surface. The three plots correspond to dimensionless times ($t^* = v_0 t/h^2$, where h is the vertical height of the enclosure) of 0.05, 0.25, and 1.25.

similarity between these experimental results and our computations as reflected by Figure 8.15 is reassuring. However, please note that an elementary model of the Rayleigh–Bénard phenomenon as depicted here cannot reproduce certain aspects of the transient behavior seen in real buoyancy-driven flows. In particular, we observe that the sense of the initial rotation is always the same in the computational procedure employed in this example. This will not necessarily be the case in an experiment, where localized heating may result in either clockwise or counterclockwise rotation of the first convection roll. This provides emphasis

for an observation we made previously: When we discretize a PDE, we are, in fact, making changes to the model itself. The gross results may be realistic, but the effects of altering the PDE, *coupled with the particular pattern* of the computational procedure, may mask (or even obliterate) some of the nuances seen in the actual physical phenomenon.

The power of the vorticity transport technique for two-dimensional problems with fluid motion may now be apparent to you, but let us extend it for final emphasis. Consider the transport of a chemical species in two dimensions governed by the continuity equation:

$$\frac{\partial C}{\partial t} + v_x \frac{\partial C}{\partial x} + v_y \frac{\partial C}{\partial y} = D\left[\frac{\partial^2 C}{\partial x^2} + \frac{\partial^2 C}{\partial y^2}\right]. \quad (8.108)$$

You will see immediately that this equation is of *exactly the same form* as the vorticity transport equation and the two-dimensional energy equation. Thus, the explicit procedure we used previously will work here as well. We now have the capability for solving two-dimensional problems involving the transport of momentum, heat, and mass, even when all three are occurring simultaneously. The only restrictive requirement is that the Reynolds number be low enough to guarantee that we have a highly ordered (laminar) flow. In a recent paper, Nikbakhti and Rahimi (2012) have solved eq. (8.102), eq. (8.105), and eq. (8.108) simultaneously for *double-diffusive* natural convection in a rectangular cavity; the convective circulation was driven by a thermally active wall *combined with* a patch of elevated concentration. This means that the local fluid density was affected by both thermal expansion and by variations in solute concentration, and both are accounted for by using the Boussinesq approximation: $\rho = \rho_0[1 - \beta_T(T - T_c) + \beta_s(C - C_1)]$. Note that the effects of heat and mass transfer are opposed. The authors have presented streamlines, isotherms, and concentration contours for different locations of the thermally active surface and some of the resulting circulation patterns are fascinating. For example, when the heated sections are placed at the top and bottom of the cavity, two strong (convection) cells arise, in the upper left and the lower right, with an offset horizontally. Under these conditions, the local velocity near the horizontal centerline is very low.

MacCORMACK'S METHOD

MacCormack's method (1969) for transient phenomena allows the analyst to solve time-dependent PDEs explicitly and for a couple of decades, it was one of the most popular techniques for the solution of high-speed compressible flow problems. For simplicity, let us contemplate a two-dimensional flow in which the density varies (i.e., we have a compressible fluid). We have two components of the Navier–Stokes equation and continuity (conservation of mass). The reader may note an inconsistency here: We said the fluid density would vary, yet we are using the constant-density Navier–Stokes equation. We will resolve this momentarily. For now, let us arrange the equations to isolate the time derivatives:

$$\frac{\partial v_x}{\partial t} = -v_x \frac{\partial v_x}{\partial x} - v_y \frac{\partial v_x}{\partial y} - \frac{1}{\rho}\frac{\partial p}{\partial x} + \nu\left[\frac{\partial^2 v_x}{\partial x^2} + \frac{\partial^2 v_x}{\partial y^2}\right], \quad (8.109)$$

$$\frac{\partial v_y}{\partial t} = -v_x \frac{\partial v_y}{\partial x} - v_y \frac{\partial v_y}{\partial y} - \frac{1}{\rho}\frac{\partial p}{\partial y} + \nu\left[\frac{\partial^2 v_y}{\partial x^2} + \frac{\partial^2 v_y}{\partial y^2}\right], \quad (8.110)$$

and

$$\frac{\partial \rho}{\partial t} = -\frac{\partial}{\partial x}(\rho v_x) - \frac{\partial}{\partial y}(\rho v_y). \quad (8.111)$$

An equation of state is used to relate pressure (p) to the fluid density (ρ) in compressible flow problems, closing the system of equations. MacCormack's approach is a predictor–corrector scheme in which the first estimates for the time derivatives are obtained with forward differences for the inertial terms and central differences for the viscous terms. We let $U = v_x$ and $V = v_y$ to reduce clutter, and we take the i and j indices to represent the x- and y-directions, respectively. Thus, using the x-component as our example and *omitting pressure*,

$$\frac{\partial U}{\partial t} = -U_{i,j}\frac{U_{i+1,j} - U_{i,j}}{\Delta x} - V_{i,j}\frac{U_{i,j+1} - U_{i,j}}{\Delta y}$$
$$+ \nu\left[\frac{U_{i+1,j} - 2U_{i,j} + U_{i-1,j}}{(\Delta x)^2} + \frac{U_{i,j+1} - 2U_{i,j} + U_{i,j-1}}{(\Delta y)^2}\right]. \quad (8.112)$$

The computed values for the time derivatives are used to predict new values for all of the dependent variables; we illustrate this with U:

$$U_{i,j}^{\text{new}} = U_{i,j}^{\text{old}} + \left(\frac{\partial U}{\partial t}\right)\Delta t. \quad (8.113)$$

The *new* (predicted) values for the dependent variables are used to find a revised estimate for the time derivatives. But in this second step, *backward* differences are used for the inertial terms:

$$\left(\frac{\partial U}{\partial t}\right)^{\text{revised}} = -U_{i,j}\frac{U_{i,j} - U_{i-1,j}}{\Delta x} - V_{i,j}\frac{U_{i,j} - U_{i,j-1}}{\Delta y}$$
$$+ \nu\left[\frac{U_{i+1,j} - 2U_{i,j} + U_{i-1,j}}{(\Delta x)^2}\right.$$
$$\left. + \frac{U_{i,j+1} - 2U_{i,j} + U_{i,j-1}}{(\Delta y)^2}\right]. \quad (8.114)$$

Now the two computed time derivatives are averaged:

$$\left(\frac{\partial U}{\partial t}\right)_{\text{average}} = \frac{1}{2}\left[\frac{\partial U}{\partial t} + \left(\frac{\partial U}{\partial t}\right)^{\text{revised}}\right], \quad (8.115)$$

and this average value is used to compute the corrected value (for each dependent variable, of course):

$$U_{i,j}^{\text{corrected}} = U_{i,j} + \left(\frac{\partial U}{\partial t}\right)_{\text{ave}}\Delta t. \quad (8.116)$$

You can see immediately why this scheme became so popular—this explicit algorithm is simple enough that the required programming logic is fairly easy to implement. However, it *cannot be applied* to incompressible flow problems without modification since for incompressible fluid continuity simplifies to $\nabla \cdot v = 0$; (no time derivative). Fortunately, MacCormack's method has been adapted for such problems by Bernard (1986, 1989). In this variation, the changes in velocity are determined exactly as shown previously *omitting pressure*. Then the pressure gradient is determined by the solution of a Poisson equation:

$$\nabla^2 p = \rho \nabla \cdot \left[\nu \nabla^2 U - \frac{dU}{dt} \right]. \quad (8.117)$$

Bernard (1989) developed a code (STREMR) for two-dimensional flows of incompressible fluids using this approach. This scheme was later extended to three-dimensional flows (MAC3D) with the ultimate intended application to turbulent flow problems.

MacCormack's method is an extremely powerful way to solve certain PDEs with time dependence. And although the technique has been around for more than four decades, it is still being applied to important practical problems and it has been focus of continued development efforts. The interested reader should see the article by Selle et al. (2008), "An Unconditionally Stable MacCormack Method."

ADAPTIVE GRIDS

At the beginning of this chapter, we noted that many problems of significance could be solved without mesh (or grid) generation. Consequently, in our preceding discussion of the numerical solution of PDEs, we normally used a square grid in which $\Delta x = \Delta y$. This meant that, in most of the examples we considered, the domain of interest was either rectangular or regular is some way such that a square array of nodal points would coincide with the important boundaries of the problem. Obviously that will not always be the case and we should expect to encounter problems involving objects placed in the field or walls or boundaries that exhibit curvature. In some cases, such problems can be handled rather easily. Let us illustrate with an elementary two-dimensional case; we will assume that we have flow in a divergent duct as illustrated in Figure 8.16.

We will let

$$\eta = x \quad \text{and} \quad \phi = \frac{y}{b(1 + cx^2)}. \quad (8.118)$$

Obviously, whenever y coincides with the upper surface of the duct, $y = y_s$ and $\phi = 1$. This transformation yields a rectangular grid in the *computational plane* since

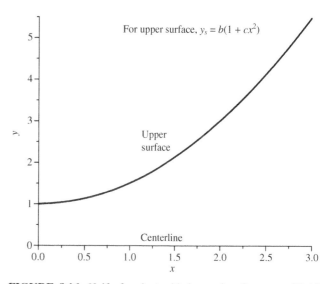

FIGURE 8.16. Half of a duct with increasing flow area. Fluid enters at the left and the origin has been placed on the centerline (the x-axis). The position of the upper surface is described by $y_s = b(1 + cx^2)$, so b is the half-width of the duct at the origin.

$$0 \le \eta \le x \quad \text{and} \quad 0 \le \phi \le 1.$$

Now we will examine a fragment of the x-component of the Navier–Stokes equation:

$$\frac{\partial v_x}{\partial t} + v_x \frac{\partial v_x}{\partial x} + v_y \frac{\partial v_x}{\partial y} \cdots. \quad (8.119)$$

Of course,

$$\frac{\partial v_x}{\partial x} = \frac{\partial v_x}{\partial \eta} \quad \text{and} \quad \frac{\partial v_x}{\partial y} = \frac{\partial v_x}{\partial \phi} \frac{\partial \phi}{\partial y} = \frac{\partial v_x}{\partial \phi} \left(\frac{1}{b(1 + cx^2)} \right). \quad (8.120)$$

What we have done in this case is *fit the coordinate system to the boundary*, and for some simple problems in which a duct wall does not correspond to our x-y grid, this can be used effectively.

Now suppose we have a transport process occurring where the size or shape of the domain changes—perhaps repeatedly. If the field variable has specified values on the boundaries, then an elliptic PDE will yield contours that will conform to the objects or intrusions. For example, consider the Laplace equation:

$$\frac{\partial^2 \psi}{\partial x^2} + \frac{\partial^2 \psi}{\partial y^2} = 0. \quad (8.121)$$

Let us examine the situation where a simple 2/3-cut step is placed in a rectangular domain. We will assign constant values to ψ at both the top and bottom, and vary ψ appropriately at the left and right boundaries (these might be

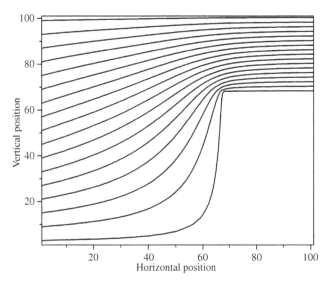

FIGURE 8.17. Contours plotted for $\psi(x, y)$ obtained from the solution of eq. (8.121).

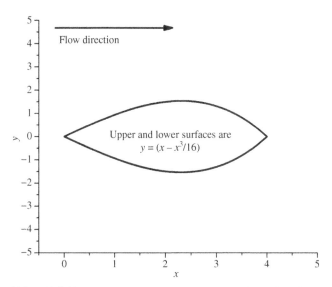

FIGURE 8.18. An object placed in a two-dimensional flow field. This is an instance where *elliptic grid generation* would be ideal. The upper and lower surfaces of the immersed object are given by $y = \pm(x - [x^3/16])$.

inflow and outflow boundaries, respectively). We solve eq. (8.121) iteratively and plot the resulting contours in Figure 8.17.

Note how the contours conform smoothly to the intrusion projecting from the bottom boundary. This suggests that we might be able to obtain a coordinate system adapted to an arbitrary shape through the solution of a Dirichlet problem. This is what we mean when we speak of *elliptic grid generation*.

A recurring scenario in fluid flow involves the situation where we have flow *around* an object—we can think of an airfoil, a bridge pier, or a heat exchanger tube. A hypothetical case in illustrated in Figure 8.18.

Of course, we could employ a square mesh and just interpolate near the boundaries, but if strong gradients exist near these surfaces (which is entirely likely), interpolation may lead to significant error. This is our motivation for finding a transformation that will yield coordinates that conform to the surface of the immersed object, corresponding to a rectangular grid in the computational plane. We will think of this process in the following way: Let the surface of the object correspond to the inner boundary—we know pairs of (x, y) values that correspond to this surface. For examples, if $x = 3/2$, $y = \pm 1.28906$, and if $x = 9/4$, $y = \pm 1.53809$. We take the outer surface be some kind of enclosing curve where once again the (x, y) values are known. If the "ends" were also known, we would have a Dirichlet problem and the connection between the x-y (physical) field and the computational plane would be governed by two elliptic PDEs:

$$\frac{\partial^2 \phi}{\partial x^2} + \frac{\partial^2 \phi}{\partial y^2} = 0 \quad \text{and} \quad \frac{\partial^2 \eta}{\partial x^2} + \frac{\partial^2 \eta}{\partial y^2} = 0. \quad (8.122)$$

Unfortunately, we must solve the *inverse* problem; we need to solve for the x-y positions corresponding to $\phi - \eta$ positions in the computational plane. When x and y are the dependent variables, we find

$$A \frac{\partial^2 x}{\partial \phi^2} - 2B \frac{\partial^2 x}{\partial \phi \partial \eta} + C \frac{\partial^2 x}{\partial \eta^2} = 0 \quad (8.123)$$

and

$$A \frac{\partial^2 y}{\partial \phi^2} - 2B \frac{\partial^2 y}{\partial \phi \partial \eta} + C \frac{\partial^2 y}{\partial \eta^2} = 0. \quad (8.124)$$

The parameters A, B, and C are

$$A = \left(\frac{\partial x}{\partial \eta}\right)^2 + \left(\frac{\partial y}{\partial \eta}\right)^2, \quad (8.125)$$

$$B = \left(\frac{\partial x}{\partial \phi}\right)\left(\frac{\partial x}{\partial \eta}\right) + \left(\frac{\partial y}{\partial \phi}\right)\left(\frac{\partial y}{\partial \eta}\right), \quad (8.126)$$

and

$$C = \left(\frac{\partial x}{\partial \phi}\right)^2 + \left(\frac{\partial y}{\partial \phi}\right)^2. \quad (8.127)$$

In principle, if we know x and y on all four boundaries, we can use eq. (8.123) and eq. (8.124) to determine the location of a mesh point in the physical $(x$-$y)$ plane relative to a location in the computational (ϕ, η) field. However, if we base the mesh generation on the pair of elliptic eq. (8.122), we often do not obtain sufficient computational points in regions where there are large gradients. Note in Figure 8.17, for example, how the contours are clustered near the step point (a convex region) but are sparsely distributed at the bottom of the step (a concave region). For this reason, production (or source) terms are often added with the purpose of adjusting the shape of the computational mesh; for example

$$\frac{\partial^2 \phi}{\partial x^2} + \frac{\partial^2 \phi}{\partial y^2} = P(\phi, \eta).$$

Now we will examine a discretization for eq. (8.123)—for simplicity, we will omit the source term and we will take $\Delta\phi = \Delta\eta = 1$:

$$A\left(x_{i+1,j} - 2x_{i,j} + x_{i-1,j}\right) - \frac{B}{2}\left(x_{i+1,j+1} - x_{i-1,j+1} - x_{i+1,j-1}\right.$$
$$\left. + x_{i-1,j-1}\right) + C\left(x_{i,j+1} - 2x_{i,j} + x_{i,j-1}\right) = 0, \qquad (8.128)$$

with

$$A = \frac{1}{4}\left[\left(x_{i,j+1} - x_{i,j-1}\right)^2 + \left(y_{i,j+1} - y_{i,j-1}\right)^2\right], \quad (8.129)$$

$$B = \frac{1}{4}\left[\begin{array}{l}\left(x_{i+1,j} - x_{i-1,j}\right)\left(x_{i,j+1} - x_{i,j-1}\right) \\ + \left(y_{i+1,j} - y_{i-1,j}\right)\left(y_{i,j+1} - y_{i,j-1}\right)\end{array}\right], \quad (8.130)$$

and

$$C = \frac{1}{4}\left[\left(x_{i+1,j} - x_{i-1,j}\right)^2 \left(y_{i+1,j} - y_{i-1,j}\right)^2\right]. \quad (8.131)$$

The equation for y, eq. (8.124) would be handled similarly, of course. Functional forms for the source terms have been put forward in the literature, and Thompson et al. (1974) suggested that $P(\phi, \eta)$ might be expressed such that it was dependent on velocity gradient or vorticity. The coordinate system would be accordingly time dependent, and the generated mesh would automatically concentrate in regions where sharp changes in the field variables were occurring. Elliptic mesh generation is an important topic in modern computational fluid mechanics, and software designed specifically for this purpose has been incorporated into many commercial CFD packages. Anderson (1995) provides a nice introduction to this mesh generation and Chung (2002) offers more detail. The reader with greater interest in body-fitted coordinate systems may also find Thompson et al. (1974) useful.

CONCLUSION

Many problems of interest in engineering and the applied sciences are governed by PDEs and only a very small number of these problems can be solved analytically. Most

will require numerical solution. Consider, for example, a simple case of heat transfer in a finite cylinder for which $L/d = 3$. Thermal energy is produced throughout the interior of the cylinder, the ends are maintained at 100° (which is also the initial temperature of the object), and the curved surface loses heat to the surroundings (with $T_\infty = 70°$) according to Newton's law of cooling. Suppose the governing equation is determined to be

$$\rho C_p \frac{\partial T}{\partial t} = k\left[\frac{\partial^2 T}{\partial r^2} + \frac{1}{r}\frac{\partial T}{\partial r} + \frac{\partial^2 T}{\partial z^2}\right] + \beta T^{3/2}, \quad (8.132)$$

where β has dimensions of (energy)/[(volume)(time)($T^{3/2}$)]. The equation is nonlinear by virtue of the production term, making analytic solution improbable. Yet the equation is very easily discretized to yield

$$\frac{T_{i,j,k+1} - T_{i,j,k}}{\Delta t} \cong \alpha\left[\frac{T_{i+1,j,k} - 2T_{i,j,k} + T_{i-1,j,k}}{(\Delta r)^2} + \frac{1}{r}\frac{T_{i+1,j,k} - T_{i-1,j,k}}{2\Delta r}\right.$$
$$\left. + \frac{T_{i,j+1,k} - 2T_{i,j,k} + T_{i,j-1,k}}{(\Delta z)^2}\right] + \frac{\beta}{\rho C_p}T_{i,j,k}^{3/2}.$$
$$(8.133)$$

We can solve this problem rapidly using the explicit technique described previously in this chapter; with the Biot modulus (hR/k) set equal to 0.596, we obtain the sequence shown in Figure 8.19.

The solution for this problem required just 35 lines of code (with no particular effort to be efficiently compact) and the programming logic was devised in less than 15 minutes. What this example reveals is that even some rather formidable nonlinear PDEs can be solved by elementary means with no more computing power than that provided by ordinary personal computers. The range of problems that the analyst can solve in this way is broad, and the transformative power the computer has exerted on routine solution of PDEs is obvious.

To provide emphasis for this last point, we want to provide the reader with an illustrative construction of a generalized elliptic PDE solver that uses SOR to handle Poisson and Laplace problems in rectangular coordinates. The code was written by the author, and it is designed to provide capability for fluid flow, heat transfer, and mass transfer with Dirichlet, Neumann, and Robin's-type boundary conditions. The governing equation has the form

$$0 = \kappa\left[\frac{\partial^2 \phi}{\partial x^2} + \frac{\partial^2 \phi}{\partial y^2}\right] + P. \quad (8.134)$$

The user is queried for the length (L) and height (H) of the rectangular region and the source term, P, can be zero, a constant, or spatially variable; for example,

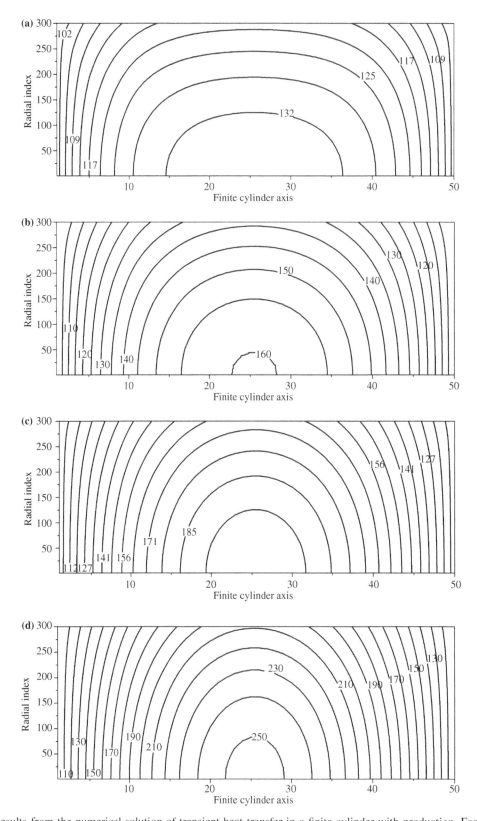

FIGURE 8.19. Results from the numerical solution of transient heat transfer in a finite cylinder with production. For these calculations, $L/d = 3$, $Bi = 0.596$, and $T_\infty = 70°$. The sequence shown is for times of 1, 2, 4, and 8 (top to bottom).

$$P = a + \frac{b}{c + (H - y)^2 + (L - x)^2}. \quad (8.135)$$

In discretized form, eq. (8.134), appropriately cast for iterative solution, appears as

$$\phi_{i,j} = \phi_{i,j} + \frac{\omega}{\beta}\left[\frac{\phi_{i+1,j} + \phi_{i-1,j}}{(\Delta x)^2} + \frac{\phi_{i,j+1} + \phi_{i,j-1}}{(\Delta y)^2} + \frac{P_{i,j}}{\kappa} - (\beta)\phi_{i,j} \right], \quad (8.136)$$

where

$$\beta = \frac{2}{(\Delta x)^2} + \frac{2}{(\Delta y)^2}.$$

The rectangular region is divided into 101 nodes in each direction, resulting in 9801 interior mesh points. The acceleration parameter for SOR, ω, is taken to be 1.85. An arbitrary measure of convergence is employed to terminate the computation (when the change in the *sixth* decimal place is less than 1). First, we will apply the program to flow of an aqueous fluid in a square microchannel, 18 μm on each side. The applied pressure gradient will be $(-)5300$ dyne/cm²/cm.

Next, we will apply the very same code to conduction of thermal energy in a slab with spatially variable production. In this case, we choose $L = 10$ cm, $H = 4$ cm, and $k = 0.01$ cal/(cm s °C). The slab's bottom and left-hand side will be insulated, but the top and the right-hand side will lose thermal energy to the surroundings. The production function has its maximum value in the center of the slab.

The examples shown in Figure 8.20 and Figure 8.21 are simply intended to illustrate how easy it is to solve a broad variety of elliptic PDEs using one basic code structure. The user merely selects the size of the rectangular domain, the numerical value of the transport coefficient, the boundary conditions, and the nature of the source term. Since the

algorithm is not constrained to a square mesh, calculations can be made for a rectangular region of any aspect ratio.

Finally, we do want to point out the existence of other useful numerical procedures for CFD that have not been described here. The first of these is the *explicit Lax–Wendroff* technique, which is appropriate for forward-marching in time. For the two-dimensional flow of an *inviscid* fluid, we would write

$$\frac{\partial \rho}{\partial t} = -\left(\rho \frac{\partial v_x}{\partial x} + v_x \frac{\partial \rho}{\partial x} + \rho \frac{\partial v_y}{\partial y} + v_y \frac{\partial \rho}{\partial y} \right), \quad (8.137)$$

$$\frac{\partial v_x}{\partial t} = -\left(v_x \frac{\partial v_x}{\partial x} + v_y \frac{\partial v_x}{\partial y} + \frac{1}{\rho} \frac{\partial p}{\partial x} \right), \quad (8.138)$$

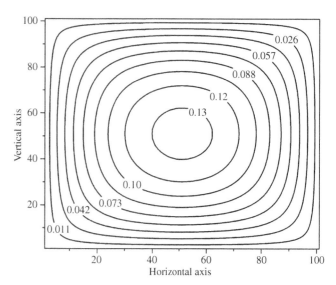

FIGURE 8.20. Velocity distribution in a square microchannel (18 μm on each side) resulting from a pressure gradient of −5300 dyne/cm²/cm. The average velocity is a little less than 0.07 cm/s.

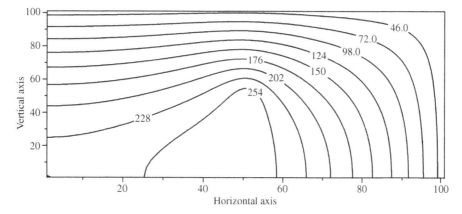

FIGURE 8.21. Temperature distribution in a two-dimensional slab with a spatially variable production of thermal energy (with maximum at the center). The bottom and left-hand side are insulated and the top and right-hand side lose thermal energy to the surroundings by Newton's law of cooling.

and

$$\frac{\partial v_y}{\partial t} = -\left(v_x \frac{\partial v_y}{\partial x} + v_y \frac{\partial v_y}{\partial y} + \frac{1}{\rho}\frac{\partial p}{\partial y}\right). \qquad (8.139)$$

Next, each of the dependent variables is expanded in a Taylor series to give the value of that quantity on the new time-step row; for example,

$$\rho_{i,j}^{\,t+\Delta t} = \rho_{i,j}^{\,t} + \left(\frac{\partial \rho}{\partial t}\right)_{i,j}^{t} \Delta t + \left(\frac{\partial^2 \rho}{\partial t^2}\right)_{i,j}^{t} \frac{(\Delta t)^2}{2} + \cdots. \qquad (8.140)$$

Note that the derivatives in eq. (8.140) are on the *previous* time-step row; if we could obtain values for these time derivatives, we could calculate $\rho_{i,j}$ on the new time-step row explicitly. The first derivative with respect to time is obtained directly from eq. (8.137) but with the *spatial derivatives on the right-hand side* rewritten as second-order central differences (all those quantities are known because they are from the previous time-step row). The estimate for the second derivative,

$$\left(\frac{\partial^2 \rho}{\partial t^2}\right)_{i,j}^{t},$$

is obtained by differentiating eq. (8.137) with respect to time. Of course, this process generates mixed derivatives, such as

$$\left(\frac{\partial^2 v_x}{\partial x \partial t}\right),$$

which are approximated by differentiating eq. (8.138) with respect to x; other mixed derivatives are handled similarly. Once again, all of the spatial derivatives are written as second-order, central differences. Although the algebra associated with Lax–Wendroff is complicated, the method is fully explicit (just like the related technique, MacCormack's method, discussed previously). For this reason, it has been popular among practitioners of CFD. Both Anderson (1995) and Chung (2002) provide additional detail for the interested reader.

The final approach we want to mention is known as the semi-implicit method for pressure-linked equations (*SIMPLE*), and it was devised by Patankar and Spalding and is very nicely explained by Patankar (1980). You may recall that we indicated previously that the major problem in CFD was computation of the pressure field (and you may also remember that we used the vorticity transport equation to circumvent this difficulty in two dimensions). In the SIMPLE procedure, the pressure field is estimated, the momentum equations are then solved to obtain the velocity vector components, the pressure field is corrected, and revised estimates are obtained for the velocities. The corrected pressure then

serves as the initial estimate and the entire process is repeated until convergence is obtained. Since its development in the 1970s, SIMPLE has been used by many fluid dynamicists and it has been incorporated into a number of commercial CFD codes for both two- and three-dimensional flows. It has been found to be divergent in some applications, and under-relaxation has been used to cure that problem. Patankar's book is recommended reading for the student wishing to know more about SIMPLE.

We have shown in this chapter that a wide variety of phenomena governed by PDEs can be modeled successfully and the solutions can be computed with relative ease. In many cases, even nonlinear phenomena can be dealt with using nothing more sophisticated than a personal computing device coupled with a spreadsheet. Most importantly, it only takes a little practice to become proficient at solving a broad spectrum of important practical problems.

PROBLEMS

8.1. Consider a slab of steel measuring 20×20 cm. The left-hand edge ($x = 0$) is maintained at $1000°$ for all time and the top surface ($y = 20$ cm) is insulated. The bottom and the right-hand side lose thermal energy to the surroundings according to Newton's law of cooling. Using the right-hand side as an example, therefore, we write

$$-k\frac{\partial T}{\partial x}\bigg|_{x=20} = h\left(T_{x=20} - T_{\infty}\right).$$

Find (and prepare a contour plot illustrating) the equilibrium temperature distribution in the slab by solving

$$\frac{\partial^2 T}{\partial x^2} + \frac{\partial^2 T}{\partial y^2} = 0.$$

The thermal conductivity of mild steel is about 0.1 cal/(g cm °C) and the heat transfer coefficient (h) may be taken as 0.033 cal/(cm² s °C).

8.2. A solid copper cylinder with a radius of 6 cm (with very large L/d) has the upper half of its (curved) surface maintained at $300°$C. The lower half of the cylinder is embedded in a medium maintained at $T = 100°$C. Find the steady-state temperature distribution in the copper cylinder. The thermal conductivity of copper is 0.93 cal/(g cm °C) and the governing equation is

$$\frac{1}{r}\frac{\partial}{\partial r}\left(r\frac{\partial T}{\partial r}\right) + \frac{1}{r^2}\frac{\partial^2 T}{\partial \theta^2} = 0.$$

8.3. A slab of anisotropic material measuring 10×10 cm lies in the x-y plane. The thermal conductivity in the x-direction is 0.000834 cal/(cm s °C), but in the y-direction, it is only 0.000361 cal/(cm s °C). If the left edge is maintained at 170°C, and the right-hand side at 25°C, find the temperature distribution in the slab. The top and bottom edges loose heat to the surroundings according to Newton's law of cooling; therefore, using the top for our example, we have

$$-k_y \frac{\partial T}{\partial y}\bigg|_{y=H} = h(T_{y=H} - T_\infty).$$

The temperature of the surrounding air is 35°C, and the heat transfer coefficient, h, is 0.0001 cal/(cm^2 s °C).

The governing equation for the interior of the slab is

$$\frac{\partial}{\partial x}\left(k_x \frac{\partial T}{\partial x}\right) + \frac{\partial}{\partial y}\left(k_y \frac{\partial T}{\partial y}\right) = 0.$$

8.4. A viscous fluid is at rest in a square duct measuring 3×3 cm. At $t = 0$, a pressure gradient of 0.7 dyne/cm^2/cm is applied in the z-direction and the fluid begins to move down the duct. The fluid's viscosity and density (μ and ρ) are 0.03 g/(cm s) and 1 g/cm^3, respectively. Solve the governing equation,

$$\frac{\partial v_z}{\partial t} = -\frac{1}{\rho}\frac{dp}{dz} + \nu\left[\frac{\partial^2 v_z}{\partial x^2} + \frac{\partial^2 v_z}{\partial y^2}\right],$$

and find the velocity at the duct's center at 1, 10, and 100 seconds. Find the ultimate centerline velocity and the expected Reynolds number.

8.5. Consider the parabolic PDE given by

$$\frac{\partial \phi}{\partial t} = 2\frac{\partial^2 \phi}{\partial x^2}.$$

Compute and plot a series of $\phi(x, t)$ curves such that the full range of dynamic behavior of this system is illuminated.

We are given the following: $0 \le x \le 4$, $\phi(0, t) = 0$, and $\phi(4, t) = 0$, with the initial condition $\phi(x, 0) = 25x$.

8.6. A long cylindrical rod with $R = 1$ (and $d = 2$) is at an initial (uniform) temperature of zero. At $t = 0$, the surface of the rod is instantaneously heated to a constant 100°. Spiegel (1971) provides the analytic solution for this problem:

$$T(r, t) = 100\left[1 - 2\sum_{n=1}^{\infty} \frac{J_0(\lambda_n r)}{\lambda_n R J_1(\lambda_n R)}\exp(-\alpha \lambda_n^2 t)\right],$$

The values for λ_n come from the roots of J_0, of course. Compute the numerical solution for this problem and compare your results at $t = 0.5$, 1.0, and 2.0 seconds with

those obtained from the previous analytic solution. The thermal diffusivity, α, has a value of 0.4. The governing PDE is

$$\frac{\partial T}{\partial t} = \alpha\left[\frac{\partial^2 T}{\partial r^2} + \frac{1}{r}\frac{\partial T}{\partial r}\right].$$

8.7. An annular fin is placed on a pipe to help dissipate heat. The fin is constructed of mild steel and it extends from $R_1 = 4$ cm to $R_2 = 8$ cm, with a thickness of 4 mm. At $r = R_1$, the pipe wall has a temperature of 200°C for all time. The temperature in the steel fin is governed by

$$\frac{\partial T}{\partial t} = \alpha\left[\frac{\partial^2 T}{\partial r^2} + \frac{1}{r}\frac{\partial T}{\partial r} + \frac{\partial^2 T}{\partial z^2}\right],$$

and the flat annular surfaces (denoted by s) lose heat to the surroundings according to

$$-k\frac{\partial T}{\partial z}\bigg|_s = h(T_s - T_\infty).$$

We can assume that (nearly) no heat is lost through the outer cylindrical edge of the fin (at $r = R_2$). We are given that h, ρ, C_p, and α (*all in centimeter-gram-second units*) are 0.00678, 7.85, 0.118, and 0.12, respectively. We want to investigate the dynamic behavior of the temperature distribution in the fin if the temperature of the surroundings (T_∞) suddenly drops from 45 to 0°C. Assume that this temperature had been 45°C long enough that equilibrium had been established prior to the change.

8.8. Find the distribution of S over the annular region, $R_1 \le r \le R_2$, where S is governed by the equation

$$\frac{\partial^2 S}{\partial r^2} + \frac{1}{r}\frac{\partial S}{\partial r} + \frac{1}{r^2}\frac{\partial^2 S}{\partial \theta^2} = 0.$$

Let $S(r = R_1) = 100$, but $S(r = R_2) = 50 + 30\sin(\theta)$, where θ varies from 0 to 2π. The inner and outer radii are 1 and 10, respectively.

8.9. A solid cylindrical rod of radius, R, is immersed in a liquid such that the surface temperature of the rod is maintained constantly at 50°. The rod has an initial uniform temperature of 50°, but at $t = 0$, thermal energy begins to be produced inside the rod by a source term, S:

$$\rho C_p \frac{\partial T}{\partial t} = k\left[\frac{\partial^2 T}{\partial r^2} + \frac{1}{r}\frac{\partial T}{\partial r} + \frac{\partial^2 T}{\partial z^2}\right] + S.$$

The production of thermal energy in the interior follows:

$$S = \beta\left(\frac{1}{a + r^2}\right).$$

The rod is 50 cm long with $0 \leq r \leq 10$ cm. We know $\beta = 40$ and $a = 2$. The properties of the material (ρ, C_p, and k) are 7, 0.22, and 0.19, respectively. Explore the dynamic behavior of $T(r, t)$ over a time period sufficient to establish the full range of thermal behavior.

8.10. The thermal conductivity of chrome steel varies with temperature between 0 and 500°C, the relationship is approximately linear: $k = 62 - 0.049T$, W/(m°C). A slab of chrome steel 40 cm thick is at a uniform initial temperature of 500°C. At $t = 0$, the front face is quickly cooled to 0°C. The back face is insulated. Find the evolution of the temperature distribution in the slab by solving the equation:

$$\rho C_p \frac{\partial T}{\partial t} = \frac{\partial}{\partial x}\left(k(T)\frac{\partial T}{\partial x}\right).$$

The density of chrome steel is about 488 lb_m/ft^3 and the heat capacity is about 0.11 Btu/(lb_m °F).

8.11. A viscous fluid in an annulus (with $R_1 = 3$ cm and $R_2 = 5$ cm) is moving in the positive z-direction under the influence of a pressure gradient. The average velocity is 4 cm/s. At $t = 0$, the inner surface (which was stationary) begins moving (in the z-direction) at a constant velocity of -10 cm/s. When will the *net* flow be exactly zero? The fluid motion is governed by

$$\rho \frac{\partial V_z}{\partial t} = -\frac{\partial p}{\partial z} + \mu\left[\frac{1}{r}\frac{\partial}{\partial r}\left(r\frac{\partial V_z}{\partial r}\right)\right].$$

8.12. Consider water, initially at a uniform temperature, flowing in the z-direction between parallel plates; for $z > 0$, both the upper and lower plates are maintained at a constant elevated temperature. The lower planar surface is located at $y = 0$ and the upper at $y = B$. For steady-state conditions, the governing equation is

$$\rho C_p V_z \frac{\partial T}{\partial z} = k\left[\frac{\partial^2 T}{\partial y^2} + \frac{\partial^2 T}{\partial z^2}\right].$$

We will assume that the fluid properties are constant, the flow is fully developed, and that the centerline (maximum) velocity is V_{max}. We want to find the bulk fluid temperature and the Nusselt number as functions of the z position. Before we begin, we should carefully consider the axial conduction term, $\partial^2 T/\partial z^2$. Under what circumstances can we expect this term to be negligible?

We will use the following specific values for our problem: $B = 3$ cm, $V_{max} = 5$ cm/s, $T_S = 150$°F, and $T_{in} = 50$ °F. Assume that the fluid properties (for water) are constant, and use values corresponding to a temperature of 80°F. Prepare a figure that shows both the bulk fluid temperature and the Nusselt number (Nu $= hd/k$) as functions of the z position.

How long must the apparatus be if the bulk fluid temperature is to be 60, 70, 80, and 90°F? Show that the bulk fluid temperature at $z = 600$ cm is 24.77 °C. Remember that the bulk fluid temperature must be determined by integration of the product, $V_z(y)T(y)$, over the flow area.

8.13. The FitzHugh–Nagumo model describes wavelike phenomena associated with nerve axons and it consists of the pair of equations

$$\frac{\partial V}{\partial t} = -R + V - \frac{V^3}{3} + \frac{\partial^2 V}{\partial x^2} + I(t) \quad \text{and}$$

$$\frac{\partial R}{\partial t} = k(V - bR + a).$$

V is the membrane potential (output of the neuron), R is a restoring effect, and $I(t)$ is the input or forcing function. For certain parametric choices (a, b, k), this model can exhibit chaotic behavior. One variant of the model that appears frequently consists of the equations

$$\frac{dV}{dt} = c(-R + V - \tfrac{1}{3}V^3 + I(t)) \quad \text{and} \quad \frac{dR}{dt} = V - bR + a.$$

Begin by solving this pair of *ordinary* differential equations with $a = 0.7$, $b = 0.8$, and $c = 10$, and let $I(t)$ be a simple periodic (sinusoidal) input with an amplitude of 1. Prepare a plot of $V(t)$ and $R(t)$ to reveal the phase-plane dynamics of this simplified system. What will the impact of the term $\partial^2 V/\partial x^2$ be on the solution of this system? Demonstrate the difference between the alternative models by solving the initial set of equations.

8.14. Solve the wave equation

$$\frac{\partial^2 u}{\partial t^2} = c^2 \frac{\partial^2 u}{\partial x^2}$$

numerically over the range $-10 < x < +10$. The initial distribution of displacement, $u(x, 0)$, is $u = \exp(-x^2)$, with zero initial velocity. Set the Courant number equal to 1 and use the results shown in Figure 8.22 as a solution guide.

8.15. Use the vorticity transport equation to compute transient two-dimensional flow over a rectangular box placed on the bottom surface between parallel walls. The height of the box corresponds to 1/2 of the vertical channel size. The fluid is initially at rest, and motion is started by sliding the upper surface at constant velocity in the x-direction. Focus initially on Reynolds numbers (using the height of the box for the characteristic length) of 15 and 25. A typical result from such a computation is shown in Figure 8.23 for the case in which Re $= 15$.

8.16. We have a two-dimensional slab that extends from $x = 0$ to $x = L$ and from $y = 0$ to $y = H$. Three sides of the slab are maintained at $100°$ and the bottom edge is insulated. Thermal energy is produced in the interior of the slab and the maximum rate of production occurs at the exact center. The governing equation is

$$k\left[\frac{\partial^2 T}{\partial x^2}+\frac{\partial^2 T}{\partial y^2}\right]+\frac{A}{\left[(x-L/2)^2+\frac{1}{4}\right]\left[(y-H/2)^2+\frac{1}{4}\right]}=0.$$

Compute the temperature distribution in the slab, $T(x,y)$, and plot the isotherms. You should see something like the result shown in Figure 8.24.

Also, evaluate the rate of heat loss from the top edge of the slab. We know $L = H = 8$, $A = 0.085$, and $k = 0.005$.

8.17. Repeat the analysis of mass transport between concentric cylinders illustrated in the section on Elementary

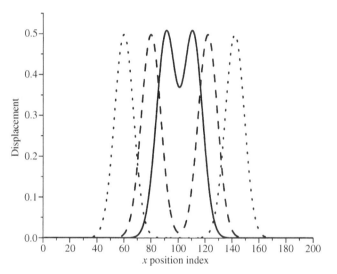

FIGURE 8.22. Note that the initial exponential distribution of displacement results in the formation of two peaks traveling in opposite directions. An x index of 101 corresponds to the center of the interval (i.e., $x = 0$). Curves shown are for times of 0.015 (solid), 0.03 (dash), and 0.06 (dot) second.

Problems with Convective Transport, pages 164 and 165, but allow the tangential component of velocity to oscillate such that

$$v_\theta = (r = R_2, t) = 24\sin\left(\frac{1}{9}t\right).$$

How much impact does the oscillation of the outer cylinder have on the radial development of the concentration plume?

8.18. Consider a cavity filled with a viscous fluid initially at rest. At $t = 0$, the upper surface of the cavity (a flat plate) begins to slide in the positive x-direction with constant velocity, V. Use the following equations to model the resulting flow:

$$\frac{\partial \omega}{\partial t}+v_x\frac{\partial \omega}{\partial x}+v_y\frac{\partial \omega}{\partial y}=\nu\left[\frac{\partial^2 \omega}{\partial x^2}+\frac{\partial^2 \omega}{\partial y^2}\right],$$

$$v_x=\frac{\partial \psi}{\partial y} \quad v_y=-\frac{\partial \psi}{\partial x},$$

and

$$-\omega=\frac{\partial^2 \psi}{\partial x^2}+\frac{\partial^2 \psi}{\partial y^2}.$$

Solve this problem for a square cavity with unit width and depth and let $V = 1$. Place the origin in the lower left-hand corner and take the kinematic viscosity to be 0.10. The Reynolds number for this flow, therefore, will ultimately be $Re = Vh/\nu = 10$. Prepare a series of plots of the stream function that illustrate the evolution of the flow. The student should note that if the Reynolds number is sufficiently small, this problem would simply require solution of the biharmonic equation, $\nabla^4\psi = 0$; that is, one could ignore the convective transport of vorticity. Chow (1979) provides an example of the solution procedure for this creeping flow problem in chapter 3 of his book, *An Introduction to Computational Fluid Mechanics*. If the upper surface is started

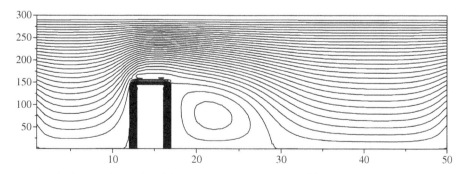

FIGURE 8.23. Computed results for two-dimensional flow over a rectangular box with $Re = 15$ (based on the box height and the midchannel velocity).

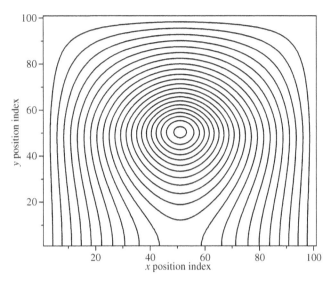

FIGURE 8.24. Temperature distribution in a slab with the production function centered.

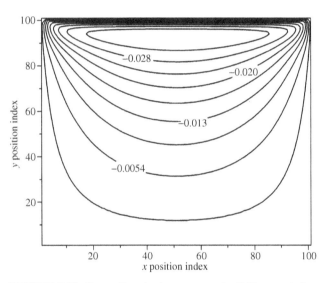

FIGURE 8.25. Streamlines in the square cavity 0.01 second after motion of the upper surface (at $V = 1$) is initiated. Note that the velocity (in the x-direction) near the bottom streamline is only about -0.0159 cm/s.

impulsively at $t = 0$ with a velocity of 1, then at very small times, the analyst should obtain streamlines similar to those shown in Figure 8.25.

8.19. An interesting variation of Problem 8.18 (see Figure 8.25) is posed by a square cavity in which the top is a *free surface* (nearly zero momentum flux) and the bottom surface slides with constant velocity, V. Solve this modified problem with the same parameters employed in Figure 8.25.

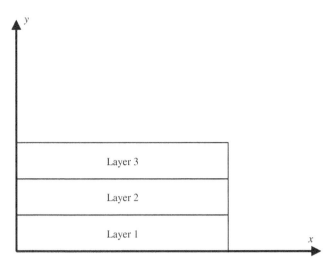

FIGURE 8.26. Three-layer solid state device arrangement.

8.20. We want to examine a solid-state device with three layers with production of thermal energy occurring in the intermediate layer (layer 2). Each layer will have a different conductivity, but all three have the same thickness (0.667 cm). The device is 3 cm long (horizontally) and 2 cm high. The general form of the equation we must solve is

$$k_n \left[\frac{\partial^2 T}{\partial x^2} + \frac{\partial^2 T}{\partial y^2} \right] + P_n = 0,$$

and the three thermal conductivities are 124/100, 52/100, and 130/100 for k_1 through k_3, respectively. We are dividing by 100 to convert from W/(mK) to W/(cmK). The production term for layer 2 has the constant value, 25 W per unit volume. A crude picture of the device is provided in Figure 8.26.

Assume that the bottom surface (the bottom of layer 1) is insulated, as is the left-hand side (vertical edge). The top loses thermal energy to the surroundings according to Newton's law of cooling: $h_{top}(T_{top} - T_{air})$. The right-hand edge also loses thermal energy to the surroundings, but the actual BC will vary with layer:

$$-k_n \left(\frac{\partial T}{\partial x} \right)_{x=L} = h_{edge}(T_{x=L} - T_{air}).$$

We are going to assume that $h_{top} = 0.75$ and that $h_{edge} = 1.5 h_{top} = 1.125$. The air temperature will be taken as 25°C. Find the temperature distribution in the interior of the device (temperature contours). Will this arrangement meet operating requirements if the thermal limit for the device is 60°C? The results of a calculation for the case in which *both vertical sides and the bottom are insulated* is shown in Figure 8.27 as a guide.

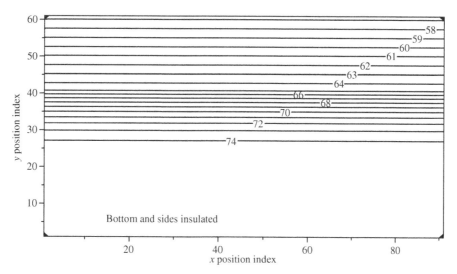

FIGURE 8.27. Temperature distribution in the three-layer device when the bottom and both sides are insulated.

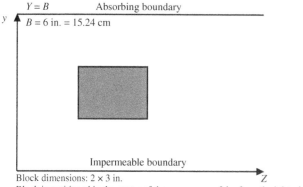

Block dimensions: 2 × 3 in.
Block is positioned in the center of the passageway, 3 in. from the inlet plane

FIGURE 8.28. Porous medium with an impermeable obstruction placed in the migration path.

8.21. Transport processes in porous media are sometimes simulated by placing impermeable obstructions in the migration pathway. Consider the situation illustrated in Figure 8.28, where an impermeable block has been placed in the center of the channel. The height of the block is 2 in., so it provides 33% occlusion of the pathway.

The contaminant, at a concentration of 1, enters the region from the left where $z = 0$. The lower boundary is impermeable and the upper boundary is absorbing; that is, the contaminant is instantaneously removed such that $C = 0$ for all t at the surface indicated by $y = B$. We assume that the mass transfer process is governed by

$$\frac{\partial C}{\partial t} = D\left[\frac{\partial^2 C}{\partial y^2} + \frac{\partial^2 C}{\partial z^2}\right].$$

Our interest is the total flow of contaminant past the trailing edge of the block where $z = 6$ in. or 15.24 cm. We will

take $D = 2 \times 10^{-3}$ cm²/s (a very large value, but this will shorten the computational time considerably). It takes almost 1 hour for the contaminant plume to reach the leading edge of the block and about 5 or 6 hours for it to reach the trailing edge. *Compare the rate at which the contaminant migrates past the block at z = 6 in. with the rate of transport past that plane with the obstruction removed.*

8.22. Consider a porous, sorbent sphere placed in a well-agitated solution of limited volume, for example, an activated carbon "particle" immersed in a beaker of water containing an organic contaminant. The contaminant (or solute) species (A) is taken up by the sphere and the concentration of A in the liquid phase is depleted. The governing equation for transport in the sphere's interior is

$$\frac{\partial C_A}{\partial t} = D\left[\frac{\partial^2 C_A}{\partial r^2} + \frac{2}{r}\frac{\partial C_A}{\partial r}\right].$$

As we have seen previously, this equation can be transformed into an equivalent problem in a "slab" by setting $\phi = C_A r$. The total amount of A in solution initially is VC_{A0}, and the rate at which A is removed from solution can be described by

$$4\pi R^2 D\frac{\partial C_A}{\partial r}\bigg|_{r=R}.$$

Therefore, the total amount removed over a time, t, can be obtained by integration of this equation. The *limiting case* is readily solved through the use of the product method and the transformation of our equation leads to

$$\frac{\partial \phi}{\partial t} = D\frac{\partial^2 \phi}{\partial r^2},$$

which is a (familiar) candidate for separation of variables:

$$C_A = \frac{A}{r}\exp(-D\lambda^2 t)\sin\lambda r.$$

The cosine term has disappeared because the concentration of solute at the sphere's center must be finite, of course. It is convenient to switch to dimensionless concentration, where

$$C = \frac{C_A - C_{Ai}}{C_{As} - C_{Ai}}.$$

It is likely that the sphere contains no solute initially, so $C_{Ai} = 0$. When the solution volume is *unlimited*, then we write

$$C = 1 + \sum_{n=1}^{\infty} \frac{A_n}{r}\exp(-D\lambda_n^2 t)\sin\lambda_n r,$$

where $\lambda_n = n\pi/R$.

Complete the analytic solution started previously for the limiting case and prepare a figure that shows M_t/M_∞ as a function of Dt/R^2. Note that M_t is the amount of solute taken up by the sphere through time, t. M_∞ is the amount taken up by the sphere after infinite time.

Solve the "limited solution" problem *numerically* and prepare similar curves for three cases: the portion of solute ultimately removed from solution (and taken up by the sphere) is 20%, 50%, and 80%.

Suppose spherical sorbent particles with $d = 4.8$ mm are used to remove benzyl alcohol from an initially saturated aqueous solution maintained at 17°C (the solubility is 4 g/100 g water). It is known that $D = 0.82 \times 10^{-6}$ cm^2/s. The number of particles employed is such that each is effectively surrounded by volume of liquid corresponding to 100 mm^3. Prepare a figure that shows the concentration of benzyl alcohol remaining in solution as a function of time, assuming the liquid phase is energetically stirred. Then, repeat this final part of the problem using a Robin's-type boundary condition at the surface with $KR/D = 0.2$.

8.23. Circumstances can arise in mass and heat transfer in which a concentration or temperature *front* can propagate through a medium. Such cases are referred to as *moving-boundary* problems and examples of how they can occur include the following:

- The diffusivity experiences a sharp, discontinuous change at a particular concentration.
- The diffusing species is immobilized at a limited number of available sites.
- A chemical reaction occurs at a reactant interface and one or more *product* species migrate through the medium.

- Heat flows through a medium in which a phase change occurs at a specific temperature (the latent heat effect may either produce or absorb thermal energy).

Suppose we have a medium in which the diffusivity is 10^{-3} cm^2/s *if* the dimensionless concentration is *above* 0.5 but is four orders of magnitude *smaller* (10^{-7}) for concentrations below 0.5. At $t = 0$, the concentration at the front face where $x = 0$ is instantaneously elevated to 1; there is initially none of the diffusing species in the medium. Compute the migration of this species into the medium and plot the resulting concentration as a function of $x/\sqrt{4D_1 t}$, where D_1 is the elevated diffusivity (10^{-3}) so that the evolution of the concentration profile is evident. You can compare your results with Figure 13.7 in Crank (1975).

8.24. Refer to Figure 8.17 (regarding elliptic grid generation). Suppose that the paucity of contours at the base of the intrusion was preventing the analyst from accurately solving a particular problem. Develop and add an appropriate source or production term to eq. (8.121) to rectify this difficulty and then demonstrate its effectiveness by computing the solution of the modified equation numerically and replotting the $\psi(x, y)$ contours.

8.25. A concrete drainage ditch is constructed in the form of a 90° "V" and it is to be used to carry 5500 gpm of wastewater away from a plant. *Assume* that the flow is governed by

$$0 \cong \mu\left[\frac{\partial^2 V_z}{\partial x^2} + \frac{\partial^2 V_z}{\partial y^2}\right] + \rho g\sin\theta.$$

The slope is 1 ft/1000 ft such that $\sin\theta = 0.001$. By computation, determine the *depth* of water in the channel. Then, repeat the analysis, but assume that the channel is a 60° "V." One concern with an open channel of this type is the possibility of particle deposition in the bottom of the "V," and if sedimentation occurs, the carrying capacity of the ditch could be reduced. Consequently, particle "scour" is extremely important; make sure that your calculations also provide an estimate of the shear stress at the bottom of the channel.

8.26. Water flows through a partially filled large pipe, 1 m in diameter, under the influence of gravity. The depth of water in the pipe (measured vertically from the bottom) is $d/3$ or $0.667R$. The velocity distribution is (approximately) governed by

$$0 = \mu\left[\frac{1}{r}\frac{\partial}{\partial r}\left(r\frac{\partial V_z}{\partial r}\right) + \frac{1}{r^2}\frac{\partial^2 V_z}{\partial\theta^2}\right] + \rho g\sin\phi.$$

The angle of declination is small, so we will take $\sin \phi = 0.0002$. The fluid flows over a section of coated pipe wall that is soluble, and we will take the concentration on the fluid side of the interface (immediately adjacent to the pipe wall) to be 1. The continuity equation for the soluble species will be

$$V_z \frac{\partial C}{\partial z} = D \left[\frac{1}{r} \frac{\partial}{\partial r} \left(r \frac{\partial C}{\partial r} \right) + \frac{1}{r^2} \frac{\partial^2 C}{\partial \theta^2} \right].$$

The coated section of wall is 1 m long and $D = 0.0002$ cm^2/s. Find the concentration distributions 200, 400, and 800 m downstream. Any contaminant that finds its way to the free surface disappears immediately (is volatilized).

REFERENCES

Anderson, J. D. *Computational Fluid Dynamics: The Basics with Applications*, McGraw-Hill, New York (1995).

Bernard, R. S. Discrete Solution of the Anelastic Equations for Mesoscale Modeling. *Report 86/E/51, GKSS Forschungszentrum*, Geesthacht (1986).

Bernard, R. S. Explicit Numerical Algorithm for Modeling Incompressible Approach Flow. *Technical Report REMR-HY-5, US Army Engineer Waterways Experiment Station*, Vicksburg (1989).

Chow, C. Y. *An Introduction to Computational Fluid Mechanics*, John Wiley & Sons, New York (1979).

Chung, T. J. *Computational Fluid Dynamics*, Cambridge University Press, Cambridge (2002).

Crank, J. *The Mathematics of Diffusion*, second Edition, Clarendon Press, Oxford (1975).

Frankel, S. P. Convergence Rates of Iterative Treatments of Partial Differential Equations. *Mathematical Tables and Other Aids to Computation*, 4:65 (1950).

James, M. L., Smith, G. M., and J. C. Wolford. *Applied Numerical Methods for Digital Computation (with FORTRAN and CSMP)*, 2nd edition, Harper and Row, New York (1977).

MacCormack, R. W. The Effect of Viscosity in Hypervelocity Impact Cratering. AIAA paper 69–354 (1969).

Nikbakhti, R. and A. B. Rahimi. Double-Diffusive Natural Convection in a Rectangular Cavity with Partially Thermally Active Side Walls. *Journal of the Taiwan Institute of Chemical Engineers*, 43:535 (2012).

Oertel, H. and K. R. Kirchartz. In: *Recent Developments in Theoretical and Experimental Fluid Mechanics* (U. Muller, K. G. Roesner, and B. Schmidt, editors), Springer-Verlag, Berlin (1979).

Patankar, S. V. *Numerical Heat Transfer and Fluid Flow*, Hemisphere Publishing, Washington, DC (1980).

Peaceman, D. W. and H. H. Rachford. The Numerical Solution of Parabolic and Elliptic Partial Differential Equations. *Journal of the Society for Industrial and Applied Mathematics*, 3:28 (1955).

Sarra, S. A. The Method of Characteristics and Conservation Laws. *The Journal of Online Mathematics and Its Applications*, Mathematical Association of America, www.maa.org (2003).

Selle, A., Fedkiw, R., Kim, B. M., Liu, Y., and J. Rossignac. An Unconditionally Stable MacCormack Method. *Journal of Scientific Computing*, 35:350 (2008).

Smith, G. D. *Numerical Solution of Partial Differential Equations*, Oxford University Press, Oxford (1965).

Spiegel, M. R. *Advanced Mathematics for Engineers and Scientists*, McGraw-Hill, New York (1971).

Thompson, J. F., Thames, F. C., and C. W. Mastin. Automatic Numerical Generation of Body-Fitted Curvilinear Coordinate Systems for Fields Contining Any Number of Arbitrary Two-Dimensional Bodies. *Journal of Computational Physics*, 15:299 (1974).

Van Dyke, M. *An Album of Fluid Motion*, Parabolic Press, Stanford, CA (1982).

9

INTEGRO-DIFFERENTIAL EQUATIONS

INTRODUCTION

Integro-differential equations (IDEs) arise in a variety of contexts; there are applications to process control and to diffusion along grain boundaries (Antipov and Gao, 2000), as well as modeling for neural networks (Jackiewicz et al., 2008), option prices (Cont and Voltchkova, 2005), and the spread of infectious diseases (Medlock and Kot, 2003). Only rarely can analytic solutions be found for such problems, so frequently, the analyst must resort to numerical methods. Note that IDEs also figure prominently in the analysis of multiphase processes where countable entities such as bubbles, drops, and particles are borne by a fluid phase. Examples include solvent extraction/emulsification, flocculation, crystallization, sedimentation, and the operation of biochemical reactors. Because so many industrial processes involve countable entities, the importance of IDEs to process engineering and the applied sciences cannot be overstated.

To provide a historical framework for our consideration of IDEs, we will explore their role in the early twentieth-century study of biological systems. Vito Volterra was an eminent Italian mathematician (1860–1940) whose career was effectively ended by his refusal to sign the oath of allegiance to the Fascist government in 1931. It is worth noting that only 12 Italian university professors refused to sign, a small number but one that is understandable given the consequences (essentially exclusion from the Italian intellectual and scientific communities). One of Volterra's principal interests was mathematical biology, particularly, the dynamic behavior of populations in conflict, which is frequently referred to as the predator–prey problem. Let us preface this part of our discussion by considering a simple system consisting of (initially) known populations of foxes (F) and rabbits (R).

Prey species usually reproduce rapidly since interactions with predators will invariably diminish their numbers. A very simple model for the prey population might be written as

$$\frac{dR}{dt} = a_1 R - a_2 RF. \tag{9.1}$$

In contrast, predators do not usually prosper by excessive breeding since they will suffer from too much competition for available prey:

$$\frac{dF}{dt} = -b_1 F + b_2 RF. \tag{9.2}$$

You will notice the similarity between the two differential equations, but with a reversal of signs, of course. This elementary model is deterministic; given values for the constants and the initial populations, the future numbers of foxes and rabbits are set for all time, t. We recognize that this cannot be correct. One obvious but trivial objection is that a fractional rabbit is not physically realizable. More importantly, there are aspects of animal behavior that are not reflected by these first-order ordinary differential equations

Applied Mathematics for Science and Engineering, First Edition. Larry A. Glasgow.
© 2014 John Wiley & Sons, Inc. Published 2014 by John Wiley & Sons, Inc.

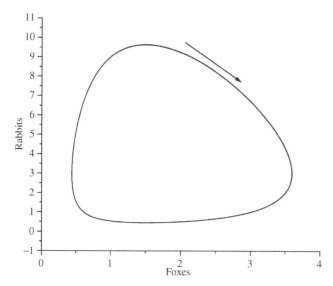

FIGURE 9.1. Phase-space portrait of the dynamic behavior of the predator–prey problem for foxes and rabbits. Motion on this limit cycle is clockwise and the parametric values employed were $a_1 = 1/2$, $b_1 = 1/4$, $a_2 = 1/3$, and $b_2 = 1/12$. The initial populations were nine rabbits and one fox. The closed cycle means that this system's behavior is *periodic*.

(ODEs); that is, some behavioral phenomena are *not deterministic*. At the same time, we may be able to learn something useful about the dynamic behavior of such simple systems that will benefit us later. We will use eq. (9.1) and eq. (9.2) to compute $R(t)$ and $F(t)$, and then we will cross plot the dependent variables to produce the phase-space trajectory.

The data provided in Figure 9.1 illustrate that an increase in the number of rabbits results in growth of the predator (fox) population. Of course, this increase in predators diminishes the number of prey animals, lack of food causes the number of predators to decrease, the prey species recovers, and so on. As we pointed out previously, the model employed is not entirely realistic. Volterra recognized that it was unreasonable to assume that the dynamic behavior of a biological system should depend *solely on its present state*. Indeed, many common physical systems exhibit dependence on past events—what Volterra referred to as "hereditary influences." An engineer might be more likely to think of a cylindrical rod repeatedly subjected to torsion, or an airliner cabin that has experienced thousands of pressurization cycles; at some point, the materials may fail due to cumulative effects of stress.

One approach to such problems is to add a time integral to the model—in essence, a term that will reflect historical influences on system behavior. We will start with the case of a single population, where an appropriate model might be written:

$$\frac{1}{y}\frac{dy}{dt} = a_1 + a_2 y + \int_{t_0}^{t} K(t, \phi) y(\phi) d\phi. \qquad (9.3)$$

The difficulty posed by the integro-differential eq. (9.3) depends mainly on the nature of the *kernel*, that is, the functional form of $K(t, \phi)$. Before we reexamine our predator–prey (populations in conflict) problem, we can carry out a small exploration of some of the effects of the addition of a simple history term. We will do this by examining an elementary control strategy that is familiar to many engineers, where part of the control action is determined by the past behavior of the system.

AN EXAMPLE OF THREE-MODE CONTROL

We now look at an example from process control that will allow us to better appreciate the significance of the addition of the integral, as illustrated by eq. (9.3); we will do so for a problem that is linear with respect to the variable that drives the controller (the error, ε). One tried and proven strategy for automatic process control is found in the *proportional-integral-derivative* (PID) algorithm. Let ε represent the error detected in an output variable (perhaps temperature, pressure, concentration, pH, etc.). P is the controller output (pressure in the case of pneumatic controllers); please note that P is in deviation form—the equilibrium value has been subtracted such that the initial (undisturbed) value for P is zero. In the time domain, our PID algorithm appears as

$$P = K_C \left(\varepsilon + \frac{1}{\tau_I} \int_0^t \varepsilon dt + \tau_D \frac{d\varepsilon}{dt} \right). \qquad (9.4)$$

K_C is the controller gain, τ_I is the integral time, τ_D is the derivative time, and ε is the measured error. We have three parameters that will affect the behavior of this system. The corrective action taken is dependent on the instantaneous error, the history (time integral) of the error, and the derivative of the error. It should be clear that the integral action will take into account the past behavior of the system and the derivative action will anticipate what is about to happen at an instant in time. For this reason, $K_C \tau_D (d\varepsilon/dt)$ is often referred to as *anticipatory control*. It is convenient to formulate a *transfer function* for eq. (9.4) using the Laplace transform—this takes us from the time domain to the s-plane:

$$\frac{P(s)}{\varepsilon(s)} = K_C \left(1 + \frac{1}{\tau_I s} + \tau_D s \right). \qquad (9.5)$$

You will observe that integration with respect to time has been replaced by division by s, and that differentiation with respect to time has been replaced by multiplication by s.

Now let us assume a particular functional form for the error; we will start with a sinusoidal error such that $\varepsilon(t) = A\sin(\omega t)$ and $\varepsilon(s) = A\omega/(s^2 + \omega^2)$. Therefore,

$$P(s) = \frac{\dfrac{A\omega K_C}{\tau_I}(\tau_I \tau_D s^2 + \tau_I s + 1)}{s(s^2 + \omega^2)}. \qquad (9.6)$$

Here is where the effect of adding integral control becomes really apparent; the order of the polynomial (in s) in the denominator has been increased by 1. This also means that in the complex plane, a pole has been placed at the origin; you may recall that a complex root with a positive real part implies unstable oscillatory behavior. In the control literature, root locus is a graphical technique used to identify roots of the characteristic equation (a polynomial in s). When complex roots exist, parts of the dynamic response of the system will be governed by terms of the form $e^{(a+ib)t} = e^{at}(\cos bt + i\sin bt)$. Thus, when a is positive, the system response will be oscillatory with *increasing* amplitude (unstable).

We can invert eq. (9.6) by partial fraction expansion, or choose to work strictly in the time domain so that our results will be immediately transparent. Our particular interest is to explore the impact of the system's history (manifested in the integral control) on its behavior. We will achieve this by varying the value of τ_I from 10 to 1 (remember, τ_I has an *inverse* impact on the integral term).

Our objective with Figure 9.2 is to learn a little bit about how the "history" term affects the solution of eq. (9.4). We see that as the importance of the time integral is increased (τ_I varies from $10 \rightarrow 1$), the oscillatory nature of the response

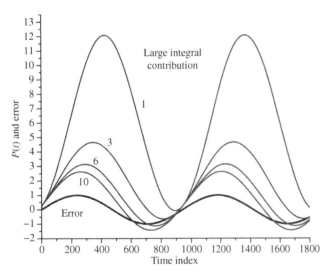

FIGURE 9.2. Illustration of the response of eq. (9.4) to sinusoidal error (the error is the heavy curve at the bottom that oscillates between -1 and $+1$). Results for four different values of τ_I are shown, 1, 3, 6, and 10.

is enhanced and the phase lag (shift of the peaks to the right) is increased. In fact, one of the more interesting features of PID control is that just about any system can be destabilized if enough integral action is added, that is, if τ_I is made small enough! As a practical matter, we note that the large response, $P(t)$, seen in Figure 9.2 would almost certainly saturate the final control element (typically a control valve). Saturation can also result from a prolonged history of error (a persistent error condition) which will cause the integral, $\int_0^t \varepsilon dt$, to continue to accumulate, resulting in what is referred to in the control literature as integral "windup." This condition is usually avoided in practice by switching from the *position form* to the *velocity form* (of the PID algorithm).

POPULATION PROBLEMS WITH HEREDITARY INFLUENCES

It will prove advantageous to return to eq. (9.3), with a slight modification, such that we have (initially) a one-variable problem:

$$\frac{dy}{dt} = a_1 y + a_2 y^2 + y\int_0^t K(t - \phi)y(\phi)d\phi. \qquad (9.7)$$

You will observe that a specific form has been chosen for the kernel in this case—this is the kernel that Volterra referred to as belonging to the *closed-cycle* class. Following Davis (1962), we approximate the kernel by writing

$$K(z) = K_0 + K_1 z + \frac{1}{2}K_2 z^2 + \cdots, \qquad (9.8)$$

then we assume that the only K that is nonzero is K_0 (we are truncating our approximation of the kernel). We substitute into eq. (9.7), divide by y, and then differentiate, resulting in

$$y\frac{d^2 y}{dt^2} = \left(\frac{dy}{dt}\right)^2 + a_2 y^2 \frac{dy}{dt} + K_0 y^3. \qquad (9.9)$$

It proves convenient to let

$$y = -\frac{a_1 u}{a_2}, \quad t = \frac{\tau}{a_2}, \quad \text{and} \quad \lambda = \frac{K_0}{-a_1 a_2},$$

which yields

$$u\frac{d^2 u}{d\tau^2} = \left(\frac{du}{dt}\right)^2 - u^2 + \lambda u^3. \qquad (9.10)$$

Davis notes that no closed-form solution is known for this ODE, so a numerical solution is required. The intriguing feature of eq. (9.10) is the profound impact that the

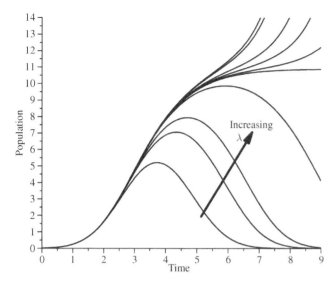

FIGURE 9.3. Solutions of the ordinary differential eq. (9.10), with increasing values of the parameter, λ. The solution in the "middle" that approaches a horizontal asymptote is often referred to as the growth, or logistic, curve.

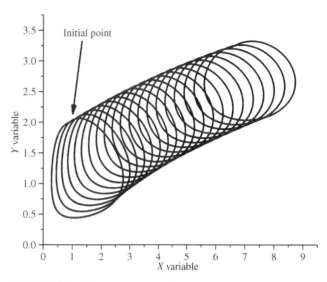

FIGURE 9.4. Phase-plane portrait for two populations in conflict with hereditary influences. Motion in this case is counterclockwise from the initial point, which is $(X, Y) = (1, 2)$. The constants K_1 and K_2 are set to -0.05 for this example.

parameter, λ, has on the solution. We will explore the solution of this ODE with increasing λs in Figure 9.3.

The principal task of interest to us now is to assess what happens when hereditary influences are added to the case of two populations in conflict. It is likely that some profound changes—relative to the results shown in Figure 9.1 for foxes and rabbits—will occur. The model we will consider takes the form

$$\frac{dX}{dt} = aX - bXY - K_1 X \int_0^t Y(\phi)d\phi \qquad (9.11)$$

and

$$\frac{dY}{dt} = -\alpha Y + \beta XY + K_2 Y \int_0^t X(\phi)d\phi. \qquad (9.12)$$

Using the approach described previously for the single-population example, we obtain the system of ODEs:

$$\frac{d^2X}{dt^2} = \frac{1}{X}\left(\left(\frac{dX}{dt}\right)^2 - bX^2\frac{dY}{dt} - K_1 X^2 Y\right) \qquad (9.13)$$

and

$$\frac{d^2Y}{dt^2} = \frac{1}{Y}\left(\left(\frac{dY}{dt}\right)^2 + \beta Y^2\frac{dX}{dt} + K_2 XY^2\right). \qquad (9.14)$$

We will confine our attention to the case for which $a = b = 2$, $K_1 = -0.05$, and $\alpha = \beta = 1$ with $K_2 = -0.05$.

The initial values for X and Y will be taken as 1 and 2, respectively. Of course, since we are now solving second-order ODEs, we need a second initial condition; we can obtain values for the first derivatives directly from eq. (9.11) and eq. (9.12) by noting that, for $t = 0$, the integrals disappear. Therefore, the initial values for X and Y can be used to obtain $(dX/dt)_{t=0}$ and $(dY/dt)_{t=0}$. We discover that the solution for this system is very different from the case reviewed in the introduction where hereditary influences were neglected. It is evident from Figure 9.4 that *the phase-plane trajectory has an evolving shape and the populations are unstable.*

There are several important lessons to be drawn from this study of populations in conflict. Foremost, the addition of hereditary influences can render the populations nonperiodic and unstable. Moreover, by making one of the Ks positive and one negative, we can also obtain extinction of one species and proliferation of the other. The reader may wish to investigate the possibilities by setting $K_1 = -0.05$ and $K_2 = +0.05$, which results in a trajectory that collapses along the y-axis with very small values for X occurring intermittently. The effects obtained with various combinations of Ks will form the basis of a student exercise at the end of the chapter.

AN ELEMENTARY SOLUTION STRATEGY

In the previous section, we converted our IDEs into second-order ODEs that we solved by methods already familiar to us. We now want to look at an IDE example to illustrate how such problems might be solved when we retain the integral but employ a simple discretization. Consider the nonlinear IDE:

$$\frac{du}{dx} = -1 + \int_0^x u^2(\phi)d\phi, \qquad (9.15)$$

for which $u(x = 0) = 0$ and $0 \leq x \leq 1$. This equation has been solved by Batiha et al. (2008) using the *variational iteration method* (VIM), so we will have a convenient comparison. We replace the derivative, du/dx, with the first-order forward difference:

$$\frac{du}{dx} \cong \frac{u(x + \Delta x) - u(x)}{\Delta x}, \qquad (9.16)$$

and the integral is approximated by the summation of rectangles each of width, Δx. We choose $\Delta x = 0.001$ and proceed from $x = 0$ to $x = 1$. Let us compare our results with those provided by Batiha et al.

x	VIM (Batiha et al., 2008)	Approximate Discretization
0.0000	0.000000	0.000
0.0938	−0.0937935	−0.094
0.2188	−0.2186091	−0.218
0.3125	−0.3117064	−0.311
0.4062	−0.4039385	−0.404
0.5000	−0.4948226	−0.495
0.6250	−0.6124315	−0.612
0.7188	−0.6969446	−0.697
0.8125	−0.7771007	−0.777
0.9062	−0.8519654	−0.852
1.0000	−0.9205578	−0.921

We can see from the tabulated results that this elementary discretization has produced a quite acceptable agreement with the published solution for this IDE. We would do well to wonder how the results of such a simple procedure would compare with a case for which the *analytic solution* can be easily determined, resulting in a more definitive test. We will do this by looking at the linear IDE,

$$\frac{du}{dx} + 2u + 5\int_0^x u(\phi)d\phi = 1, \quad \text{with } x \geq 0 \text{ and } u(0) = 0. \qquad (9.17)$$

We observed previously that a linear equation of this type can be solved readily through the use of the Laplace transform, and we now illustrate this process. We begin by taking the Laplace transform of each term in the equation (noting that $u(x = 0) = 0$):

$$su(s) + 2u(s) + 5\frac{u(s)}{s} = \frac{1}{s}. \qquad (9.18)$$

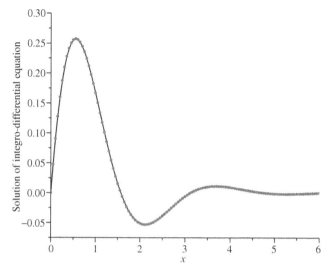

FIGURE 9.5. Comparison of the analytic solution (solid curve) of the integro-differential eq. (9.17) with the approximate solution obtained by discretization (filled circles).

We isolate $u(s)$:

$$u(s) = \frac{1}{s^2 + 2s + 5}, \qquad (9.19)$$

then by partial fraction expansion:

$$\frac{1}{(s+1+2i)(s+1-2i)} = \frac{A}{s+1+2i} + \frac{B}{s+1-2i}, \quad (9.20)$$

and we find that $A = (1/4)i$ and $B = -(1/4)i$. Thus, we can now invert our transform:

$$\begin{aligned}
u(x) &= \frac{1}{4}i\exp((-1-2i)x) - \frac{1}{4}i\exp((-1+2i)x) \\
&= \frac{1}{2}\exp(-x)\sin(2x).
\end{aligned} \qquad (9.21)$$

To provide a comparison, we will employ the very same discretization scheme we used for the first (nonlinear) example in this section, using a step size (Δx) of 0.005. Both solutions are provided in Figure 9.5. The analytic case is shown as the solid (black) curve and the approximate computed values are represented by the small filled circles. The typical discrepancy between the two solutions is on the order of about 0.4%. For all but the most exacting applications, this would be completely satisfactory.

VIM: THE VARIATIONAL ITERATION METHOD

Previously in this section, we cited some results obtained using *VIM*. Pioneering work by He (1999, 2000, 2007) led to rapid exploitation of the technique, and the efforts of Batiha et al. (2008), He and Wu (2007), and many others indicated that VIM might be especially well suited to the solution of IDEs. It is, therefore, appropriate that we deviate from our course a little to provide a discussion of VIM. So that we better understand the technique, we must spend a few moments discussing Lagrange's multiplier method, which we saw for the first time in Chapter 2.

Let us suppose that we have two variables that are to be combined such that

$$x_1{}^2 + 2x_2{}^3 = 30. \tag{9.22}$$

We are seeking values for the two variables that cause the product $(x_1 x_2)$ to be a maximum, subject to the constraint, eq. (9.22). These values can be identified through use of the Lagrange multiplier, λ. We write

$$F = x_1 x_2 + \lambda(x_1{}^2 + 2x_2{}^3 - 30). \tag{9.23}$$

This expression is differentiated with respect to each of the *three* variables, and the partial derivatives are set equal to zero:

$$\frac{\partial F}{\partial x_1} = x_2 + 2\lambda x_1 = 0 \tag{9.24}$$

$$\frac{\partial F}{\partial x_2} = x_1 + 6\lambda x_2{}^2 = 0 \tag{9.25}$$

$$\frac{\partial F}{\partial \lambda} = x_1{}^2 + 2x_2{}^3 - 30 = 0. \tag{9.26}$$

These three equations are solved, and the reader may wish to show that $x_1 = 4.24264$, $x_2 = 1.81712$, and $\lambda = -0.21415$; accordingly, the maximum product subject to the constraint is $x_1 x_2 = 7.70939$. Next, we examine how the Lagrange multiplier technique is modified for application to an *ODE*. Suppose we have the ODE,

$$\frac{du}{dx} + u^2 = 2, \tag{9.27}$$

with $u(0) = 1/4$ and $0 \le x \le 1$. In this case, the ODE can be solved yielding

$$\frac{\sqrt{2} + u}{\sqrt{2} - u} = \exp\left(2\sqrt{2}(x + 0.126326)\right). \tag{9.28}$$

We will calculate a few values to have a comparison readily available:

x	$u(x)$
0	0.25
0.13495	0.50
0.29129	0.75
0.496899	1.00
0.85885	1.25

Now we are set; let our ODE, eq. (9.27), be rewritten as $u_u' + u_n{}^2 - 2$. The successive iterations are determined from

$$u_{n+1} = u_n + \int_0^x \lambda(u_n' + u_n{}^2 - 2)ds. \tag{9.29}$$

The Lagrange multiplier, λ, can be identified from the stationary condition(s), and for linear ODEs, λ *can be determined exactly*. For the nonlinear eq. (9.27), we will presume $\lambda = -1$ and proceed. We take $u_0 = 1/4$ (satisfying the initial condition) and find by integration $u_1 = (1/4) + (31/16)x$. Therefore,

$$u_2 = \frac{1}{4} + \frac{31}{16}x - \int_0^x \left(\frac{31}{32}s + \frac{961}{256}s^2\right)ds, \tag{9.30}$$

which yields

$$u_2 = \frac{1}{4} + \frac{31}{16}x - \frac{31}{64}x^2 - \frac{961}{768}x^3. \tag{9.31}$$

If we let $x = 0.29129$, we find $u \approx 0.74235$; setting $x = 0.4969$ produces $u \approx 0.9396$. These two results have corresponding errors of about 1% and 6%. For many applications, this level of error would be tolerable, but additional iterations will be necessary if we require improved accuracy.

Let us consider one more example to ensure our familiarity with this very powerful technique. Suppose we have the ODE,

$$\frac{du}{dx} = -\frac{x^3}{(u+1)^2}, \quad \text{with } u(x=0) = 1. \tag{9.32}$$

The analytic solution is known:

$$3x^4 + 4(u+1)^3 = 32, \quad \text{or alternatively, } u = \left[\frac{32 - 3x^4}{4}\right]^{1/3} - 1. \tag{9.33}$$

Our succession of estimates is obtained from

$$u_{n+1} = u_n - \int_0^x \left(u_n' + \frac{s^3}{(u_n+1)^2}\right)ds. \tag{9.34}$$

If we start with $u_0 = 1$, we can very easily find $u_1 = 1 - (x^4/16)$. Therefore,

$$u_2 = 1 - \frac{x^4}{16} - \int_0^x \left[-\frac{s^3}{4} + \frac{s^3}{\left(2 - \frac{1}{16}s^4\right)^2} \right] ds, \quad (9.35)$$

and

$$u_2 = 3 + \frac{64}{x^4 - 32}. \quad (9.36)$$

We can compare these estimates with the analytic solution in the following table:

$x=$	0.50	1.00	1.25	1.50	1.75
$u(x)$ analytic	0.996086	0.935438	0.834018	0.613829	−0.011526
$u_1(x)$	0.996094	0.93750	0.847412	0.683594	0.413818
$u_2(x)$	0.996086	0.935484	0.834809	0.624130	0.170782

Except for higher values of x, the second iteration does a remarkable job of representing the solution of this ODE.

We want to facilitate the application of VIM more broadly and this will require that we make use of some elements from the *calculus of variations*. Because our goal is to use VIM as a tool, we will pursue the approach described by Ji-Huan He in his interview in *ScienceWatch*, July 2008: We want to present the technique in such a way that "anyone who knows nothing of variational theory in mathematics can apply the method." For the reader who is curious about other applications of the calculus of variations, an introduction is provided at the end of this text in Chapter 11. In addition, Robert Weinstock's (1974) book is very useful and there are very good online primers available including James Nearing's (University of Miami) work, for example. There is also a nice appendix in Kenneth Huebner's (1975) book that was specifically written for engineers.

We generalize this part of our discussion by rewriting the ODE we wish to consider as

$$Lu + Nu = g(x), \quad (9.37)$$

where L and N are linear and nonlinear operators, respectively. Now we construct a *correction functional*:

$$u_{n+1}(x) = u_n(x) + \int_0^x \lambda[Lu_n(\phi) + N\tilde{u}_n(\phi) - g(\phi)]d\phi. \quad (9.38)$$

The nonlinear terms are to be removed by substituting $\tilde{u}$, a restricted variation such that $\delta(\tilde{u}) = 0$, and the Lagrange multiplier, λ, is identified using variational theory. To find the optimal value for λ, we write

$$\delta u_{n+1}(x) = \delta u_n(x) + \delta \int_0^x \lambda[Lu_n(\phi) + N\tilde{u}_n(\phi) - g(\phi)]d\phi \quad (9.39)$$

and make the correction functional *stationary*. We can illustrate the process with an example (the Lienard equation) provided by Matinfar et al. (2008). Suppose we have the nonlinear ODE,

$$\frac{d^2u}{dt^2} + au + bu^3 + cu^5 = 0, \quad (9.40)$$

where a, b, and c are real constants. We now write the correction functional as before:

$$u_{n+1}(t) = u_n(t) + \int_0^t \lambda\left[u_n''(\phi) + au_n(\phi) + b\tilde{u}_n^3(\phi) + c\tilde{u}_n^5(\phi)\right]d\phi. \quad (9.41)$$

The restricted variations are used on the nonlinear terms so they can be eliminated, resulting in

$$\delta u_{n+1}(t) = \delta u_n(t) + \delta \int_0^t \lambda[u_n''(\phi) + au_n(\phi)]d\phi. \quad (9.42)$$

The integral is handled by integration by parts, which yields three stationary conditions for identification of the optimal λs:

$$\lambda = \phi - t, \quad (9.43)$$

$$\lambda = \frac{1}{\sqrt{-a}} \sinh(\sqrt{-a}(\phi - t)), \quad (9.44)$$

and

$$\lambda = \sin(\phi - t). \quad (9.45)$$

Any one of the three choices will work—they will only differ by the speed of convergence of the correction formula. If we choose the first of the trio of λs, our iterative process will be based on:

$$u_{n+1}(t) = u_n(t) + \int_0^t (\phi - t)[u_n'' + au_n + bu_n^3 + cu_n^5]d\phi. \quad (9.46)$$

An initial approximation for $u_{n=0}$ is all that is required to get started with the iterative process. Matinfar et al. used

MATLAB for their computations with a, b, and c set equal to $-1, 4,$ and -3, respectively. For the initial (trial) function, they used

$$u_0 = u(t = 0) + \left(\frac{du}{dt}\right)_{t=0}(t),$$

and their results were in excellent accord with the closed-form solution. It should be noted that there are many examples of the application of He's method in the recent literature; just a few of the differential equations that have been solved this way include the Kawahara equation by Ganji et al. (2007), the Laplace equation by Jassim (2012), the nonlinear oscillator by He and Wu (2007), and Sturm–Liouville equations by Altintan and Ugur (2012). VIM has even been applied to model contamination (the spread of pollution) in a system of interconnected lakes (see Merdan, 2009).

Of course, our real objective in this chapter is the solution of IDEs. The procedure is wholly analogous to our treatment of ODEs earlier, and to begin, let us take

$$\frac{du}{dx} = f(x) + \int_0^x g(\phi, u(\phi), u'(\phi))d\phi, \qquad (9.47)$$

where $f(x)$ is the source term. We build a *correction functional* just as before:

$$u_{n+1}(x) = u_n(x)$$
$$+ \int_0^x \lambda \left[u_n'(s) - f(s) - \int_0^s g(\phi, \tilde{u}(\phi), \tilde{u}'(\phi))d\phi\right]ds. \qquad (9.48)$$

Our task now is to choose the optimal Lagrange multiplier, λ. Remember that the *tilde* in eq. (9.48) denotes a restricted variation; therefore,

$$\delta u_{n+1}(x) = \delta u_n(x)$$
$$+ \delta \int_0^x \lambda(s)\left[(u_n)_s(s) - f(s) - \int_0^s g(\phi, u(\phi), u'(\phi))d\phi\right]ds. \qquad (9.49)$$

The integration is performed and the stationary conditions are used to identify λ, which in this case is -1. Therefore, the iteration formula is

$$u_{n+1}(x) = u_n(x)$$
$$- \int_0^x \left[u_n'(s) - f(s) - \int_0^s g(\phi, u(\phi), u'(\phi))d\phi\right]ds. \qquad (9.50)$$

We will now apply this technique to a nonlinear problem considered by Batiha et al. (2008):

$$\frac{du}{dx} = 1 + \int_0^x u(\phi)\frac{du}{d\phi}d\phi, \qquad (9.51)$$

given that $(0 \leq x \leq 1)$ and $u(x = 0) = 0$. Remember that we select a trial function such that the boundary conditions are satisfied, and in this case, an obvious function is $u_0(x) = x$. Once again, $\lambda = -1$, so the iteration formula is

$$u_{n+1}(x) = u_n(x) - \int_0^x \left[\frac{du_s}{ds} - 1 - \int_0^s u(\phi)\frac{du}{d\phi}d\phi\right]ds. \qquad (9.52)$$

Batiha et al. reported an incorrect sequence of results in the original paper, and the errors have been corrected here:

$$u_1(x) = x + \frac{1}{6}x^3 \qquad (9.53)$$

and

$$u_2 = x + \frac{1}{6}x^3 + \frac{1}{30}x^5 + \frac{1}{504}x^7. \qquad (9.54)$$

We will also employ discretization with eq. (9.51) to provide a comparison:

x	$u(x)$ Discretized	$u(x)$ VIM Approx.
0	0.0000	0.0000
0.0938	0.0940	0.09394
0.2188	0.2207	0.22056
0.3125	0.3178	0.31769
0.4062	0.4178	0.41774
0.5000	0.5220	0.52189
0.6250	0.6693	0.66894
0.7188	0.7879	0.78729
0.8125	0.9156	0.91416
0.9062	1.0548	1.05159
1.0000	1.2086	1.20198

In this example, Batiha et al. (2008) used three VIM iterations, but the values reported in their original table are incorrect. The corrected values for the second iteration are shown in the preceeding table, and you will immediately note that these values are in remarkable agreement with the results computed using discretization.

We now look at an example with a higher-order derivative provided by He and Wu (2007), a fourth-order IDE:

$$\frac{d^4y}{dx^4} = x(1 + e^x) + 3e^x + y(x) - \int_0^x y(s)ds. \qquad (9.55)$$

We are given that $y(0) = 1$, $y(1) = 1 + e = 3.71828$, $y''(0) = 2$, and $y''(1) = 3e = 8.15485$. Of course, we might think about discretizing this equation and trying to solve it just as we did in some of the previous examples. However, the combination of the fourth derivative and the split boundary conditions renders this approach quite unappealing. We will be better served by applying VIM to this IDE.

For this linear case, the exact solution is known, $y = 1 + x \exp(x)$, facilitating comparison. He and Wu suggest a trial function,

$$y_0(x) = 1 + \exp(x)[a + bx + cx^2 + dx^3], \quad (9.56)$$

and clearly $a = 0$ so that $y(0) = 1$. One of the most important contributions of the He and Wu (2007) paper is that appropriate iteration formulae are given for a wide variety of problems, including this one. For example, given the equation

$$\frac{d^2u}{dx^2} + f(u, u', u'') = 0, \quad (9.57)$$

the first-order approximation is given by

$$u_1 = u_0(x) + \int_0^x (s-x)f(u, u', u'')ds. \quad (9.58)$$

We can simplify matters for our example by letting

$$f(x) = \int_0^x \int_0^\xi \left[s(1+e^s) + 3e^s + y(s) - \int_0^s y(s)ds \right] d\xi dx. \quad (9.59)$$

The iteration formula can now be written more compactly and applied to obtain y_1:

$$y_1(x) = y_0(x) + \int_0^x (s-x)f(s)ds. \quad (9.60)$$

He and Wu found that

$$y_1 = (0.0803x^3 - 0.2113x^2 + 2.302x - 6.9014)e^x + 0.0164x^4 + 0.7162x^3 + 2.3418x^2 + 5.5516x + 7.9014. \quad (9.61)$$

Please be advised that the first-order approximation as given by He and Wu contains an error (their equation number 37); the coefficient provided for x^2 is *not* 10.3418 (it should be 2.3418 as indicated previously). In this case, we can easily compare the analytic solution with the first-order approximation; some values are provided in the following table and the agreement is excellent—generally within about 0.5%.

x	$1 + x \exp(x)$	Approximation, $y_1(x)$
0.0	1.00000	1.0000
0.2	1.24428	1.2390
0.4	1.59673	1.5891
0.6	2.09327	2.0867
0.8	2.78043	2.7770
1.0	3.71828	3.7183

VIM provides us with an extraordinarily powerful tool for the solution of *both* ODEs and IDEs.

INTEGRO-DIFFERENTIAL EQUATIONS AND THE SPREAD OF INFECTIOUS DISEASE

IDEs figure prominently in deterministic models of epidemics. Let us begin by considering a simple model for a population consisting of two types of individuals: those who are susceptible to infection (S) and those already infected (I). We assume the disease is not fatal, so there is no death rate, and we exclude birth during the time period of interest for simplicity. An elementary model for this situation might consist of two ODEs:

$$\frac{dS}{dt} = -\beta SI \quad (9.62)$$

and

$$\frac{dI}{dt} = +\beta SI. \quad (9.63)$$

The meaning of this model construct is that the number of susceptibles decreases as a result of interaction with infected individuals, and the number of infectives increases due to interaction. For this example, we will take the rate constant, β, to be positive. Since there is neither birth nor death, the total population size is constant; that is, $S + I = C_1$. Therefore,

$$\frac{dI}{dt} = \beta I(C_1 - I). \quad (9.64)$$

We will take the total population to be 100 individuals and the initial number of infected to be 3; we will use different values for the rate constant to carry out a numerical exploration of eq. (9.64). The results we obtain are predictable: No matter what positive value is selected for the parameter, β, the infection will spread to the entire population and the total number of infectives will rapidly approach 100. This behavior is revealed in Figure 9.6.

These data reveal that the model given by eq. (9.64) does not produce very realistic results; for one thing, the infection spread throughout the population because the interaction

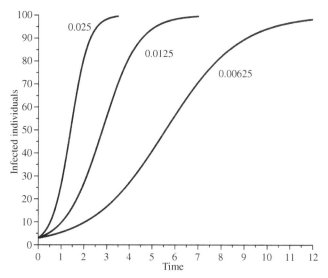

FIGURE 9.6. Increase in the number of infected individuals for βs of 0.00625, 0.125, and 0.025. As β increases, we see an increasingly rapid approach to a population that is 100% infected.

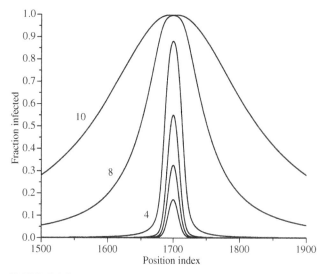

FIGURE 9.7. Solutions of eq. (9.60) for a fixed time using values for K ranging from 0.1 to 10 (specifically 0.1, 1, 2, 4, 8, and 10).

between susceptibles and infectives was mandated by the differential equation. It completely failed to account for one of the conditions of modern life that has significantly affected the transmission of disease: ease of travel. The very simple model we described earlier does not take into account the movement of infected individuals—and the intercontinental spread of AIDS revealed how very important this factor is. So, what might we do to incorporate the movement of infected individuals into the modeling?

"Diffusion" approximations have been used to try to model the spread of epidemics through what is often referred to as the Fisher–Kolmogorov–Petrovskii–Piskunov (FKPP) equation:

$$\frac{\partial I}{\partial t} = D\frac{\partial^2 I}{\partial x^2} + KI(1-I). \qquad (9.65)$$

We will think of I in this case as the fraction of infected individuals. D is a kind of diffusion coefficient and K is a rate constant. The FKPP model is tractable, but it displays an unphysical characteristic that we will now demonstrate. Suppose we have an initial cluster (spike) of infectives at a particular spatial location. Our plan is to solve eq. (9.65) employing different values for K to explore the effect of the rate constant on the solution. All other parameters are fixed.

The data in Figure 9.7 reveal the failure of the FKPP model as applied to epidemics; the speed of propagation increases as the rate constant, K, increases. In fact, the diffusion model suggests that the rate of spread of disease will become infinite if K is allowed to increase without bound. Fedotov (2001) points out that this behavior is unphysical and that the origin of this problem is due to the parabolic scaling associated with the FKPP equation. He notes that the

correct scaling for a propagating front (in our case, the interface between infectives and susceptibles) must be *hyperbolic*. Thus, a different modeling approach is required.

We will take $M(x)$ to be the rate at which infected individuals leave position x and travel to a new location in the spatial domain. We will also assume that M is a positive constant. Our model now has the form

$$\frac{\partial I}{\partial t} = \beta I(C_1 - I) - MI + M\int K(x,y)I(y,t)dy. \qquad (9.66)$$

$K(x, y)$ is the kernel; it is a density function for the fraction of infected individuals who *leave position* y *destined for* x. This density function has the characteristics

$$K(x,y) \geq 0 \quad \text{and} \quad \int K(x,y)dx = 1. \qquad (9.67)$$

Medlock and Kot (2003) assumed that the kernel was of the convolution type

$$K(x,y) = k(x-y) = k(u) = \frac{1}{2\alpha}\exp\left(-\frac{|u|}{\alpha}\right), \qquad (9.68)$$

where u is the separation between the x and y positions. They set $M = 1$, $\beta = 1$, and $\alpha = 0.7698$; I, the fraction of infected individuals, was taken to be 1 in the center of the domain. The problem thus posed by eq. (9.66) is formidable. The repeated evaluation of the integral is computationally expensive, so Medlock and Kot used a Runge–Kutta scheme for the time derivative and then the fast Fourier transform (FFT) for the integral. The solution of this problem takes the form of a traveling wave as illustrated in Figure 9.8. At small ts,

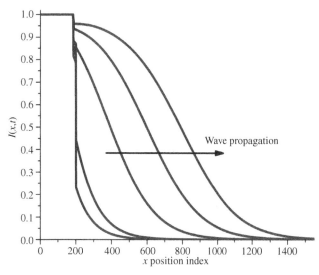

FIGURE 9.8. Emergence of the traveling wave from the initial block of infectives. The wave continues to move to the right with increasing time. This model is far more realistic than the one posed by the FKPP equation.

FIGURE 9.9. Floc formation in Couette device in which the outer cylinder is rotating (inner cylinder at rest). The large aggregates seen in this image consist of clay (kaolin) particles, polymer flocculant, and water trapped in the interstitial spaces. Measurements of the density of these aggregates reveal that the overwhelmingly dominant component is the trapped water.

the traveling wave emerges from the original block of infectives and then moves to the right (positive x-direction) with increasing time. The results shown in this figure were obtained using a Euler scheme on the time derivative (with a small time step) and the trapezoidal rule for the integral. This numerical procedure is inefficient computationally, but very simple to apply and execute.

The behavior we see in Figure 9.8 is very different from the FKPP model we discussed previously. The transmission of disease with the movement of infected individuals is much more wavelike; the boundary between infectives and susceptibles travels in the x-direction, but the characteristics of the wave are essentially unchanged as it emerges from the initial block of infectives.

EXAMPLES DRAWN FROM POPULATION BALANCES

We observed at the beginning of the chapter that many important industrial processes are multiphase, often consisting of a continuous fluid phase carrying countable entities. These dispersed objects may be particles, droplets, bubbles, cellular entities, or possibly some combination of several of these. Furthermore, these countable entities are commonly experiencing birth, death, growth (aggregation), breakage (or comminution), interphase transport, and so on. These processes may result in changes to extensive variables such as number, mass, or volume, and such changes are often crucial with respect to process performance and control. An illustration of such a process is shown in Figure 9.9; here,

very small (colloidal) clay particles are aggregating in a concentric-cylinder Couette device to form very large flocs (agglomerates). There is particle growth in this system, but little or no breakage due to the highly ordered nature of the hydrodynamic environment. One of the interesting features of this particular system is that elements of the fluid phase become trapped in the interstices of the aggregate such that the apparent volume of the dispersed phase *increases significantly* during aggregation. Thus, neither volume nor mass would be conserved in a system such as that illustrated in Figure 9.9.

One of the formidable challenges offered by problems of this type is the incredible range seen in property values (such as entity size). When colloidal clay particles are flocculated in a more typical environment (inhomogeneous turbulence), we find entity sizes ranging all the way from about 1 μm to several millimeters (the difference, therefore, is three or even four orders of magnitude). To compound the problem, the number densities for very small particles can be 5–10 orders of magnitude larger than those for very large aggregates formed in the coagulation. This is exactly the situation illustrated by the comparison shown in Figure 9.10.

It will be useful for us to preface our continuing discussion of the application of population balances to multiphase processes by spending a little time considering entity *birth* and *death*. Often when we have a population of individuals (countable entities), the appearance of new individuals (birth) may occur. In biological systems, this might be by reproduction, cell division, and so on. In crystallization, new entities—though *very small ones*—appear by nucleation. But at a finite specified size, crystals are actually born by the breakage of larger parents. Of course, at any

FIGURE 9.10 (a,b) Batch flocculation of colloidal kaolin in a stirred 1-L reactor. The disparity in entity size is revealed by this comparison; we see mainly *primary* clay particles in the dispersion shown on the left, and mostly large aggregates on the right. The number density for the small particles on the left is on the order of 1×10^7 per cm^3, while for the case on the right, one might find only 1–10 large aggregates per cubic centimeter.

instant in time, some existing individuals may disappear (death). Generally, both birth and death may depend in some manner on the size of the existing population—large numbers may lead to many births and deaths. The reader may find chapter 8 in Bailey (1964) to be a valuable introduction to this topic, should he/she wish to know more. Let us suppose that the probability that a population of individuals will total n at time t is $p_n(t)$. And we will assume that the probability that a given entity will produce a new one in the time interval, Δt, is $\lambda \Delta t$. Now, how could we arrive at a population of n entities at time $t + \Delta t$ taking only birth processes into account? We could come from the $n - 1$ state by a birth, or we could simply remain at the n-state if no births occurred over the time interval. Let us write this down:

$$p_n(t + \Delta t) = p_{n-1}(t)\lambda(n-1)\Delta t + p_n(t)(1 - \lambda n \Delta t). \quad (9.69)$$

We subtract $p_n(t)$ and divide by Δt, taking the limit as $\Delta t \rightarrow 0$; the result is a *differential-difference equation*,

$$\frac{dp_n}{dt} = \lambda(n-1)p_{n-1} - \lambda n p_n. \quad (9.70)$$

Since we start with a nonzero population, say, n_0, at $t = 0$, the initial condition for eq. (9.70) is $p_{n_0}(t=0) = 1$. This gives us a set of equations that we can solve in succession, beginning with

$$\frac{dp_{n_0}}{dt} = -\lambda n_0 p_{n_0}. \quad (9.71)$$

The model described earlier is often referred to as the Yule–Furry birth process. We say that such a birth process is homogeneous because the transition probability is constant—it does not depend on time. We might ask ourselves how birth processes are changed if the rate of birth depends on the population to the second power, that is, when the birth rate per individual depends on population size (we are requiring interaction between individuals). Under these circumstances, the total population grows according to

$$\frac{dn}{dt} = \lambda n^2, \quad (9.72)$$

resulting in

$$n = \frac{1}{\frac{1}{n_0} - \lambda t}. \quad (9.73)$$

The consequence of this model is divergent (explosive) growth, where $n \rightarrow \infty$ as $\lambda t \rightarrow 1/n_0$. Of course, neither the Yule–Furry process nor the explosive growth scenario accounts for the disappearance of individuals (death). Suppose we now add the assumption that individuals die at a rate that is proportional to their number—let the probability that an individual will die during a time interval, Δt, be $\mu \Delta t$. Once again, we think about how we can get to the n-state (i.e., where we have n individuals). This could occur due to a birth (coming from $n - 1$) or a death (coming from $n + 1$), or by having *no* births or deaths over the time interval, Δt (and thus remaining at n). Just as before, this model leads to a differential-difference equation:

$$\frac{dp_n}{dt} = \lambda(n-1)p_{n-1} - (\lambda+\mu)np_n + \mu(n+1)p_{n+1}. \quad (9.74)$$

The initial condition is that $n = n_0$ at $t = 0$, such that $p_{n_0}(t=0) = 1$. Note that if the death rate exceeds the birth rate in the combined model, extinction is guaranteed. Among other possibilities in population problems are immigration and emigration. In the case of the latter, the effects of exiting individuals can be handled by simply adjusting the death rate upward. For immigration, the simplest possibility is that the rate of influx is *independent* of the present population, $n(t)$, and therefore, the likelihood of a new individual appearing over the time interval, Δt, will be written as $\nu\Delta t$. Thus, the probability that the total number of individuals will *increase* by one over Δt is

$$(\lambda n(t) + \nu)\Delta t. \quad (9.75)$$

Once again, we consider how a population can arrive at the n-state over a time interval, Δt, now taking into account birth, death, and immigration: Of course, we can grow from $n - 1$ by birth or immigration, we could remain at n, or we could decline in number from $n + 1$. And, just as we saw previously, the result is a (set of) differential-difference equation(s); we should point out that Ji-Huan He has observed that the VIM, as described previously in this chapter, can also be used to advantage for this problem type.

In the preceding discussion of birth and death processes, our objective was the determination of probability that a population would total n individuals. But in the process industries, our concern is more likely to be centered on the distribution of entity volume, size, age, and so on, and how those distributions evolve with time in different physicochemical environments. Thus, we seek a framework that we might employ to model such phenomena in view of their enormous practical importance. The reader with interest in this area is urged to consult the book *Population Balances: Theory and Applications to Particulate Systems in Engineering* by Ramkrishna (2000). We will begin with the simplest possible one-dimensional problem type, and we will adopt Ramkrishna's notation.

Suppose we have an initial population of particles that are distributed homogeneously in space. Furthermore, let us assume that we have *only particle growth* occurring and that the growth rate at a given size, x, is $\dot{X}(x,t)$. Our focus is the particular size range, $a \le x \le b$; if the number density function, $f_1(x, t)$, changes in this size interval, then it will be due to either growth into (a, b) from below or growth out of (a, b) at the upper end. We can write a description for the dynamic behavior of this growth process:

$$\frac{d}{dt}\int_a^b f_1(x,t)dx = \dot{X}(a,t)f_1(a,t) - \dot{X}(b,t)f_1(b,t). \quad (9.76)$$

This dynamic balance is equivalent to

$$\frac{\partial f_1(x,t)}{\partial t} + \frac{\partial}{\partial x}\left[\dot{X}(x,t)f_1(x,t)\right] = 0. \quad (9.77)$$

We will take the number density function to be (initially)

$$f_1 = x\exp(-x), \quad (9.78)$$

and we will assume that the growth rate is described by

$$\dot{X} = \frac{C_1}{C_2 + x^2}. \quad (9.79)$$

Of course, the latter eq. (9.79) means that the growth rate is strongly damped by large particle size. We will make a sequence of calculations to see how the number density function will evolve with time under these conditions.

The data provided in Figure 9.11 show how the growth rate limitation at larger entity size is constraining the movement of the number density with respect to larger x. The population is becoming increasingly concentrated at intermediate sizes. If we lessen the impact of the size-limited growth rate (e.g., let $\dot{X} = C_1/(C_2 + x)$), we obtain more rapid movement of the number density function toward larger sizes; this is illustrated in Figure 9.12.

Of course, the model we have been considering is unrealistic in that it only describes entity growth. In many systems, breakage is a real possibility, and this leads us to rewrite eq. (9.77) with the addition of a combined birth and

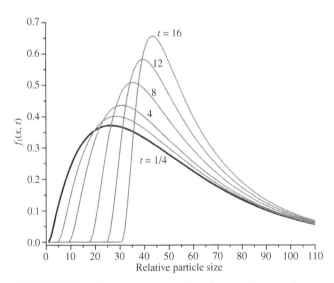

FIGURE 9.11. Change in the number density function for the growth-only case. The heavy black curve is for $t = 1/4$. Note how the small sizes are disappearing as the density function becomes increasingly narrow.

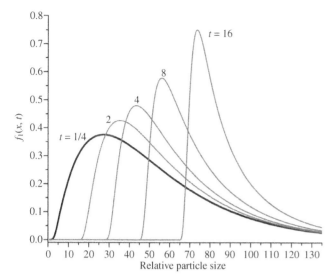

FIGURE 9.12. Change in the number density function (growth only) using a growth rate expression with less large-size attenuation.

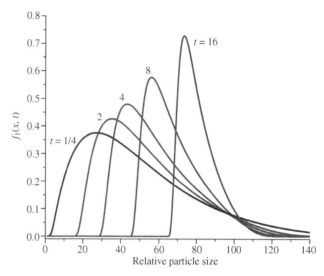

FIGURE 9.13. Growth process with breakage imposed at a threshold size (corresponding to a relative particle size index of 100). Distributions are shown for $t = 1/4$, 2, 4, 8, 16, and 24. Note how the distributions at $t = 16$ and 24 are being constrained at the upper end.

death term to account for the net generation of particles in the size range of interest:

$$\frac{\partial f_1(x,t)}{\partial t} + \frac{\partial}{\partial x}\left[\dot{X}(x,t)f_1(x,t)\right] = h(x,t). \quad (9.80)$$

Let us spend a moment thinking about particle breakage. If the breakage rate is dependent on hydrodynamic conditions, then there may exist a size threshold below which no breakage occurs. For example, in turbulent flows, we can use the dissipation rate per unit mass, ε, to assess the strain rate at different eddy scales (sizes). We usually argue that a disruptive eddy must have a scale comparable to the particle or entity size (much larger eddies simply entrain the particle, and very small eddies do not affect it at all). Naturally, larger eddies carry more energy, but they occur at a fixed location less frequently. One way to establish a breakage condition is to propose a balance between disruptive hydrodynamic energy and the restoring force (this could be surface tension, e.g., in the case of a suspended droplet). It may now be apparent to you how difficult the characterization of $h(x, t)$ really is: We will need a breakage criterion that takes into account the energetics, we will need a rate at which these energetic events occur, and we will need to know the number and sizes of the fragments produced by the disintegration. Since we are talking about process elements that are almost always both nonlinear and stochastic, these characterizations will not be easy. A major challenge in the construction of population balance models is to produce a result that works *and* is actually grounded in sound physics.

Let us explicitly describe some of the difficulties that arise in the characterization of $h(x, t)$. If, at a given size, x,

entities disappear by breakage, then $h(x, t)$ is a "sink" term consisting of a breakage rate, $b(x, t)$, multiplied by the density function: $b(x, t)f_1(x, t)$. We will replace the right-hand side of eq. (9.80) with this product and impose a size threshold where the breakage rate becomes large, thereby limiting the movement of the density function to the right (toward larger sizes). Some computed results are provided in Figure 9.13.

Normally, when breakage is occurring, larger entities are increasingly likely to break and some of those events will generate new entities at size x. This phenomenon represents a "source" of particles at size x. Now things are a good deal more complicated. When one of these larger entities breaks, some number of daughter particles will be formed, and we will represent the average number of fragments with $\nu(x', t)$, where $x' > x$, of course. Clearly, ν must be at least 2 (this is what we call binary breakage), but under sufficiently energetic circumstances, it could be much larger. Obviously, this also means that we must know where in x-space these fragments will be distributed. We will let $P(x|x')$ be the probability that a larger entity at size x' creates a fragment at size x. The disintegration of *any* larger entity might produce a fragment (or daughter particle) at size x, so we must take *all larger particles* into account:

$$\frac{\partial f_1(x,t)}{\partial t} + \frac{\partial}{\partial x}\left[\dot{X}(x,t)f_1(x,t)\right]$$

$$= \int\limits_x^\infty \nu(x')b(x')P(x|x')f_1(x',t)dx - b(x)f_1(x,t). \quad (9.81)$$

There are some clear restrictions placed on the conditional probability, $P(x|x')$, including $P(x|x') \geq 0$ and $P(x|x') = 0$ if $x > x'$. In models of this type where both growth and breakage are taken into account, an important concern is that eq. (9.81) is consistent with an appropriate conservation principle. For example, if the total mass in the system is constant, then the first moment of the density function cannot vary; that is,

$$\mu_1 = \int\limits_0^\infty x f_1(x,t) dx \qquad (9.82)$$

must not be a function of time. Let us use a very simple illustration to highlight some of the problems encountered with conservation. Suppose we are using mass to characterize entity size and that we have two particles, each of mass M_1, that combine to form a single new particle. Then, $M_1 + M_1 = M_2$ and, of course, $M_2 = 2M_1$.

Now suppose we have a liquid–liquid system where the coalescence of two identical droplets occurs; we also assume we are vitally interested in droplet size and volume: $V_1 + V_1 = V_2$. Of course, $R_2^3 = [R_1^3 + R_1^3]$, which implies $d_2 = 1.26d_1$.

Next, assume two small aggregates formed by flocculation in an aqueous system collide and affiliate. These flocs are typically very irregular, and each structure contains a significant amount of interstitial water; moreover, when they combine, even more fluid will be trapped in the porous spaces between the contacting aggregates, and neither mass nor volume will be conserved. Let us assume that the additional fluid volume amounts to 75% of the volume of the colliding flocs (this percentage is actually on the small side):

$$V_2 = (1.75V_1 + 1.75V_1) = 3.5V_1. \qquad (9.83)$$

Under this scenario, we find $d_2 \cong 1.52d_1$. Naturally, when breakage of such an entity occurs, some of that trapped, interstitial fluid is released, so both volume and mass will be lost. It is effectively impossible to enforce a conservation principle in a system of this type, so there is little benefit to working with mass or volume. Therefore, in the following discussion, we will turn our attention to a straight number balance.

Particle Size in Coagulating Systems

Let us begin by providing a general verbal description of the balance we wish to formulate, and we will consider systems of the types illustrated by Figure 9.9 and Figure 9.10. We will develop a number balance and let (a, b) be the particular size range of interest.

The rate of change of population in the interval (a, b) will be determined by

- production in (a, b) by the growth of smaller entities
- depletion in (a, b) by growth of entities in the interval (beyond b)
- production of fragments in (a, b) by breakage of larger entities
- depletion in (a, b) by breakage of entities from the interval.

Remember that these terms are specific to flocculating systems of the type we cited earlier. In more general situations, particle growth could occur by surface reaction, coating, deposition, and so on. We observed previously that birth in crystallization processes might appear to be the result of nucleation; however, we must remember that when saturation is exceeded, the nuclei that are formed initially are extraordinarily small. Thus, entity birth at a *finite size* in crystallization may actually be the result of breakage of larger particles.

Continuous crystallizers, flocculators, and so on, may operate at steady state such that the number density of entities is virtually constant. A batch process, in contrast, may yield a significant change in number density. For example, in the batch coagulation of colloidal clay particles, the number density may begin at 10^7 or 10^8 per cm^3 (for primary particles), and then decline by several orders of magnitude during the flocculation process. This is exactly what Figure 9.10 reveals.

If particle growth occurs solely by aggregation, then particle–particle contact will be necessary and a mechanistic description will require a model for the collision rate between entities. For fluid-borne particles, collision can result from Brownian motion, laminar or turbulent shear, particle inertia, and differential sedimentation. The functional forms for these collision rates are

$$\beta_{i,j} = \frac{2kT}{3\mu}\left[\frac{1}{V_i^{1/3}} + \frac{1}{V_j^{1/3}}\right]\left(V_i^{1/3} + V_j^{1/3}\right) \quad \text{Brownian motion}$$
$$\qquad (9.84)$$

$$\beta_{i,j} = \frac{4}{3}(R_i + R_j)^3 \frac{dU}{dy} \quad \text{Laminar shear} \qquad (9.85)$$

$$\beta_{i,j} = 1.3\left(\frac{\varepsilon}{\nu}\right)^{1/2}(R_i + R_j)^3 \quad \text{Isotropic turbulence} \qquad (9.86)$$

$$\beta_{i,j} = 5.7(R_i^3 + R_j^3)|\tau_i - \tau_j|\frac{\varepsilon^{3/4}}{\nu^{1/4}} \quad \text{Turbulent inertia} \quad (9.87)$$

$$\beta_{i,j} = \pi\alpha(R_i + R_j)^2(V_{Si} - V_{Sj}) \quad \text{Differential sedimentation.} \qquad (9.88)$$

In these rate expressions, R is entity radius and V is volume. We observe that the last two collision rates (eq. 9.87

and eq. 9.88) require *disparity in particle size* to be important. For turbulent inertia, there must be a difference in characteristic times for the participating particles, where $\tau = $ (mass of particle)$/(6\pi\mu R)$. For differential sedimentation, there must be a difference in settling velocities, $V_{Si} - V_{Sj}$. In a batch process where we start with all primary particles (monodisperse conditions), these two collision mechanisms will not be important at small times.

Note that in eq. (9.85), dU/dy is the velocity gradient (strain rate) associated with a highly ordered laminar flow; in eq. (9.86), ε is the dissipation rate per unit mass and ν is the kinematic viscosity of the fluid phase. Each of these collision rates has the dimensions L^3/t. Thus, when the expressions are multiplied by the number densities of the participating particles, n_i and n_j, we obtain the number of collisions per unit volume per unit time. We will look at an example for particle collisions in isotropic turbulence: Suppose we have two classes of spherical particles, $d_i = 5$ µm and $d_j = 6$ µm, and each class has a number density of 10^5 particles per cubic centimeter. We let the dissipation rate per unit mass be 400 cm^2/s^3 and the fluid medium be water ($\nu = 0.01$ cm^2/s). We want to calculate the collision rate produced by isotropic turbulence:

$$\beta_{i,j}n_in_j = (1.3)\left(\frac{400}{0.01}\right)^{1/2}(2.5\times10^{-4} + 3.0\times10^{-4})(10^5)(10^5)$$

$$= 24.66 \text{ collisions/(cm}^3\text{s)}. \qquad (9.89)$$

Now we will compare with the rate produced by laminar shear with $dU/dy = 400$ 1/s:

$$\beta_{i,j}n_in_j = \left(\frac{4}{3}\right)(2.5\times10^{-4} + 3.0\times10^{-4})^3(400)(10^5)(10^5)$$

$$= 887.33 \text{ collisions/(cm}^3\text{s)}. \qquad (9.90)$$

One of the characteristics of the highly ordered shear field created by the Couette device depicted in Figure 9.9 is that there is virtually no breakage of aggregates. In such cases, we only need to model particle growth. Therefore, an appropriate balance can be written:

$$\frac{dn(V)}{dt} = \frac{1}{2}\int_0^V \beta(V-u, u)n(V-u)n(u)du$$

$$- n(V)\int_0^\infty \beta(V, u)n(u)du, \qquad (9.91)$$

where $u < V$. The first term on the right-hand side accounts for production by collision of particles smaller than V; the second term on the right accounts for a loss of entities of size V due to additional growth by contact with other particles.

Solutions for eq. (9.91) are often sought through discretization; for example, Farley and Morel (1986) used the following set of equations:

$$\frac{dn_i}{dt} = \frac{1}{2}\sum_{i+j=k}\alpha(i, j)\beta(i, j)n_in_j - n_k\sum_{i=1}^m\alpha(i, k)\beta(i, k)n_i,$$

$$(9.92)$$

where $\alpha(i, j) = 1$ if $i \neq$ j, but 2 if $i = j$. Discretization of a population balance in which growth and breakage are occurring raises an obvious question: How many partitions should be used and how wide should they be? A popular choice is geometric spacing (multiples of 2); if aggregation is occurring and mass is conserved, then the affiliation of two particles of class 1 will result in an entity with mass doubled: $M_1 + M_1 = 2M_1 = M_2$. Ramkrishna (2000) points out that a coarse discretization makes it impossible for the model to be internally consistent. In other words, a collision event will not necessarily produce an entity in the next larger class or bin. One method that has been used to accommodate this problem is to employ weighting fractions so that only a portion of *i-j* affiliations produce an entity in the next larger bin. Therefore, we should probably regard a set of discrete balances (eq. 9.92) as semi-quantitative; the problem thus posed is more easily solved as it involves merely a set of simultaneous ODEs, but compromises have been made. We will illustrate this by simulating an aerosol system.

Let us assume we have an aerosol for which particle growth occurs solely by Brownian motion, but particles are lost from the control volume due to sedimentation. We will use eight particle classes with bins centered about the diameters 0.375, 0.75, 1.5, 3, 6, 12, 24, and 48 µm. We will solve eight simultaneous ODEs of the form given by eq. (9.92) but with the addition of a loss term based on the settling velocity (which is size-dependent). The collision rate will be described by eq. (9.84). For case 1, we will initially populate the first four (smallest) bins with 1×10^7 particles per unit volume each. For case 2, we will heavily populate the smallest (0.375) bin, but the others will have *much smaller* number densities. This will cause a profound change in the dynamic behavior of the numbers of small particles. In Figure 9.14a,b, the number density in the smallest (0.375) bin is shown with a heavy black curve.

Application of the Population Balance to a Continuous Crystallizer

In this section, we look at a continuous crystallizer operated at steady state. Liquor containing the solute is fed to a well-mixed vessel and the suspension containing crystals is withdrawn from the apparatus at the same rate. We write the balance for crystals of size s:

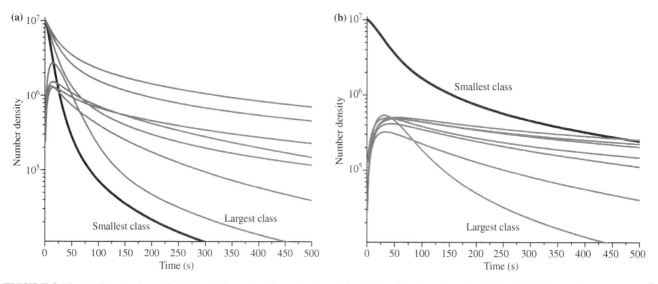

FIGURE 9.14. (a) Case 1: dynamic behavior from the discretized model with the first four (smallest) bins initially populated at 1×10^7 particles/cm^3. Because of the increased opportunity for small particle collisions, the 0.375 class decreases rapidly. (b) Case 2: dynamic behavior from the discretized model with only the 0.375 bin heavily populated. The opportunity for small particle collisions has been severely reduced.

$$\frac{d}{ds}(Ln) = -\frac{n}{\theta} + B - D, \qquad (9.93)$$

where B and D are birth and death functions, respectively. n is the number density of crystals, θ is the average holding time, and L is the linear crystal growth rate. Death (disappearance of crystals of a given size) is the result of breakage, and D is assumed to have the functional form

$$D(s) = kn(s)s^m, \qquad (9.94)$$

where k and m are empirical constants. The "birth" of crystals of a given size is assumed to occur by breakage of larger entities, and it is *convenient to assume* that the breakage process is binary (i.e., the breakage of a parent particle results in the production of two daughter particles). The birth function is written as

$$B(s) = \int_0^1 P(\varepsilon)\left[D\left(\frac{s}{\varepsilon}\right) + D\left(\frac{s}{1-\varepsilon}\right)\right]d\varepsilon. \qquad (9.95)$$

$P(\varepsilon)$ is a probability density and $P(\varepsilon)d\varepsilon$ is the likelihood that the breakage of a crystal of size s will produce a crystal in the size range, εs to $(\varepsilon + d\varepsilon)s$. Obviously, ε is a fractional quantity ranging from zero to 1. Among the idealizations used in developing this model are the assumption that the breakage process is strictly binary (a bit of a stretch) and that the linear growth rate (L) is a constant. The latter assumption is referred to as McCabe's law, and it is based on the idea that crystal growth is more likely to be dependent

on supersaturation than on crystal size. Randolph and Larson (1988) note that this is often observed to be true. The birth and death functions can be inserted into eq. (9.93), and once a functional form for $P(\varepsilon)$ is chosen, a solution may be sought. The simplest case, $P(\varepsilon) = 1$, corresponds to uniform binary breakage, and it is convenient to recast the model in dimensionless form by setting

$$x = \frac{s}{\theta L}, \quad y = \frac{n(s)}{n(0)}, \quad \text{and} \quad k_1 = k\theta(L\theta)^m. \qquad (9.96)$$

The result is the IDE:

$$\frac{dy}{dx} + (1 + k_1 x^m)y = k_1 \int_0^1 \left[\left(\frac{x}{\varepsilon}\right)^m y\left(\frac{x}{\varepsilon}\right) + \left(\frac{x}{1-\varepsilon}\right)^m y\left(\frac{x}{1-\varepsilon}\right)\right]d\varepsilon. \qquad (9.97)$$

Of course, x is the dimensionless crystal size and y is the dimensionless crystal population density; since y has been normalized with $n(0)$, $y(x = 0) = 1$. This formidable problem has been solved by Singh and Ramkrishna (1977) using the *method of weighted residuals* (MWR). They constructed their trial functions with Laguerre polynomials. We can gain some limited appreciation for the behavior of $y(x)$ by assuming that the birth rate is zero. Then, with the right-hand side of eq. (9.97) gone,

$$\frac{dy}{dx} = -(1 + k_1 x^m)y, \qquad (9.98)$$

and a solution is easily found:

$$y = C_1 \exp\left[-\left(x + \frac{k_1}{m+1}x^{m+1}\right)\right].$$

Singh and Ramkrishna used this result (with $C_1 = 1$) as the weight function for their solution by MWR, and they obtained the numerical results shown in the following table. The values selected for the parameters were $m = 4$ and $k_1 = 0.006173$.

x	Y
0.0	1.0000
0.2	0.8191
0.5	0.6085
1.0	0.3743
1.5	0.2332
2.0	0.1461
2.5	0.8972×10^{-1}
3.0	0.5141×10^{-1}
3.5	0.2545×10^{-1}
4.0	0.9701×10^{-2}
4.5	0.2529×10^{-2}
5.0	0.3700×10^{-3}
5.4	0.4565×10^{-4}
6.0	0.5746×10^{-6}

Ramkrishna (2000) reviews the use of MWR as a solution technique for population balance equations in chapter 4 of his book.

CONCLUSION

IDEs appear in many important problems in the applied sciences, ranging from biology to separation processes. For linear IDEs, the Laplace transform can be extremely useful and it can produce exact solutions. However, many of the IDEs of interest to us are nonlinear, and rarely can these equations be solved analytically. Our purpose in this chapter was to provide some alternative solution techniques so that students confronted by an IDE have options. In this vein, the VIM developed by He and coworkers is especially useful and *highly* recommended. The reader interested in applying VIM may find the recent article by Prajapati et al. (2012) very useful. They solved Abel's integral equation using four different initial guesses for the trial function, $y_0(x)$, and they were able to show that all four produced convergent series solutions with small absolute errors. Their paper contains sufficient detail to be a valuable aid to the analyst who is new to VIM.

However, by no means is VIM the only viable alternative; for example, for nonlinear Fredholm IDEs of the form

$$\frac{dy}{dt} = f(t) + \int_0^1 k(t,s)F(y(t))ds, \qquad (9.99)$$

with $0 \le t \le 1$ and $y(t = 0) = y_0$, a few of the solution methods that have appeared in the literature are homotopy perturbation, rationalized Haar functions, hybrid functions and collocation, and the Tau method with Chebyshev and Legendre bases. For an example of application of the latter, see Pour-Mahmoud et al. (2005).

There are other prominent IDEs that have been intensively studied; two examples where several alternative solution procedures have been devised include the neutron transport equation and the equation for radiative heat transfer. In the case of the latter, for a medium that absorbs, emits, and scatters (anisotropically),

$$\frac{dI}{ds} + \beta I = \kappa_a I_b + \frac{\kappa_s}{4\pi}\int_{4\pi} I(s,\Omega')\phi(\Omega',\Omega)d\Omega'. \quad (9.100)$$

I is the intensity of the radiation and Ω is the radiation (vector) direction, and for eq. (9.100), $0 \le s \le L$ and $I = I_0$ for $s = 0$. Zhao and Liu (2007) investigated this IDE and reported that many numerical methods exhibit unphysical oscillations in this particular application. They transformed eq. (9.95) to obtain a second-order differential equation that was of the diffusion type (with predictable behavior when solved numerically). You will observe that this is an approach that we have illustrated several times in this chapter; where feasible, this practice can often make the solution of such problems a good deal easier.

PROBLEMS

9.1. We want to solve the IDE,

$$\frac{du}{dt} - 3u + 6\int_0^x u(\phi)d\phi = t,$$

with $u(t = 0) = 0$. Find solutions by both the approximate discretization technique and the method of Laplace transform. Verify that your results correspond to the behavior shown in Figure 9.15 for ($0 \le t \le 4$).

9.2. Find the solution for the IDE,

$$\frac{dy}{dt} + 5\int_0^t \cos[2(t-u)]y(u)du = 10,$$

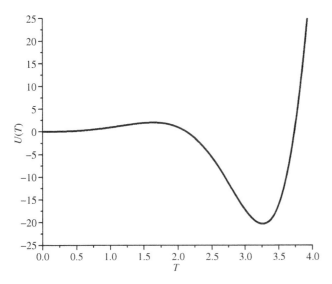

FIGURE 9.15. Behavior of $u(t)$ for the IDE in Problem 9.1.

x	Approximate $y(x)$
0.05	0.95333
0.10	0.91019
0.15	0.87116
0.20	0.83585
0.25	0.80390
0.30	0.77499
0.40	0.72517
0.50	0.68438
0.60	0.65099
0.70	0.62366
0.75	0.61192
0.80	0.60129
0.85	0.59167
0.90	0.58298
0.95	0.57511
1.00	0.56799

Determine how accurate these results are.

given that $y(t = 0) = 2$. Compare your result with

$$y(t) = \frac{1}{27}(24 + 120t + 30\cos(3t) + 50\sin(3t)).$$

9.3. Find the solution for the IDE,

$$\frac{dy}{dx} = 1 + 2x - y + \int_0^x x(1 + 2x)\exp[s(x - s)]y(s)ds,$$

given that $y(x = 0) = 1$. Verify that $y(1.75) = 21.3809$.

9.4. Using two methods, find the solution of the IDE,

$$\frac{dy}{dt} = 1 - \int_0^t y(s)ds,$$

given that $y(0) = 0$ and $0 \le t \le 1$. We are particularly interested in obtaining an accurate value for $y(0.5555)$.

9.5. The IDE,

$$\frac{dy}{dx} + y = \int_0^x \exp(s - x)y(s)ds,$$

has been solved for $0 \le x \le 1$ with $y(0) = 1$.
Some of the results obtained are given in the following table:

9.6. One of the examples worked in the text as part of the discussion of VIM was the ODE,

$$\frac{du}{dx} = -\frac{x^3}{(u + 1)^2}, \quad \text{with } u(x = 0) = 1.$$

The analytic solution is known,

$$3x^4 + 4(u + 1)^3 = 32, \quad \text{or alternatively, } u = \left[\frac{32 - 3x^4}{4}\right]^{1/3} - 1.$$

With VIM, the succession of estimates is obtained from

$$u_{n+1} = u_n - \int_0^x \left(u_n' + \frac{s^3}{(u_n + 1)^2}\right)ds.$$

Since $u_0 = 1$, we can very easily find: $u_1 = 1 - (x^4/16)$. Therefore,

$$u_2 = 1 - \frac{x^4}{16} - \int_0^x \left[-\frac{s^3}{4} + \frac{s^3}{\left(2 - \frac{1}{16}s^4\right)^2}\right]ds$$

and

$$u_2 = 3 + \frac{64}{x^4 - 32}.$$

Find u_3 for this problem and compare its performance with the table of values that accompanied the example.

9.7. Let us repeat the approximate analysis of two populations in conflict by solving the simultaneous second-order ODEs:

$$\frac{d^2X}{dt^2} = \frac{1}{X}\left[\left(\frac{dX}{dt}\right)^2 - bX^2\frac{dY}{dt} - K_1X^2Y\right]$$

and

$$\frac{d^2Y}{dt^2} = \frac{1}{Y}\left[\left(\frac{dY}{dt}\right)^2 + \beta Y^2\frac{dX}{dt} + K_2XY^2\right].$$

Set $a = b = 2$ and $\alpha = \beta = 1$, and take $K_1 = -0.05$ and $K_2 = +0.05$. Use the same initial point that we employed previously, $(X, Y) = (1, 2)$. Remember, we can obtain the correct initial values for the first derivatives from the undifferentiated equations by setting $t = 0$. How does the phase-plane trajectory behave in this case? How will it differ if both Ks are positive, $+0.05$?

9.8. We saw that the FKPP equation was an unrealistic diffusion model for the influence of travel on the spread of infectious disease. An alternative is the model proposed by Medlock and Kot (2003) in which $M(x)$ is the rate at which infected individuals leave position x and travel to a new location in the spatial domain. We take M to be a positive constant. The model has the form

$$\frac{\partial I}{\partial t} = \beta I(C_1 - I) - MI + M\int K(x, y)I(y, t)dy.$$

$K(x, y)$ is the kernel; it is a density function for the fraction of infected individuals who leave position y destined for x. This density function has the characteristics

$$K(x, y) \geq 0 \quad \text{and} \quad \int K(x, y)dx = 1.$$

Medlock and Kot (2003) assumed that the kernel was of the convolution type,

$$K(x, y) = k(x - y) = k(u) = \frac{1}{2\alpha}\exp\left(-\frac{|u|}{\alpha}\right),$$

where u is the separation between the x and y positions. They set $M = 1$, $\beta = 1$, and $\alpha = 0.7698$; I, the fraction of infected individuals, was taken to be 1 in the center of the domain. The problem thus posed by this model is formidable. The repeated evaluation of the integral is computationally expensive, so Medlock and Kot used a Runge–Kutta scheme for the time derivative and the FFT for the integral. We will let our domain extend from $x = 0$ to some very large x, and we will place a block of infected individuals at the origin. Our goal is to explore the propagation of the "wave front" of infected people. Let the initial block for I occupy 5% of the

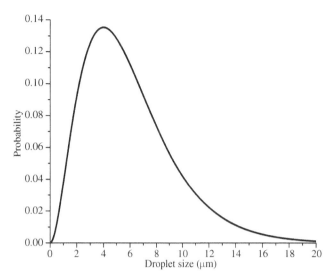

FIGURE 9.16. Droplet (number) density function for Problem 9.8.

domain. Use the Medlock–Kot parameters and compute the evolution of the "infected" wave. How long will it take the center portion of the wave to travel five times the width of the initial block of infectives?

9.9. In an emulsification process, data show that the droplet (number) density function is approximately described by

$$f(d) = \frac{1}{16}d^2\exp(-d/2).$$

Therefore, when $d = 4$ μm, $f(d) = 0.1353$; for $d = 11$ μm, $f(d) = 0.0309$. The probability density is shown as a function of droplet size in Figure 9.16.

Measurements show that the total number density is 1×10^7 droplets per cubic centimeter. Therefore, for droplets with diameters between 8 and 12 μm, the cumulative probability is about 17%; that is, there are about 1.7×10^6 droplets *in this size range* per cubic centimeter. For many applications, the probability distribution for *volume* is more useful than $f(d)$ given in Figure 9.16. Convert the data shown previously to $f(V)$. Where (at what volume) is the maximum in the distribution now located?

9.10. Solve the dynamic balance for a pure growth process,

$$\frac{\partial f_1(x, t)}{\partial t} + \frac{\partial}{\partial x}\left[X(x, t)f_1(x, t)\right] = 0,$$

assuming that the initial density function is $f_1 = x\exp(-x)$, but that the growth rate is given by

$$\dot{X} = \frac{C_1}{C_2 + x^3}.$$

Let $C_1 = 0.75$ and $C_2 = 2.0$. Prepare a figure that illustrates the evolution of the number density function.

9.11. Solve the discretized model, eq. (9.87), for an aerosol with eight particle classes using the example in the text as your guide. Center the size bins on 0.375, 0.75, 1.5, 3, 6, 12, 24, and 48 μm. Neglect sedimentation, but compare the results obtained with *three different* collision mechanisms: Brownian motion (as illustrated in Figure 9.14a), laminar shear with $dU/dy = 300$ s^{-1}, and isotropic turbulence with $\varepsilon = 1000$ cm^2/s^3. Which mechanism results in the most rapid reduction in number density of small particles? Let the initial concentrations be 1×10^7 particles per cubic centimeter in each of the three smallest classes (let n_i for the five larger classes be zero initially).

9.12. Yüzbasi and Sezer (2012) studied a linear Volterra IDE with a weakly singular kernel:

$$y''(x) - x^2 y'(x) = 6x - \frac{1}{2}x^4 - \frac{16}{5}x^{5/2}$$
$$+ \int_0^x \frac{y'(t)}{\sqrt{x-t}}\,dt - \frac{5}{4}\int_0^x xty''(t)\,dt,$$

where $y(0) = 1$, $y'(0) = 0$, with $0 \leq x$ and $t \leq 1$. They note that the exact solution for this problem is $y(x) = x^3 + 1$. Solve this Volterra IDE using the method of your choice and compare your solution with the exact result.

9.13. Consider the IDE

$$\frac{dy}{dt} = \int_0^t y(u)\cos(t-u)\,du,$$

with $y(0) = 1$. Your colleague has developed his own method for solving such equations, and he reports that $y(t = 1.5) = 2.135$ and that $y(t = 3) = 4.499$. Use the method of your choice to confirm or refute the results of your colleague.

9.14. In 2013, an online discussion occurred in which solution strategies were sought for the *partial* IDE,

$$\frac{\partial u}{\partial t} = \frac{\partial^2 u}{\partial x^2} - u^3 + u + B\left(u_0 - \frac{1}{L}\int_0^L u\,dx\right),$$

given that $L = 4$, $B = 10$, and $u_0 = \frac{1}{2}$. For this problem, $0 \leq x \leq 4$ and the initial distribution of u consists of a block centered at $x = 2$ with a width of $\frac{1}{2}$ and an amplitude of 1; u is 0 elsewhere. Try the following approach: Discretize the equation (with respect to x) to get a system of n-ODEs of the form

$$\frac{du}{dt} \cong \frac{u_{n+1} - 2u_n + u_{n-1}}{(\Delta x)^2} - u_n^3 + u + B\left(u_0 - \frac{1}{L}\sum u_n \Delta x\right);$$

divide L into 40 pieces, and then solve the set of ODEs to find the time evolution of the nonlinear system. It has been reported that this problem exhibits diffusive behavior. Is that borne out by your results?

REFERENCES

Altintan, D. and O. Ugur. *Variational Iteration Method for Sturm–Liouville Differential Equations*, Institute of Applied Mathematics, Middle East Technical University, Ankara, Turkey (2012).

Antipov, Y. A. and H. Gao. Exact Solution of Integro Differential Equations of Diffusion along a Grain Boundary. *The Quarterly Journal of Mechanics and Applied Mathematics*, 53:645 (2000).

Bailey, N. T. J. *The Elements of Stochastic Processes*, John Wiley & Sons, New York (1964).

Batiha, B., Noorani, M. S. M., and I. Hashim. Numerical Solutions of Nonlinear Integro-Differential Equations. *International Journal of Open Problems in Computer Science*, 1:34 (2008).

Cont, R. and E. Voltchkova. Integro-Differential Equations for Option Prices in Exponential Levy Models. *Finance and Stochastics*, 9:299 (2005).

Davis, H. T. *Introduction to Nonlinear Differential and Integral Equations*, Dover Publications, New York (1962).

Farley, K. J. and F. M. M. Morel. Role of Coagulation in the Kinetics of Sedimentation. *Environmental Science and Technology*, 20:187 (1986).

Fedotov, S. Front Propagation into an Unstable State of Reaction-Transport Systems. *Physical Review Letters*, 86:926 (2001).

Ganji, D. D., Nourollahi, M., and M. Rostamian. A comparison of Variational Iteration Method with Adomian's Decomposition Method in Some Highly Nonlinear Equations. *International Journal of Science and Technology*, 2:179 (2007).

He, J. H. Variational Iteration Method—A Kind of Nonlinear Analytical Technique: Some Examples. *International Journal of Nonlinear Mechanics*, 34:699 (1999).

He, J. H. Variational Iteration method for Autonomous Ordinary Differential Systems. *Applied Mathematics and Computation*, 114:115 (2000).

He, J. H. Variational Iteration Method—Some Recent Results and New Interpretations. *Journal of Computational and Applied Mathematics*, 207:3 (2007).

He, J. H. and X.-H. Wu. Variational Iteration Method: New Development and Applications. *Computers and Mathematics with Applications*, 54:881 (2007).

Huebner, K. H. *The Finite Element Method for Engineers*, John Wiley & Sons, New York (1975).

Jackiewicz, Z., Rahman, M., and B. D. Welfert. Numerical Solution of a Fredholm Integro-Differential Equation Modeling θ-Neural Networks. *Applied Mathematics and Computation*, 195:523 (2008).

Jassim, A. M. A Modified Variational Iteration Method for Schrodinger and Laplace Problems. *International Journal of Contemporary Mathematical Sciences*, 7:615 (2012).

Matinfar, M., Hosseinzadeh, H., and M. Ghanbari. A Numerical Implementation of the Variational Iteration Method for the Lienhard Equation. *World Journal of Modeling and Simulation*, 4:205 (2008).

Medlock, J. and M. Kot. Spreading Disease: Integro-Differential Equations Old and New. *Mathematical Biosciences*, 184:201 (2003).

Merdan, M. He's Variational Iteration Method Applied to the Solution of Modeling the Pollution of a System of Lakes. *DPU Fen Bilimleri Dergisi*, 18:59 (2009).

Pour-Mahmoud, J., Rahimi-Ardabili, M. Y., and S. Shahmorad. Numerical Solution of the Volterra Integro-Differential Equations by the Tau Method with Chebyshev and Legendre Bases. *Applied Mathematics and Computation*, 170:314 (2005).

Prajapati, R. N., Mohan, R., and P. Kumar. Numerical Solution of Generalized Abel's Integral Equation by Variational Iteration Method. *American Journal of Computational Mathematics*, 2:312 (2012).

Ramkrishna, D. *Population Balances: Theory and Applications to Particulate Systems in Engineering*, Academic Press, Inc., San Diego (2000).

Randolph, A. D. and M. A. Larson. *Theory of Particulate Processes*, 2nd edition, Academic Press, Inc., San Diego (1988).

Singh, P. N. and D. Ramkrishna. Solution of Population Balance Equations by MWR. *Computers and Chemical Engineering*, 1:23 (1977).

Weinstock, R. *Calculus of Variations with Applications to Physics and Engineering*, Dover Publications, New York (1974).

Yüzbasi, S. and M. Sezer. A Numerical Method to Solve a Class of Linear Integro-Differential Equations with Weakly Singular Kernel. *Mathematical Methods in the Applied Sciences*, 35:621 (2012).

Zhao, J. M. and L. H. Liu. Second-Order Radiative Transfer Equation and its Properties of Numerical Solution using the Finite-Element Method. *Numerical Heat Transfer, Part B*, 51:391 (2007).

10

TIME-SERIES DATA AND THE FOURIER TRANSFORM

INTRODUCTION

At the very beginning of this book, we pointed out that some problems are not deterministic. Many real processes include nonlinear, stochastic components that defy conventional modeling efforts. And yet, it is essential that scientists and engineers be able to interpret and analyze such phenomena so that some level of confidence with respect to outcomes can be achieved. The purpose of this chapter is to provide a few tools that can be used in such scenarios.

In the applied sciences, it is common to record experimental observations as a function of time, and we refer to such a sequence of data as a time series. We might collect these data to support a model or a hypothesis, or we might obtain them with the hope that a suitable model could be identified later. It is possible that such data might reveal particular functional behavior that could be periodic. Alternatively, the data might contain multiple periodicities or, in the worst case, exhibit fluctuations without a discernible period; that is, the phenomenon under study might be chaotic. Let us examine some meteorological data from the Bradford Station (United Kingdom) in Figure 10.1.

In the form shown in this illustration, these data are not very revealing. However, if we change to a line plot and expand the horizontal (time) axis, a very different picture emerges.

Of course, we see the expected behavior in Figure 10.2; over a time period of 100 months, there are about 8+ cycles in the data set; the annual variation is now completely obvious. The peak (and trough) temperatures occur at 12-month intervals.

Annual (local) climate variations are simple enough that no tools are necessary to see and comprehend periodic behavior of the kind shown in Figure 10.2. We do not have to look far, however, to find greater challenges. Let us consider the Dow Jones Industrial Average (DJIA) and focus on the 22-year span from 1960 to 1982. The DJIA entered the 1960s over 600 and during 1982 was as low as about 770. If those numbers are indicative of the stock prices of interest to us (assuming we were long in our positions), then we were certain to be disappointed since the annual increase in value was only about 1.3%! If we look more closely at the data, however, we find quite a different picture. For example, in late 1974, the DJIA was as low as 577, but in 1976 and 1977, it reached 1000, a 73% increase. We will look at the monthly averages over this two-decade plus span of time to get a visual understanding of the dynamic behavior.

Fluctuations in the DJIA are apparent in Figure 10.3, but there does not seem to be any obvious regular, periodic behavior. This is important because, if we were able to identify a periodicity (or periodicities) in these data, we could anticipate the peaks and valleys and become very, very rich. What we need is a tool, or a method, by which we could identify periodicities in time-series data when they exist. *Harmonic* (or Fourier) analysis provides us with a means to do this—we can assess fluctuations in a time series by comparison with sinusoids. Even more useful is *spectrum analysis*, which allows us to identify the tendency for oscillations of a given frequency to appear.

Applied Mathematics for Science and Engineering, First Edition. Larry A. Glasgow.
© 2014 John Wiley & Sons, Inc. Published 2014 by John Wiley & Sons, Inc.

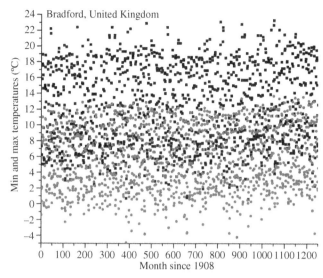

FIGURE 10.1. A scatter plot of the minimum and maximum (average) temperatures recorded at the Bradford Station (United Kingdom) for every month since 1908. The maximum temperatures are the filled black squares and the minimum temperatures are the filled circles. These data were obtained from www.metoffice.gov.uk.

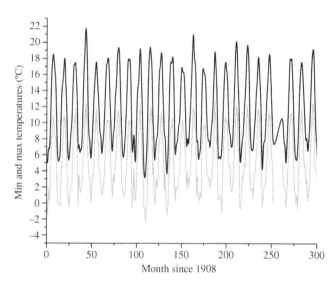

FIGURE 10.2. Line plot of temperature data from the Bradford Station with an expanded time axis. The black curve represents the average maximum temperatures, and the lighter curve the average minimum temperatures.

A NINETEENTH-CENTURY IDEA

The detection of solar and lunar periodicities by man dates back many thousands of years. But during the eighteenth and nineteenth centuries, scientists and mathematicians began to look at many other phenomena with the idea that

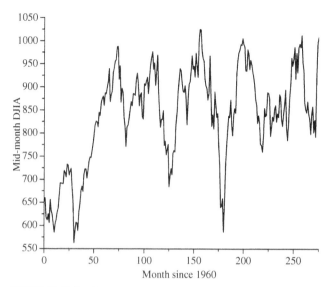

FIGURE 10.3. The Dow Jones Industrial Average (mid-month) from 1960 through 1982. These data were obtained from www.davemanuel.com. These are closing values for the middle of each month.

some of them—such as earthquakes—might occur periodically. Knott (1897), for example, concluded that there was *probably* a connection between lunar tides and earthquakes by means of a Fourier series. Schuster (1897), however, argued that one should consider the *relative magnitudes* of the Fourier coefficients compared with those obtained if the events in question were perfectly random.

Schuster proposed the following assessment: Suppose one has a time series, $y(t)$, and computes the integrals,

$$A = \int_t^{t+T} y \cos \kappa t\, dt \quad \text{and} \quad B = \int_t^{t+T} y \sin \kappa t\, dt, \quad (10.1)$$

where T is a given time interval. We now define R as

$$R = \sqrt{A^2 + B^2}. \quad (10.2)$$

If $y \approx \cos \kappa t$, then R will increase as the time interval, T, increases. However, if $y \approx \cos \lambda t$, where $\lambda \neq \kappa$, then R will fluctuate about some constant value as T increases. We will explore this proposal using a function, $y(t)$, that consists of a limited number of sinusoids (Figure 10.4).

Our plan is to fix the time interval, T, and then to compute R for five different frequencies with 34π in the center of the range tested. We can then increase T and repeat the calculation as many times as we wish.

We find through the curves shown in Figure 10.5 that the test recommended by Schuster is capable of identifying periodicities in a given data set. As a practical tool, however, it is severely limited; it would be computationally expensive

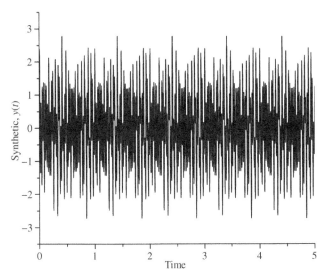

FIGURE 10.4. A synthetic signal constructed from four sinusoids with radian frequencies: 30π, 34π, 58π, and 84π. The fact that there are a limited number of characteristic frequencies is obvious.

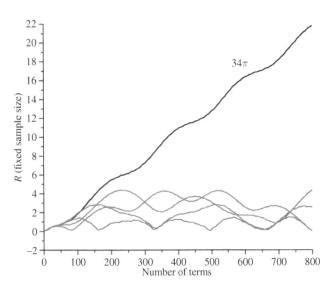

FIGURE 10.5. Application of Schuster's test to the data shown in Figure 10.4. The calculations were made for a single value of T using five radian frequencies: 29π, 31.5π, 34π, 36.5π, and 39π. One of the characteristic frequencies of the data set shown in Figure 10.4 has been clearly identified; the curve for 34π is growing without limit.

to test a time series for every possible frequency and in the Victorian age—when the test was devised—it would have been utterly impossible.

In the middle of the twentieth century, the importance of *spectrum estimation* in electrical engineering applications (such as radio, radar, and electronic communications)

drove an intensified interest in the Fourier transform. This culminated in the Cooley–Tukey algorithm and variants, which are broadly known as the *fast Fourier transform* (FFT). James Cooley (1987) provided a wonderful first-hand account of the development of the FFT and he pointed out that Gauss had grasped the essential ideas in the nineteenth century. Of course, the pace at which the use of the discrete Fourier transform (DFT) has expanded was driven by the development and widespread use of digital computers in the twentieth century. Before we discuss the FFT, however, we need to explore some other aspects of fluctuating signals.

THE AUTOCORRELATION COEFFICIENT

Suppose we observe a signal, $y(t)$, that fluctuates about some constant mean value (such a process is said to be statistically stationary). For the synthetic data shown in Figure 10.4, it is clear that the mean is approximately zero. We let these observations be represented by

$$y(t) = \bar{y} + y', \tag{10.3}$$

where $\bar{y}$ is the mean and y' is the fluctuation. We will define an *autocorrelation coefficient* in the following way:

$$\rho(\tau) = \frac{\overline{y'(t)y'(t+\tau)}}{\overline{y'^2}}. \tag{10.4}$$

The quantity in the denominator is referred to as the *mean-square* fluctuation; naturally, the square root of this quantity is the *rms* fluctuation. If the time offset, τ, is 0, then the autocorrelation coefficient is equal to 1: $\rho(\tau = 0) = 1$. How $\rho(\tau)$ behaves depends for larger τ, of course, on the nature of $y(t)$. If y is periodic (or consists of a set of periodic functions), then $\rho(\tau)$ will show strong correlation at distinct values of τ; in such a situation, the oscillatory correlation is said to be *ringing*. If, on the other hand, the signal under observation is random or chaotic, then the fluctuations should become uncorrelated as the offset (or time delay) becomes large and $\rho(\tau \rightarrow \infty) = 0$.

Historically, the correlation coefficient has been used extensively in fluid mechanics and especially in the study of turbulence. The development of the hot-wire anemometer in the twentieth century produced large sets of time-series data that were observed in real time with oscilloscopes and processed by analog instruments such as spectrum analyzers. It is important, though, that we recognize that the autocorrelation has a significant limitation: It is *not* reversible; that is, since one *cannot retrieve* the original data from the autocorrelation, the process of obtaining $\rho(\tau)$ automatically entails a loss of information. Consequently, the autocorrelation is

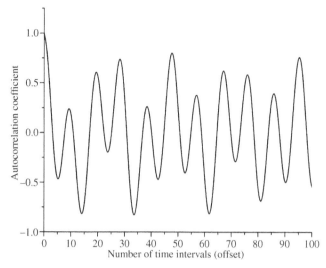

FIGURE 10.6. Autocorrelation coefficient for synthetic signal composed of four sinusoids. The expected *ringing* behavior is apparent.

of little value for tasks associated with signal processing and image manipulation.

Nevertheless, we will begin exploration of the data set shown in Figure 10.4 by computing the autocorrelation coefficient. Since these data were constructed from a limited set of sinusoids, we should expect an *oscillatory* $\rho(\tau)$.

The data we used to compute the autocorrelation coefficient shown in Figure 10.6 were constructed from four periodic functions added together and previously illustrated in Figure 10.4. The computation of $\rho(\tau)$ reveals that there were *regular fluctuations* in the time-series data; note the strong recurring peaks in the correlation at (number of time intervals) 19, 28, 38, 48, 57, 67, and so on. This suggests that something with a period of about 9 to $10\Delta t$ was present in the time series, corresponding to a frequency range of about 250–279 rad/s. Of course, we know that there was a very strong sinusoidal component present at 263.9 (84π) rad/s—right in the middle of the indicated band of frequencies!

In simple cases of this type (where a regular oscillation in the time series is obvious), one might consider developing a model by fitting a periodic function(s) to the data directly. For example, suppose through our examination of the data we concluded that a function of the type

$$y(t) = A\sin(\omega t) + B\cos(\omega t) \tag{10.5}$$

might adequately represent the behavior. Assuming we were able to estimate the radian frequency directly, we could then minimize the sum of the squares of the deviations by setting

$$F = \sum (y - A\sin(\omega t) - B\cos(\omega t))^2 \tag{10.6}$$

and setting the derivatives equal to zero:

$$\frac{\partial F}{\partial A} = 0 \quad \text{and} \quad \frac{\partial F}{\partial B} = 0. \tag{10.7}$$

Therefore, we could use our data to seek solutions for the following equations:

$$-2\sum (y - A\sin(\omega t) - B\cos(\omega t))\sin(\omega t) = 0 \tag{10.8}$$

and

$$-2\sum (y - A\sin(\omega t) - B\cos(\omega t))\cos(\omega t) = 0. \tag{10.9}$$

This least-squares estimation technique is fine, as long as we have a sharply defined periodic behavior where the radian frequency (or a very small number of radian frequencies) is obvious. In such cases, we are merely trying to identify the amplitude and the phase. Of course, we could add the radian frequency to the determination if necessary. For many problems of interest to us, however, the observed behavior will be altogether too complicated to let direct model development be a realistic alternative.

A FOURIER TRANSFORM PAIR

We saw previously that the autocorrelation coefficient, $\rho(\tau)$, though not reversible, could reveal much information about the time-series data under study. In fact, the autocorrelation and the *power spectral density* (or spectrum), $S(\omega)$, contain the same information and they are related through the Fourier transform pair:

$$\rho(\tau) = \int\limits_{-\infty}^{+\infty} \exp(i\omega\tau)S(\omega)d\omega \quad \text{and}$$

$$S(\omega) = \frac{1}{2\pi}\int\limits_{-\infty}^{+\infty} \exp(-i\omega\tau)\rho(\tau)d\tau. \tag{10.10}$$

Since the autocorrelation coefficient is an even function, and since negative frequencies do not hold any physical meaning for us, it is standard practice to write the *one-sided* spectrum:

$$S_1(\omega) = \frac{1}{\pi}\int\limits_{0}^{\infty} \cos(\omega\tau)\rho(\tau)d\tau. \tag{10.11}$$

We have already computed $\rho(\tau)$ for the data shown in Figure 10.4, so why not take this result and use it to evaluate the spectrum using eq. (10.11)? We will do this, but we must keep in mind two important limitations:

1. We will not be able to learn anything about fluctuations occurring more rapidly than the *Nyquist cutoff frequency*. For a sample interval of Δt,

$$f_{Nyq} = \frac{1}{2\Delta t}. \qquad (10.12)$$

For the data shown in Figure 10.4, $\Delta t = 0.0025$ second, so $f_{Nyq} = 200$ Hz or 1256 rad/s.

2. We cannot find out anything about fluctuations that occur so slowly that one cycle does not fit within our total time of observation. For infrequent (very long-period) oscillations, we must acquire a *very lengthy* set of time-series data.

The spectrum shown in Figure 10.7 was computed by repeated integration of the estimated autocorrelation coefficient, after $\rho(\tau)$ had been obtained from the experimental data by calculation of the mean, the fluctuation for each observation, and the mean-square fluctuation. Although this example demonstrates that such a procedure will work, it is computationally expensive and for many applications, simply too time-consuming. We need a process that will allow us to obtain the spectrum more rapidly, and the approach we will employ is the FFT.

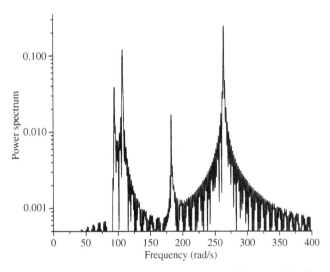

FIGURE 10.7. Computed spectrum for data of Figure 10.4. The important radian frequencies in the data set were approximately 94.2, 106.8, 182.2, and 263.9 rad/s. All four appear as dominant features in the spectrum shown here. The oscillation at 263.9 rad/s had the largest amplitude of the four sinusoids. Make particular note of the broadening apparent at the sides of the spectral peaks; we will discuss that further a little later in this chapter.

THE FAST FOURIER TRANSFORM

Brigham (1988) describes the FFT as *ubiquitous*, and indeed, he provides a list of about 75 applications (many of which appear to be quite unrelated). To highlight just a few, consider that the FFT has been used for studies of aircraft wing flutter, acoustic imaging, diagnosis of airway obstruction, echo or reverberation elimination, speech recognition, chromatography, phased-array antenna analysis, video bandwidth compression, and image restoration. Quite recently, the FFT has been used in the study of microseismic events triggered by hydraulic fracturing in gas-bearing shales; see Warpinski et al. (2012) for example spectra.

We mentioned previously that the rapid growth of applications for the FFT was driven by the digital computer and by the paper published by Cooley and Tukey (1965). Their algorithm is widely used around the world and is referred to colloquially by a variety of names including the *cool-turkey FFT*. Bingham et al. (1967) helped to introduce the technique to a broader audience, and since the 1960s, many versions of the algorithm have been developed for specific applications. Several contributors to the literature have pointed out that Gauss actually originated the idea in the early years of the nineteenth century. However, the reality is that James Cooley, Richard Garwin, and IBM (their employer) were mainly responsible (assisted by the concurrent development of fast analog-to-digital converters) for the explosive growth in the number of applications of the FFT. Cooley (1987) and Rockmore (1999) have also noted that some of the initial urgency was driven by the Cold War due to the need for seismic monitoring (the detection of nuclear explosions) and acoustic detection and identification of submarines.

If we have N observations of a fluctuating signal, $y(t)$, then we can think of the Fourier transform as an interpretation of that signal in terms of sinusoids of different frequencies. Specifically, as Bloomfield (1976) points out, we can write

$$A_0 = \frac{1}{N}\sum_{n=0}^{N-1} y_n = \overline{y}, \qquad (10.13)$$

$$A_j = \frac{2}{N}\sum_{n=0}^{N-1} y_n \cos \omega_j t, \qquad (10.14)$$

and

$$B_j = \frac{2}{N}\sum_{n=0}^{N-1} y_n \sin \omega_j t. \qquad (10.15)$$

If the sequence of observations consists of an *even* number, then

$$A_{N/2} = \frac{1}{N}\sum_{n=0}^{N-1} y_n \cos\omega_{N/2}t = \frac{1}{N}\sum_{n=0}^{N-1}(-1)^n y_n. \quad (10.16)$$

The time-series data can now be represented, if N is even, with

$$y_n = A_0 + \sum_{j<N/2} A_j \cos(\omega_j t) + B_j \sin(\omega_j t) + (-1)^n A_{N/2}. \quad (10.17)$$

A compact, shorthand representation of the DFT for sampled data (in the form of $X(j)$, a complex vector of length N) can be written as

$$S(k) = \sum_{j=0}^{N-1} X(j)W_N^{jk}, \text{ where } W_N = \exp\left(\frac{2\pi i}{N}\right). \quad (10.18)$$

The main problem here is that the number of arithmetic operations for a time series with N observations scales as N^2. Thus, if one had 512,000 data points (the reason for choosing this number will be made clear in a moment), about 2.6214×10^{11} operations would be required. Modern PC processors are capable of about 5×10^9 to 1×10^{10} flops (floating point operations per second); thus, very roughly, $(2.6214 \times 10^{11})/(7 \times 10^9) \approx 37.5$ seconds. While not a ridiculous amount of time, this is far too long for processes requiring near-instantaneous feedback.

However, John Tukey had shown that if the N observations can be written as a product, $N = ab$, then the Fourier series could be written as an a-term series of b-term (each) subseries. This meant that the scaling for the required number of computations would change from N^2 to $(a + b)N$. Richard Garwin relayed this information to James Cooley and persuaded him to work on algorithm development and programming. The result, now known as the Cooley–Tukey algorithm for the FFT, actually reduces the scope of the computations to $N \log(N)$; if N is 512,000, then the computational effort is reduced to 2.92×10^6 operations. The comparison is striking; for a modern PC, we would now have about $(2.92 \times 10^6)/(7 \times 10^9) \approx 0.0004$ second for the required processing time. As Cooley (1987) later described, the development of the FFT algorithm meant that the Fourier transform could now be extended to very large problems. An additional motivation for his work was the interferometer data brought to Cooley by Janine Connes; her husband, the astronomer Pierre Connes, had constructed an interferometer that could produce a time series (a data record) consisting of (the then incredibly large number) 512,000 points. The power of the Cooley–Tukey algorithm was such that it made the processing of lengthy time series of this type of routine.

For the analyst who must perform FFTs on time-series data, the options are many. There are literally hundreds of FFT codes available in the public domain, written in high-level languages including FORTRAN, BASIC, PASCAL,

and C. For convenient working examples, see chapter 8 in Brigham (1988). There are also numerous websites devoted to the Fourier transform. Moreover, many commercial software packages have FFT capabilities including Mathcad™, MATLAB™, MS-Excel™, and Origin™. Thus, just about any collection of time-series data can be analyzed and the signal content in frequency space can be determined.

We should look at an illustration to underscore this point. Suppose we have a signal consisting of four sinusoids added together with a random fluctuation (z):

$$x(t) = \sin\frac{\pi t}{2} + \sin\frac{\pi t}{10} + \sin\frac{\pi t}{5} + \sin\frac{2\pi t}{3} + z. \quad (10.19)$$

Take note of the radian frequencies employed here: $\pi/10$, $\pi/5$, $\pi/2$, and $2\pi/3$. We will allow t to assume integer values from 0 to 1023 such that we have 2^{10} samples; then we will determine the Fourier transform, first by using Mathcad. We will use the fast DFT function, $fft(v)$. This is a radix-2 transform program; by radix-2, we mean that the number of data points, N, can be written as $N = 2^m$ data points (where m is an integer).

Implementation of Mathcad fft(v):

$$i := 0 .. 1023$$

$$x_i := \sin\left(3.1416\frac{i}{2}\right) + \text{rnd}(1) + \sin\left(i \cdot \frac{3.1416}{10}\right)$$
$$+ \sin\left(i \cdot \frac{3.1416}{5}\right) + \sin\left(i \cdot \frac{3.1416}{1.5}\right)$$

$$c := fft(x)$$

$$N := \text{last}(c)$$

$$N := 511$$

$$j := 0 .. N$$

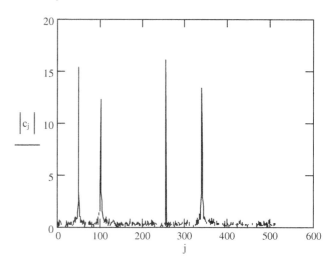

Note where the peaks are located (the j index values) in this spectrum: approximately 50, 101, 255, and 340. The

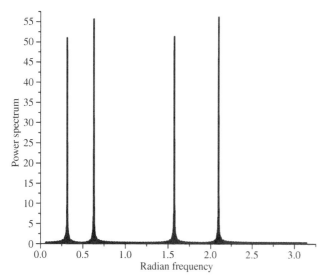

FIGURE 10.8. Power spectrum for the signal produced by four sinusoids (eq. 10.19). In this case, the autocorrelation coefficient, $\rho(\tau)$, was computed directly from the data and the Fourier transform was found by integration. Note that $\pi/10 = 0.314$, $\pi/5 = 0.628$, $\pi/2 = 1.571$, and $2\pi/3 = 2.094$.

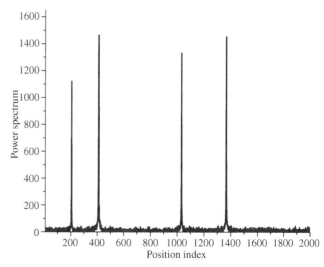

FIGURE 10.9. Spectrum from a modified version of Brigham's BASIC DFT (radix-2) program. Frequency is determined from the position index by $j\pi/v$. Therefore, the second spectral spike in Figure 10.9 corresponds to about $405\pi/2048$ or roughly 0.2π rad/t. Recall that the signal that was transformed consisted of four sinusoids with radian frequencies of 0.1π, 0.2π, 0.5π, and 0.667π.

corresponding radian frequencies are determined from $j\pi/512$.

Now we repeat the analysis of the time series, computing the Fourier transform by integration of the autocorrelation coefficient, using eq. (10.11) as illustrated previously. You will note that the two spectra shown (one above and one below) are essentially identical, differing mainly due to the normalization employed. Although the results are the same, the processing speed was radically different; the computational result shown in Figure 10.8 required *several seconds*, while the execution of the *fft* function in Mathcad appeared to be almost instantaneous. This is the reason the FFT has become so important—it is now possible to compute Fourier transforms for large time-series data sets and to do so almost in real time. Of course, this speed is crucial to signal processing applications.

Finally, we will also test a modified version of the BASIC DFT code provided by Brigham (1988) on the very same data sequence and present the result in Figure 10.9.

In the preceding example, the spectra were determined in three ways: using the built-in capability, *fft(v)*, in Mathcad, by computing the autocorrelation coefficient, $\rho(\tau)$, directly from the time-series data and then integrating the results, and by using a radix-2 DFT code that was modified from a version of the program supplied by Brigham (1988). The resulting spectra are very nearly identical and all three procedures have clearly identified the four dominant frequencies present in the time-series data.

The example that we just explored did not present much challenge for the FFT/DFT procedure. Let us look at a more

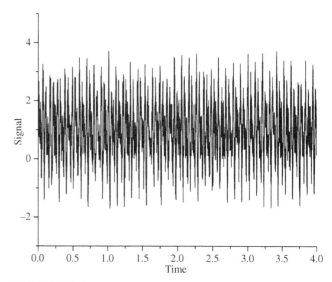

FIGURE 10.10. Synthetic signal consisting of 11 sinusoids with a random contribution added.

complicated signal (one with 11 distinct frequencies) that also includes a random contribution. To further complicate things, we will truncate the time series at 2800 and append 1296 zeros to reach 2^{12} samples when we compute the spectrum. The raw data appear as shown in Figure 10.10.

Now we will compute the Fourier transform to see if all of the different elements of this signal have been captured. Keep in mind that we have added a random component *and* filled out the data set (to reach 4096, or 2^{12}, observations)

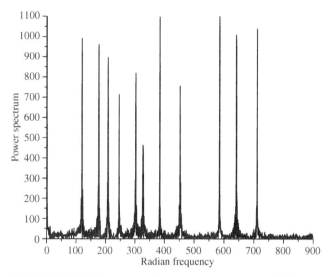

FIGURE 10.11. The computed spectrum has captured all 11 sinusoidal contributions; the original frequencies were 119.38, 175.93, 207.3, 245, 301.59, 326.73, 383.28, 452.4, 584.34, 640.89, and 710 rad/s.

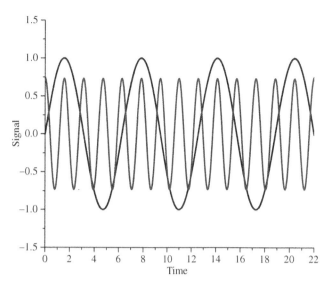

FIGURE 10.12. A plot illustrating $\sin(\omega t)$ and $0.75 \cos(4\omega t)$, with $\omega = 1$.

with zeros. We know that the kinds of abrupt changes in the data that are generated by adding a random contribution and by appending zeros to obtain 2^m total samples will produce additional contributions in the spectrum that may complicate identification of the crucial periodicities.

The computed transform shown in Figure 10.11 reveals how powerful the DFT really is; we constructed a synthetic signal that was an absolute mess, and the algorithm has successfully identified all 11 of the contributing sinusoids. And this is despite the facts that the data series was truncated (filled out with zeros) and that a strong random contribution was added to the signal.

ALIASING AND LEAKAGE

We described the Nyquist cutoff frequency previously, noting that if the sample interval, Δt, was *too large*, we would not be able to obtain any useful information about a high-frequency (oscillating) waveform. More generally, too large a Δt produces *aliasing*, which yields a distorted spectrum. Suppose we were interested in the *sum* of the two sinusoids shown in Figure 10.12.

If we had the poor judgment to begin sampling at $\pi/2$, and then further to employ an interval of 2π, the sampling of the *sum of the two waveforms* would consist only of the sequence 1.75, 1.75, 1.75, and so on, and absolutely none of the important dynamic behavior would be revealed. It is critical that we pick a sampling Δt that can capture any oscillating behavior that may be occurring. Naturally, if we compute a DFT for time-series data sampled too infre-

quently, we will not get the correct spectrum. Suppose, for example, that we sampled the *sum* of the two waveforms illustrated in Figure 10.12 using a time interval of $\pi/4$; we would see the sequence, 0.75, -0.0429, 1.75, -0.0429, 0.75, -1.457, -0.25, -1.457, 0.75, -0.0429, 1.75, -0.0429, 0.75, and so on. These equally spaced samples could be represented by *several* sinusoids; the consequence is that the spectrum folds, or overlaps. If *aliasing* is suspected, the easiest fix is to halve the sample interval, to compute the spectrum, and to note any changes, and then to repeat the process until the spectrum does not change.

Leakage appears in a DFT spectrum in the form of what are called "sidelobes." This feature of computed spectra results when a particular harmonic component leads to nonzero transform values at *other* frequencies. It can be exacerbated by the fairly common practice of filling out an incomplete data set with zeros to execute a radix-2 DFT; that is, if we had 1750 observations, we might add 298 zeros to get to 2048 (2^{11}). Any phenomenon that produces sharp (abrupt) changes in the time domain will result in additional spectral components in the frequency domain. Let us illustrate this point by generating time-series data with the simple function

$$y(t) = \sin(120t) + 0.85\cos(180t). \qquad (10.20)$$

We apply the DFT with the result shown in Figure 10.13a. Next, we truncate the time series and append 1320 zeros to provide a total sample of 4096 for Figure 10.13b. Finally, we add a strong contribution to the time series utilizing the random number generator and include the 1320 zeros at the end to generate the spectrum shown in Figure 10.13c.

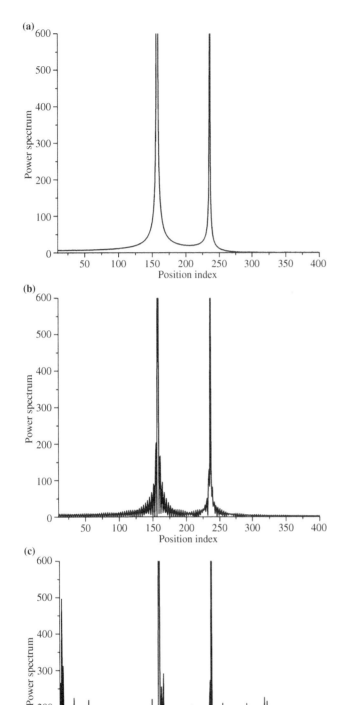

FIGURE 10.13. (a–c) Comparison of spectra obtained by DFT for time-series data generated by modifications to $\sin(120t) + 0.85 \cos(180t)$. The frequencies are obtained from $j(500\pi)/2048$, such that $j = 155$ corresponds to 120 and $j = 235$ corresponds to 180.

The first thing we should note about Figure 10.13a is that, although the time-series data were generated with the sum of sine and cosine terms (only), the resulting spectrum is not exactly what we would expect had we obtained a *continuous* transform. A pure sinusoid in time should yield a δ function in frequency space. Next, when we truncated the time series by taking 2776 sampled values with 1320 zeros appended, we produced sharp oscillations on the shoulders of the two main peaks (Figure 10.13b). And finally, in Figure 10.13c, we have used the random number generator to create discontinuities in the time domain which have, in turn, produced many additional frequency components in the spectrum. The important conclusion that we should take from this example is that discrete sampling in the time domain can yield discontinuities that, upon DFT, yield other contributions in frequency space. These contributions appear as additional peaks or sidelobes in the spectra that are computed. One remedy, which is pretty obvious but usually impractical, is to eliminate all of the abrupt changes that may appear in the time domain.

Another common source of difficulty is *time-domain truncation*, which occurs when the number of discrete sample data points is limited. In some cases, our ability to obtain time-series data may be constrained (perhaps the hardware has limited storage capacity). This can adversely affect the accuracy of the DFT, and we will illustrate this problem using a signal composed of two clean sinusoids:

$$y(t) = \cos\left(\frac{2\pi}{5.99}t\right) + \sin\left(\frac{2\pi}{2.88}t\right). \qquad (10.21)$$

We will sample this function using integer spacing ($\Delta t = 1$) and we will begin with a data record length of 1024. We will then compute the DFT for $N = 64$ and repeat using $N = 16$.

The spectrum shown in Figure 10.14a ($N = 1024$) is a close approximation of the (continuous) Fourier transform. In this case, the spectral peaks are located at j index values of 172 and 357, corresponding to frequencies of $(172/512)\pi = 0.336\pi$ and $(357/512)\pi = 0.697\pi$; these values are very close to the actual frequencies employed, 0.334π and 0.694π. But as we decrease the number of samples (time-domain truncation), we get an increasingly poor representation of the two δ functions we expected to see. Note how the spectra are broadened and the peak amplitudes diminished. In the case of Figure 10.14c ($N = 16$), we have connected the discrete points with spline fitting, but even so, our estimate of the spectrum is of little value. Although the best remedy for time-domain truncation is to increase N, there are alternatives for leakage control and we will discuss one of them next.

In the preceding examples, we saw that abrupt, or sharp, changes in the time domain, or time-domain truncation, might produce unwanted features in the computed spectra.

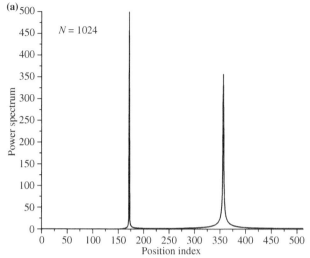

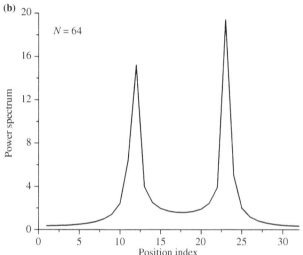

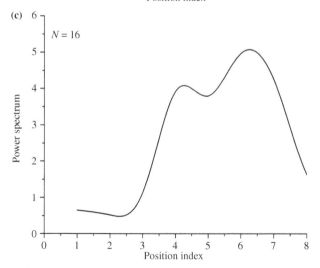

FIGURE 10.14. (a–c) Comparison of spectra computed by DFT for record lengths of 1024, 64, and 16 samples. The test signal was

$$y(t) = \cos\left(\frac{2\pi}{5.99}t\right) + \sin\left(\frac{2\pi}{2.88}t\right).$$

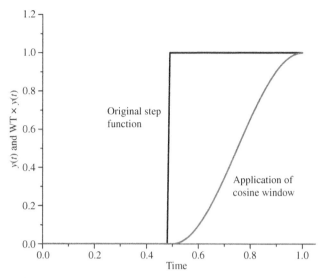

FIGURE 10.15. Application of the cosine window to a step change in the time-series data.

If we could use a data window (sometimes referred to as a *fader*) to smooth out the shoulders, we might be able to eliminate some undesirable features from computed spectra. One possibility is to use a *taper* on the time-series data and *then* to transform the tapered data. An example given by Bloomfield (1976) is the window

$$W_t = \frac{1}{2}\left[1 - \cos\left(\frac{2\pi(t + \frac{1}{2})}{n}\right)\right], \qquad (10.22)$$

where t corresponds to the discrete times in the sampled data (consisting of n points), $t = 0, \ldots, n - 1$. Please note that in this context, when we speak of n points, we actually mean $n\Delta t$. Suppose, for example, that our original time-series data form a step with an amplitude of 1. When we apply the *cosine window* to the data, we get a smooth transition that is "bell" shaped, as shown in Figure 10.15.

It is clear from inspection of Figure 10.15 that the windowed (or tapered) data are significantly different from the original step. In many cases, we would rather limit the influence of the taper so as to preserve *most* of the character of the original data. Tukey (1967) suggested a modification of the data window to

$$\frac{1}{2}\left[1 - \cos\left(\frac{\pi(t - \frac{1}{2})}{m}\right)\right], \quad t = 0, \ldots, m - 1 \quad (10.23)$$

$$W_t = 1, \quad t = m, \ldots, n - m - 1 \quad (10.24)$$

$$\frac{1}{2}\left[1 - \cos\left(\frac{\pi(n - t + \frac{1}{2})}{m}\right)\right], \quad t = n - m, \ldots, n - 1. \quad (10.25)$$

Remember that m and n are to be interpreted as $m\Delta t$ and $n\Delta t$. With this approach, the fraction of the data that will be tapered is $2m/n$; a smaller fraction of windowed data will naturally yield *less* improvement in leakage control. Tukey recommended that m be chosen such that $2m/n \approx 20\%$.

We want to look at a detailed example that will first demonstrate the problems created by leakage and then reveal how we might diminish them using tapering. Suppose, first, that we have time-series data in the form of a square wave (a rectangular window) with a maximum amplitude of 1. We will then apply a sinusoidal taper to remove both sharp edges, but leave the center intact. Specifically, we will shave off one-third of the rectangular window at both shoulders and leave the center third intact with its original value of 1. A comparison of the original rectangular window with the modified shape (where the sharp edges have been removed) is shown in Figure 10.16.

Our plan is to apply the DFT to both the original square wave and the tapered wave form so we can see the differences in the spectra; the results appear in Figure 10.17.

The leakage produced by the original square wave is significant; however, we immediately see that trimming one-third of the pulse off (at each shoulder) results in far less leakage while preserving the really important characteristics of the spectrum. It will be left as a student exercise to explore the effects of more aggressive tapers on the results.

SMOOTHING DATA BY FILTERING

We saw previously how time-series data could be windowed (or tapered) to minimize the impact of sharp irregularities. We can also use digital filtering to smooth very irregular (noisy) data. Let us return to eq. (10.3) and let

$$y(t) = \overline{y} + y'. \tag{10.26}$$

Since the mean $\overline{y}$ is nearly constant (or at least varies only very slowly), we could run a moving, three-point average through the data set, replacing $y(t)$ with z:

$$z(t) = \frac{1}{3}\left[y(t-1) + y(t) + y(t+1)\right]. \tag{10.27}$$

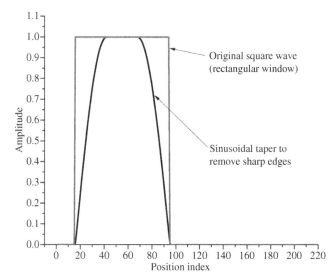

FIGURE 10.16. The initial square wave (rectangular window) is shown and the tapered waveform is in black. The tapered edges correspond to one-third (at each top edge) of the original rectangle. The center third is left alone with an amplitude of 1.

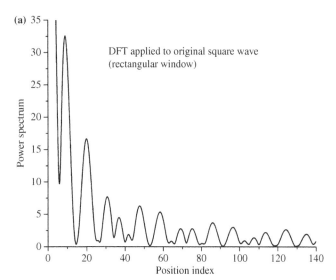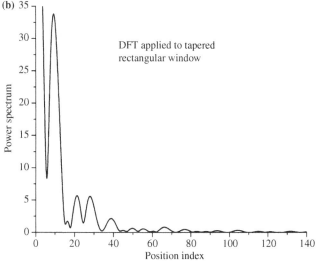

FIGURE 10.17. Spectra obtained from the original time-series data (a) and from the tapered data set in which the sharp-edged shoulders have been trimmed off (b).

Naturally, we cannot apply this to the endpoints of the data. Let us construct a synthetic data set to see how well this works; we will use a random number generator to produce noise, which we superimpose on a single sinusoid. We will also filter the data with the moving, three-point average to gauge its impact on the noisy data; the process is illustrated in Figure 10.18.

Our main interest, of course, is how the filtering affects the computed spectrum. Figure 10.19a shows the spectrum for the unfiltered (noisy) data and Figure 10.19b shows the effects of two passes of the linear filter (the moving, three-point average) on the computed spectrum.

The filter that was employed to produce the results seen in Figure 10.19b was the simple moving average. There are many other and probably better options. One that is quite effective at attenuating noise but preserving the essential character of the oscillations is the digital triangular filter. The idea is to place the greatest weight on the central point and to decrease weight on more distant data points; one example of a triangular filter is

$$z(t) = \frac{1}{9}[y(t-2) + 2y(t-1) + 3y(t) + 2y(t+1) + y(t+2)].$$

$$(10.28)$$

Note that the area of the triangle is $\frac{1}{2}bh = (\frac{1}{2})(6)(\frac{3}{9}) = 1$. Of course, other weighting schemes—and a commonly used one is exponential—can be employed as well. We can get a better sense of how effectively eq. (10.28) works by applying it to a test set of data (Figure 10.20).

In the preceding example, we looked at the effects of data filtering in a practical way by examining spectra computed

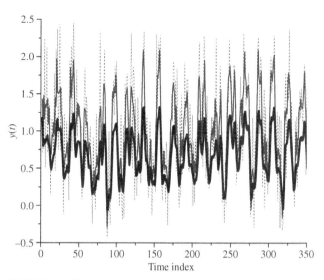

FIGURE 10.18. Original noisy data (dotted), filtered once (continuous), and filtered twice (heavy continuous black). The time index has been expanded to better reveal the behavior of $y(t)$.

from the original noisy data and then from the filtered data. There is another very convenient approach to assessing the effects of filter application. Suppose we have a simple signal in the form of a sinusoid, which we write as

$$y(t) = A\cos(\omega t + \phi),   (10.29)$$

which is the real part of $A\exp[i(\omega t + \phi)]$. For the linear filter (the three-point moving average), we have

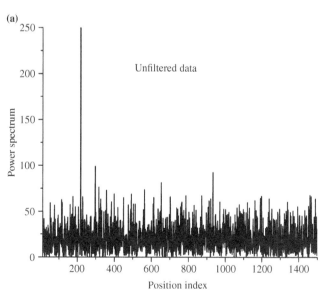

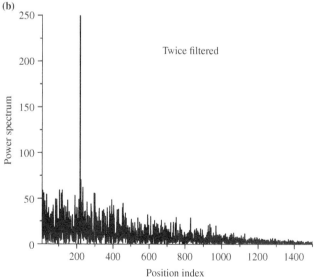

FIGURE 10.19. Comparison of the spectra for the original, noisy data (a) and then computed for the data set after two applications of the linear filter (b). Many of the additional frequency components that were introduced by the abrupt fluctuations have been attenuated or even eliminated.

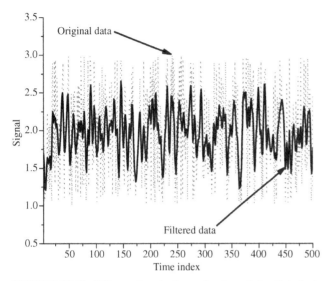

FIGURE 10.20. Effects of a *single* application of the digital triangular filter to a noisy data set. For severely noisy signals, the triangular filter (eq. 10.28) can be run through the data set repeatedly.

$z(t) = \frac{1}{3}[y(t-1) + y(t) + y(t+1)]$

$\quad = \frac{1}{3}A[\cos(\omega t - \omega + \phi) + \cos(\omega t + \phi) + \cos(\omega t + \omega + \phi)]$

$\quad = \frac{1}{3}A[\exp(i(\omega t - \omega + \phi)) + \exp(i(\omega t + \phi))$

$\quad\quad + \exp(i(\omega t + \omega + \phi))].$ (10.30)

This expression can be written more compactly as

$\quad = \frac{1}{3}A\exp[i(\omega t + \phi)][\exp(-i\omega) + 1 + \exp(i\omega)].$ (10.31)

Of course, *cos* is an even function, so we can write the real part of eq. (10.31) as

$\quad = \frac{1}{3}A\cos(\omega t + \phi)(1 + 2\cos\omega).$ (10.32)

When we compare this output with the original signal, eq. (10.29), it is clear that we are modifying it with the factor, $1/3(1 + 2\cos\omega)$. We see immediately that a signal component with $\omega = 2\pi/3 = 2.0944$ will be *completely removed* by the filter (since $\cos(2\pi/3) = -1/2$).

MODULATION (BEATS)

We have now seen on numerous occasions that the FFT/DFT is capable of revealing periodicities in time-series data. Let us begin this part of our discussion by considering a synthetic signal consisting of

$y(t) = z + \cos(82t) + \cos(89t) + \sin(196t) + \sin(207.3t).$ (10.33)

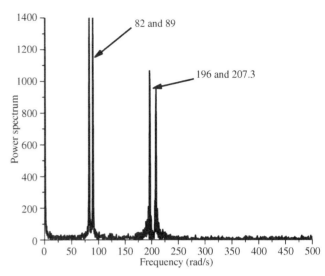

FIGURE 10.21. DFT computed for the simple sum of four periodic functions with frequencies of 82, 89, 196, and 207.3 rad/s.

Notice that the *two pairs* of radian frequencies are close together. If we compute the DFT for these data, we see four concentrated peaks in the spectrum, just as expected.

Now we will carry out the DFT computation again, but this time, we use the *product* of the two cosine contributions added to the *product* of the sine contributions. Notice in Figure 10.21 that the pair of cosine contributions is separated by 7 rad/s and the sine contributions are separated by 11.3 rad/s. This will prove to be very significant.

When the DFT is applied to the sum of the cosine–cosine product and the sine–sine product, new spectral components appear at about 7 and 12 rad/s. These are *beat* frequencies that arise from amplitude modulation. If we multiply $\cos(\omega_1 t)$ $\cos(\omega_2 t)$, where the two radian frequencies ω_1 and ω_2 are fairly close together, then we produce a modulated wave with fluctuating amplitude—notice the peak in Figure 10.22 located at 7 rad/s. You will also notice that the spectral content at 82–89 rad/s has disappeared entirely. Clearly, harmonic analysis (the DFT) has failed to reveal all of the periodicities of the original data. What is needed is what is called *complex demodulation*, and we will illustrate it with an elementary example. Let us assume we have a function described by

$y(t) = A(t)\cos(\omega t + \phi(t)).$ (10.34)

We will allow both the amplitude function, $A(t)$, and the phase, $\phi(t)$, to vary but at frequencies significantly lower than ω, in particular,

$A(t) = 3\sin\left(\frac{\omega}{5}t\right) \quad \text{and} \quad \phi(t) = 4\sin\left(\frac{\omega}{6}t\right).$ (10.35)

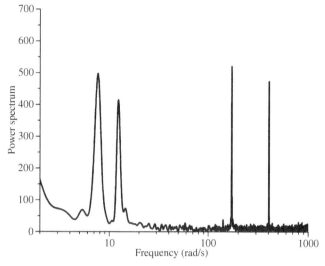

FIGURE 10.22. DFT computed for two *products* added together, the cosine pair and the sine pair. The frequency axis has been changed to a log scale to better reveal the two low-frequency components.

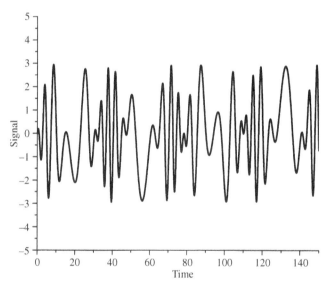

FIGURE 10.23. An oscillating function with frequency, ω, for which both the amplitude and the phase vary sinusoidally but at much lower frequencies.

The resulting signal is illustrated in Figure 10.23. If we perform a harmonic analysis on these data, what would we see?

We will take eq. (10.34) and rewrite it as

$$y(t) = \tfrac{1}{2}A(t)[\exp(i(\omega t + \phi)) + \exp(-i(\omega t + \phi))]. \quad (10.36)$$

Now suppose we form a new signal from $y(t)$ by writing

$$z(t) = y(t)\exp(-i\omega t);$$

the result is

$$z(t) = \tfrac{1}{2}A(t)[\exp(i\phi) + \exp(-i(2\omega t + \phi))]. \quad (10.37)$$

Note that the first term on the right-hand side varies slowly—remember that

$$\phi(t) = 4\sin\left(\frac{1}{6}\omega t\right).$$

The frequency for the second term in eq. (10.37) is 2ω, so it varies rapidly. If we apply a low-pass filter to the signal to remove all content with frequencies larger than ω, we are left with

$$z_f(t) = \left[\tfrac{1}{2}A(t)\exp(i\phi(t))\right]_f. \quad (10.38)$$

The subscript f indicates that the signal has been smoothed by application of the filter.

Now the amplitude is just the magnitude of z_f, and we merely divide eq. (10.38) by $A(t)$ to isolate $\phi(t)$.

SOME FAMILIAR EXAMPLES

Turbulent Flow in a Deflected Air Jet

Previously, we alluded to the fact that time-series data are generated frequently in engineering and scientific investigations. One area where this is particularly important is the study of turbulent flows, where a number of instruments have been developed that can monitor fluid velocities at a point in space. Examples include laser Doppler velocimetry (LDV) and hot-wire anemometry (HWA). These devices produce signals as functions of time, and in the case of HWA, the signal is usually voltage. The response time for a hot-wire anemometer is short, so high-frequency fluctuations can be observed without difficulty. Figure 10.24 shows the output from a hot-wire anemometer used to measure point velocity in a high-speed deflected air jet. The volumetric flow rate of air was 200 L/min, and the round jet was deflected off of a concave surface producing a *very thin* shear layer. This arrangement yields velocities mainly between 10 and 100 m/s. The measurement point for the data shown in Figure 10.24 was near the centerline (axis) of the deflected jet, so the velocities are relatively large.

The approximate range of velocities illustrated by the data in Figure 10.24 is about 10 to well over 100 m/s, and the data were recorded for a duration of about 1/8 second. Thus, the *minimum* frequency that can be observed would be about 50 rad/s or 8 Hz. A digital storage oscilloscope was used to record these data, and the device had a built-in capability for computing power spectra for time-series data. The

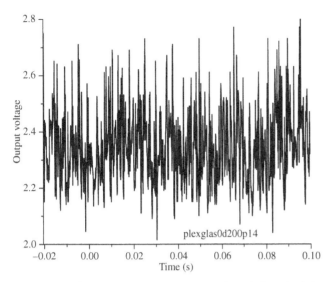

FIGURE 10.24. Data obtained from HWA in a deflected, high-speed air jet. The sampling interval was $\Delta t = 0.0001$ s, so the Nyquist frequency is 5000 Hz. The mean voltage is approximately 2.36 V, which corresponds to an air velocity of about 55 m/s.

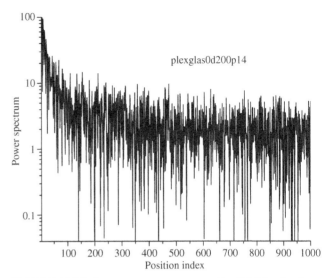

FIGURE 10.26. Power spectrum computed by DFT for the time-series data shown in Figure 10.24. There were 1200 data points in the record, so 848 zeros were appended to reach 2048 (or 2^{11} observations).

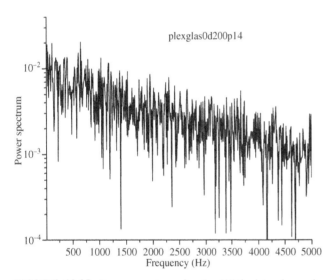

FIGURE 10.25. Power spectrum for the HWA data shown in Figure 10.24. The Fluke™ DSO computed values between 8 Hz and 5 kHz.

Bubbles and the Gas–Liquid Interface

When gas is introduced into a liquid, whether it be through a plain orifice, by jet aeration, or by liquid drops impacting the surface, a number of processes are set in motion that are capable of producing acoustic noise. Examples of these phenomena include detachment from the orifice, followed by oscillations about spherical shape, then bubble breakage and coalescence, and finally bubble disengagement at the free surface. Small bubbles are generally spherical, but other shapes are also common, depending on size, which affects buoyancy and drag forces, and so on. An image of air bubbles (Figure 10.27) produced by jet aeration in water illustrates the variety of shapes seen in two-phase (gas–liquid) processes.

Notice how the small bubbles shown in Figure 10.27 are spherical, while larger bubbles tend to be ellipsoidal. The spherical bubbles oscillate with a characteristic frequency determined by surface tension (τ), the density of the surrounding liquid (ρ_1), and bubble size (R):

$$s^2 = \frac{12\tau}{\rho_1 R^3}. \tag{10.39}$$

Thus, for a 0.5-mm air bubble in water,

$$s^2 = \frac{(12)(72)}{(1)(0.025)^3} = 5.53 \times 10^7 \text{ rad}^2/\text{s}^2, \text{ or } 1183 \text{ Hz}.$$

Of course, this frequency is in the acoustic range and these oscillations can produce detectable sound.

A number of experiments were carried out in which air was introduced through a sieve plate into column of a

spectrum provided by the recording device is shown immediately as follows.

The spectra shown in Figure 10.25 and Figure 10.26 are certainly not identical nor would we expect them to be. As we discussed previously, appending zeros to a time series to use a radix-2 DFT algorithm will result in a sharp discontinuity and the result will be additional spectral components in frequency space.

FIGURE 10.27. Characteristic shapes of air bubbles in water produced by jet aeration. The very large population of bubbles at the top of the image at the gas–liquid interface affords many opportunities for coalescence to occur prior to bubble rupture.

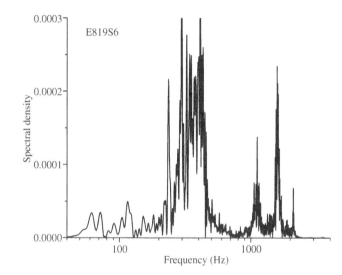

FIGURE 10.28. Spectrum for the recorded acoustic noise at the top of the column where bubble disengagement processes are the dominant sound producers.

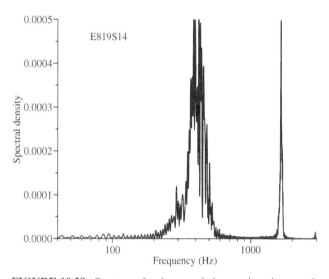

FIGURE 10.29. Spectrum for the recorded acoustic noise associated with bubble formation, detachment, and oscillation at the bottom of the column.

liquid medium consisting of water and glycerol. The objective was to learn more about energetic phenomena associated with bubbles in aqueous media of higher viscosities and the possible impact of those energetic motions on entrained entities (perhaps cells, or cells being grown on microcarriers). A microphone was used to monitor sounds produced both at the top of the column (above the interface) and at the bottom at the same level as the sieve plate sparger. Glycerol slightly decreases the surface tension when added to aqueous solutions, but the effect is small. At the top of the column, the sounds produced are mainly associated with disengagement processes (coalescence, bubble breakage, cavity collapse, and droplet ejection and impact). At the bottom of the column, the noises mainly come from bubble formation, detachment, and shape oscillations. Figure 10.28 shows how the signal energy was dis-

tributed with respect to frequency at modest gas (air) rates as detected at the top of the column. Figure 10.29 shows the distribution of signal energy at the bottom of the column adjacent to the sieve plate sparger.

The spectrum for disengagement processes at the top of the column (Figure 10.28) shows significant signal energy at 60, 70, 115, 220, 300–500, 1100, 1750, and 2020 Hz. In contrast, the spectrum shown for the bottom of the column (Figure 10.29) indicates a broad band of important frequencies between 300 and 600 Hz, with a very sharp, narrow contribution centered at 1605 Hz. Clearly, these detected acoustic signals do not merely represent spherical bubble oscillations; after all, we know that 400 Hz would correspond to the fundamental oscillation of a spherical bubble with $R = 0.0515$ cm. Most of the bubbles formed at the sieve plate in this aqueous glycerol solution were much larger than that. Furthermore, when bubbles arrive at the free surface, some may coalesce before film thinning produces an unstable interface followed by rupture. When large bubbles break, they typically throw small liquid "film droplets" laterally. When small bubbles break, the cavity created at the liquid surface due to the higher pressure in the bubble's interior will collapse; this collapse can eject larger "jet" droplets usually with a nearly vertical trajectory. And often, these breakage processes at the free surface subsequently produce much smaller bubbles capable of generating higher-frequency sounds (the next time you pour a soft drink onto ice, listen carefully at the top of the glass).

Shock and Vibration Events in Transportation

When cargo is moved by ship, rail, or truck, it is subjected to shock and vibration. If that cargo is delicate, such phenomena may lead to damage and loss; consequently, much effort has been expended in measuring strains and accelerations in transport environments. For example, in the case of railroad transport, low-frequency motions arise from irregular track surfaces; a low spot opposite a high spot will lead to rail car rocking. Shust (2007) found that rocking and vertical bounce from spring deflection occurred at frequencies mainly under 4 Hz. On the other hand, vibrations measured in a locomotive bearing box revealed important energy content at 440, 500–540, 607, and 978 Hz. A source of violently abrupt motions in railroad transport is car coupling. Magnuson and Wilson (1977) measured acceleration spectra for coupling events occurring at an impact velocity of 5.25 mph (about 99% of car coupling impacts occur at speeds less than 11 mph). One of the more interesting features of the coupling shock event is that accelerations of significant energy are produced at *higher* frequency. Figure 10.30 illustrates a spectrum redrawn from data presented by Magnuson and Wilson.

Shust (2007) notes that a common railroad car design specification (the ability to withstand 4 g accelerations) is frequently violated by the accelerations produced at higher frequencies generated by coupling events.

Truck transport also results in shock events that occur through travel over pavement irregularities, potholes, railroad crossings, and so on. Magnuson and Wilson (1977) reported spectra from over-the-road tests performed with seven different tractor–trailer combinations. They included measurements for all three axes (vertical, transverse, and longitudinal) in their study and, as one might expect, vertical accelerations were the largest by a significant margin. Spectra that have been transcribed and redrawn from their report are shown in Figure 10.31.

Large vertical accelerations are evident in Figure 10.31, but unlike shocks experienced by railcars during coupling events, these data do not show the same degree of transfer of energy to higher frequencies. In fact, the peak vertical

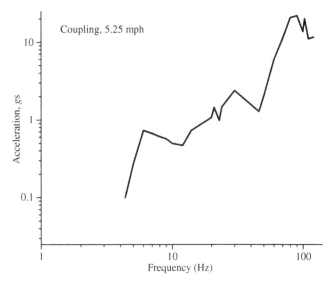

FIGURE 10.30. Spectrum of measured accelerations resulting from car coupling at 5.25 mph. Note the large measured accelerations occurring at about 100 Hz.

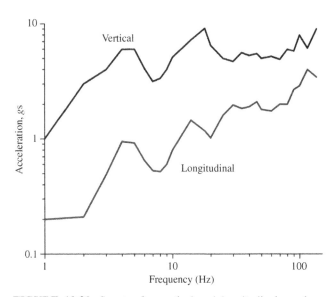

FIGURE 10.31. Spectra for vertical and longitudinal accelerations measured in trucks traveling over irregular surfaces (redrawn because of the poor quality of the original figure).

accelerations occur at about 18 Hz where the maximum is nearly 10 gs.

Time-series data collected in rail transport scenarios reveal that large strains (and perhaps very large accelerations) are especially likely to be generated by shock events such as coupling or, even worse, derailment. In some cases, the accelerations may exceed design specifications for the vehicle or car. This can pose a serious problem when hazardous materials are being transported; a ruptured tank car, for example, might disrupt revenue service and even lead to evacuations (and liability) in populated areas. Recent tragic accidents in Casselton, North Dakota, where 400,000 gal of crude oil were spilled, and the devastating explosion and fire in Lac-Megantic, Quebec, resulting from tank car brake failure have highlighted the need for reexamination of tank car design. The FFT/DFT is absolutely indispensible in investigations of this type, where improved car design criteria can be developed to improve transportation safety.

CONCLUSION AND SOME FINAL THOUGHTS

Harmonic analysis has been a standard tool in engineering practice for decades, but it may be difficult for the student new to this area to appreciate the significance. For final emphasis, let us consider an engineering study in aviation that gives us one more example of what the role of the Fourier transform is in the context of public safety. In 1968, Slusher authored FAA Report NA-68-27, "Reciprocating Engine and Exhaust Vibration and Temperature Levels in General Aviation Aircraft." Safety, of course, is a critical concern in general aviation and one problem area—known to have caused fatal accidents—is the failure of exhaust system integrity. The environment for exhaust systems on reciprocating aircraft engines is incredibly hostile due to the combination of thermal stresses, engine vibration, and chemical attack by combustion products. A mechanical failure of the exhaust system could result in engine fires or carbon monoxide infiltration into the passenger cabin (both have occurred). Slusher found, and not surprisingly, that vibration levels were highest during takeoff when the engines

were operating at maximum power. Furthermore, since the study focused on horizontally opposed reciprocating engines, the largest amplitude vibrations were lateral (the direction of both piston travel and valve motion). The data revealed that the largest lateral accelerations (at high speed, 2331 rpm) were about ± 20 gs, and many of the spectra showed important signal contributions at 1500–2000 Hz. Slusher noted that the time-series data (engine vibration waveforms) were periodic but consisted of a complex and continuous frequency spectrum with some very energetic content in certain frequency ranges. The study demonstrated that the materials that were being used for exhaust systems on general aviation aircraft in the 1960s (often types 321 and 347 stainless steels) were not ideal for the application. Though prosaic, this is just one more example of how the Fourier transform has been used in engineering practice to enhance public safety.

In recent years, the Fourier transform (and specifically the FFT) has become enormously important to modern life through signal processing, communications, analytic chemistry, seismic event detection and interpretation, collision warning systems for automobiles, process control, and so on. The list of applications of the FFT/DFT is incredibly long, and many of the technological marvels we take for granted depend on very fast determination of the Fourier transform of time-series data. Let us think a little bit about the future with a final example we had previously mentioned in passing: speech recognition. Two sets of time-series data are compared in Figure 10.32a,b, obtained from the recordings of a human voice saying "yes" and "no."

Figure 10.32a shows that the "y" in "yes" is nearly a distinct frequency, but the "e" in the middle and the "s" at

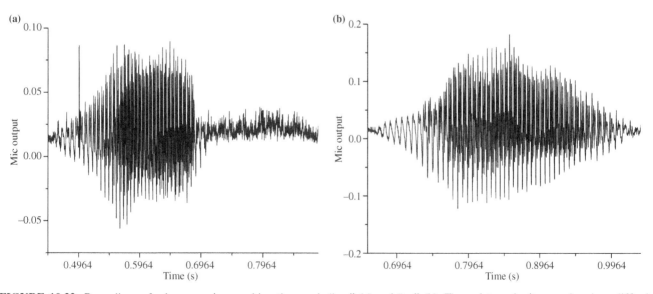

FIGURE 10.32. Recordings of a human voice speaking the words "yes" (a) and "no" (b). These data make it very clear how difficult speech recognition really is due to the fact that so many sounds involve combinations of frequencies.

the end are significantly different. It is clear the "s" portion of "yes" has much smaller amplitude but much higher frequency. In fact, the "s" is one unambiguous characteristic that distinguishes Figure 10.32a from Figure 10.32b. In the case of the spoken "no" (Figure 10.32b), the "n" sound (about the first eight cycles) is sharply defined at about 127 Hz. In contrast, it is apparent that the "o" sound involves several frequencies. Now imagine that we wished to issue instructions to a computer verbally and have those instructions—in near real time—be translated into executable code. It will be necessary for the machine to recognize the individual phones (a phone is a single speech sound) and to assemble them into an intelligible instruction. If you think about how people actually speak, often running words together, failing to enunciate certain letter combinations clearly, and so on, you can understand what an incredibly difficult problem speech recognition really is; and to make it worse, in critical applications, the system would have to be 100% reliable. The data in Figure 10.32 suggest that distinguishing between a spoken "yes" and a spoken "no" would be easy, but interpreting an entire sentence flawlessly would be far more difficult! We will have an opportunity to explore the use of the FFT in the detection of speech patterns in a student exercise where we will examine four words, all with at least some similar sounds.

PROBLEMS

10.1. Apply Schuster's test to the oscillating function

$$y(t) = \sin(39\pi t) + \sin(51\pi t) + \sin(59\pi t) + \sin(79\pi t).$$

Examine the discrete frequencies 57, 58, 59, 60, and 61π and prepare a plot similar to Figure 10.5.

10.2. Consider a set of sampled values consisting of 1024 observations. These data are uniformly zero until $n = 85$, where a ramp occurs with amplitude increasing at a rate corresponding to 0.05 per sample. This "waveform" then has an amplitude of 1 until $n = 175$; for $n > 175$, $y(t) = 0$, that is, the amplitude reverts to zero. Compute the DFT for this waveform without tapering. How severe is the leakage?

10.3. Apply the DFT to a square wave (a rectangular window) with an amplitude of 1, then apply a split cosine (bell-shaped) window to the data, leaving the center 80% of the data unaffected. Finally, allow the window to shrink so that the center amplitude falls below 1.0 to 0.75. What impact does this have on the DFT? Is leakage completely eliminated?

10.4. By an inviscid analysis (see Lamb, 1945), it can be shown that bubbles and (immiscible) droplets immersed in

liquids have a characteristic frequency of oscillation that is given by

$$s^2 = n(n+1)(n-1)(n+2)\frac{\tau}{[(n+1)\rho + n\rho_1]R^3}.$$

For a 0.2-cm diameter water droplet surrounded by air, τ (the surface tension) is about 72 dyne/cm, $\rho = 1$ g/cm and $\rho_1 \approx 0$. If the most important mode of vibration is that for which $n = 2$, then

$$s^2 = \frac{(24)(72)}{(3)(0.1)^3} = 576,000 \text{ rad}^2/\text{s}^2, \quad \text{and therefore,}$$

$$f \cong 120 \text{ Hz}.$$

If, on the other hand, an air bubble of the same size is surrounded by water, the frequency of vibration is approximately

$$s^2 = \frac{(12)(72)}{(1)(0.1)^3} = 864,000 \text{ rad}^2/\text{s}^2, \quad \text{and thus, } f \cong 148 \text{ Hz}.$$

If bubbles were formed at a single orifice at a rate of 79 per minute and if their sizes ranged from 0.9 mm (diameter) to 2.39 mm, which dominant frequencies would you expect to see in a computed spectrum (data obtained from a hydrophone or a sensitive pressure transducer)?

10.5. Dodge (1971) carried out a study of slosh suppression in LOX tanks; of course, sloshing is critical to stability and control of rockets using cryogenic liquids (oxygen and hydrogen) for propulsion. The work focused on the use of lightweight, flexible plastic baffles for dampening of the periodic forces created by the moving liquid. In the experimental work, a shake table was used to oscillate a 76-cm diameter tank containing liquid nitrogen. The natural slosh frequency in this tank was cited as 1.05 Hz, and the amplitude obtained was a little less than 4 cm. A load cell was used to record the decay of the liquid motion within the tank. *Suppose* the data record obtained from such an experiment appeared as shown in Figure 10.33.

Use the DFT to see if you can identify all of the frequencies appearing in this data set. The data will be provided to you separately in electronic format. Does the rapid attenuation cause any significant problems? The sample interval, Δt, is 0.02 second.

10.6. Baseball bats, golf clubs, tennis racquets, and the like vibrate on impact with a ball. In some cases, the resulting stinging sensation (transmitted to the hands) can be severe, and in baseball, the amplitude of the vibration can be such that the batter's hand(s) may even lose contact with the bat handle. This is a phenomenon that has seen much study by

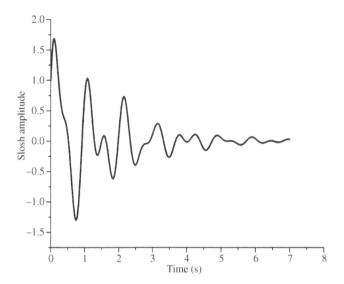

FIGURE 10.33. Slosh amplitude for cryogenic liquids in tanks.

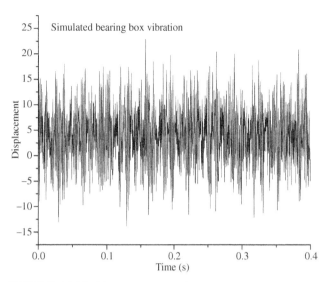

FIGURE 10.35. Simulated locomotive bearing box vibrations, with $\Delta t = 0.00025$ second. Therefore, $f_{Nyq} = 1/(2\Delta t) = 2000$ Hz.

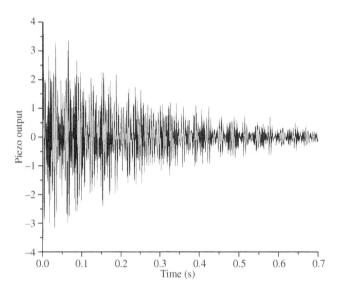

FIGURE 10.34. Data of the type obtained from a piezoelectric sensor mounted on the barrel of a wooden baseball bat.

physicists including Brody (1986) at the University of Pennsylvania and Cross (1998) at the University of Sydney. It turns out that wooden baseball bats typically have a couple of fundamental frequencies of about 500–600 Hz and about 150–200 Hz. Naturally, the precise vibration frequencies depend on where the ball is contacted on the bat, how the bat is secured, and where the piezoelectric vibration sensors are mounted. *Suppose* a ball impact test on a wooden Louisville Slugger™ produced the data set illustrated in Figure 10.34.

Perform a harmonic analysis of these data and determine the fundamental frequencies produced by the impact of a ball on the wooden bat. The data will be supplied to you

separately in electronic format. The sample interval in this case was 0.0005 second (0.5 ms).

10.7. Miles et al. (1999) studied both bending and torsional vibrations occurring in the crankshaft of a 2-L diesel engine (with four cylinders) running under load at both 600 and 750 rpm. They used laser light scattered off of the face end of the crankshaft pulley to detect the motion at the end of the crankshaft. Because of the proximity of the measurement point to the number one cylinder, one could expect firing in that cylinder to have a major impact on the motion. Each segment of piston travel corresponds to 180° of rotation on the crankshaft, so the power pulses on the number one cylinder are separated by (4)(180) = 720°. Therefore, if the engine is running at 600 rpm (10 revolutions per second), power pulses on the number one cylinder should occur about every 0.2 second. This is the pattern revealed by figure 17 of the Miles et al. paper. Their computed spectrum is shown in figure 18 and it exhibits important signal content at *many discrete frequencies including* 9, 32, 42, 55, 87, 211, 300, 352, 367, 413, 432, 454, 471, and 644 Hz, among others. *Construct a set of sinusoids with appropriate amplitudes and frequencies* such that the DFT produces a spectrum comparable to figure 18 of Miles et al. Which components of the oscillating signal must have the largest amplitudes for the two spectra to be comparable?

10.8. Shust (2007) studied shock and vibration in railroad operations. In one part of the study, vibrations in an operating locomotive bearing box were measured. Some simulated data with similar periodicities are reported in Figure 10.35. Perform a DFT on these data (supplied separately) and identify the important contributions to the spectrum. What are the most significant frequencies in the data set?

10.9. The failure of the Tacoma Narrows (TN) bridge in November 1940 led to multidisciplinary efforts to better understand oscillations occurring in suspended structures and how those vibrations might lead to structural failures. Although *vortex shedding* has frequently been identified as the cause of the TN bridge deck's oscillations, this explanation is incorrect. At the time of the failure, Professor Farquharson (of the University of Washington) observed that the destructive *torsional* oscillation was occurring at 0.12 Hz—a frequency much lower than that arising from vortex shedding from a suitably sized bluff body (with a wind speed of 42 mph). The aeroelastic torsion that doomed the bridge was nothing at all like the benign vertical "galloping" that was observed even while the bridge was still under construction. The failure of the TN bridge had an important positive result, however; it forced bridge designers and engineers to better understand (and design for) aeroelastic oscillations.

Abdel-Ghaffar and Scanlan (1985) studied vertical, torsional, lateral, and longitudinal motions of the center span of the Golden Gate bridge in response to normal excitations (wind, vehicular traffic, and wave action). In Figure 10.36, a spectrum obtained from span station 3 has been reconstructed from the published paper (the quality of the original figure was poor).

Construct a set of sinusoids that, when Fourier-transformed, will reproduce the essential characteristics of this spectrum. Use a random number generator to produce some fine-scale structure in the spectrum and compare your results with Figure 10.36.

10.10. Recordings of a human voice speaking the words "example," "execution," "expansion," and "exponent" will be provided to you separately (and illustrated in Figure 10.37). Apply the FFT to all four data sets. Are the spectral characteristics different enough to allow you to distinguish between the words? What are the essential differences, and are you confident that, based on your spectra, you could identify all four words without error?

10.11. In recent years, the problem of concussions resulting from collisions in dynamic sports like football has been revealed to be a much more serious problem than most observers thought. It has become all too evident that even mild traumatic brain injuries (TBIs) may have cumulative effects leading to chronic traumatic encephalopathy (CTE), which may manifest itself years later in memory loss, behavioral changes, and even suicide. For many years, the standard mechanism for assessment of head trauma has been the Gadd severity index (GSI), which is determined from the linear acceleration (at the headform center of gravity) and the duration of the event:

$$\text{GSI} = \int_0^T a^{5/2} dt.$$

Throughout the literature of head trauma, the threshold value for the severity index has been cited as 1000, with a typical event duration of about 15 ms (0.015 second). An alternative, but closely related, measure is the head injury criterion (HIC):

$$\text{HIC} = (t_2 - t_1) \left[\frac{1}{t_2 - t_1} \int_{t_1}^{t_2} a(t) dt \right]^{5/2}.$$

Studies have shown that concussions in professional football (the National Football League) typically result from impact velocities on the order of 9.3 ± 1.9 m/s (note that this corresponds to about 20.8 mph!). This raises an interesting question: Is it possible that significant energy is being produced at higher frequencies as a result of football-related collisions? You may recall that, near the end of this chapter (see Figure 10.30), we saw exactly that phenomenon occurring in railroad car-coupling events, where measured spectra revealed large accelerations at much higher frequencies (e.g., close to 100 Hz). This transfer of energy was obtained at much lower velocities (about 5.25 mph) and Shust (2007) further noted that energy transfer in car-coupling events routinely resulted in the 4 *g* design specification being

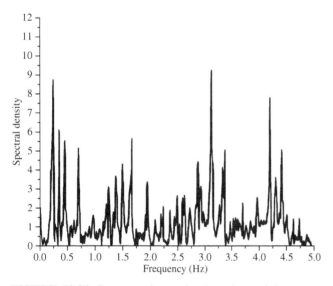

FIGURE 10.36. Spectrum for torsional motions of the center span. The frequency scale is arithmetic (hertz), and the two highest peaks in the right-half of the spectrum occur at 3.1 and 4.18 Hz. The major contribution at the left edge is centered at approximately 0.225 Hz.

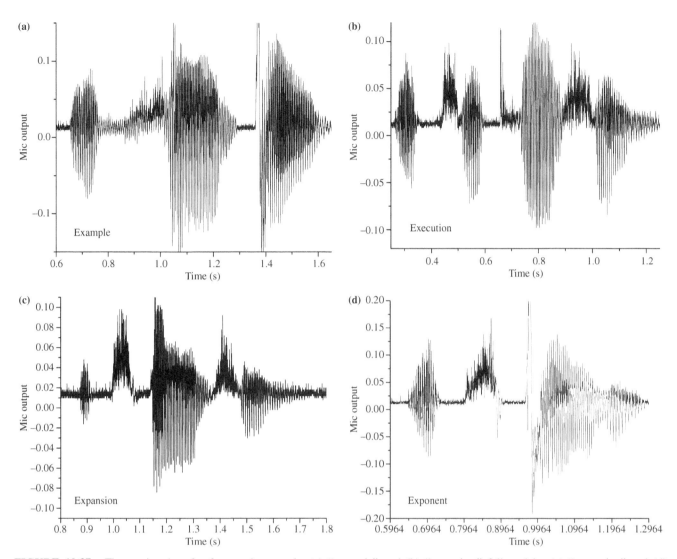

FIGURE 10.37. Time-series data for four spoken words: (a) "example" and (b) "execution," followed by (c) "expansion" and (d) "exponent."

exceeded. As of this writing (the end of 2013), the are few such (spectral) data available for helmet impacts, although this area is seeing very intensive work with the objective that CTE might be minimized or even eliminated. Search the latest literature to see what spectral data have appeared for helmet impacts. Do the spectra show the type of energy transfer seen in railroad car-coupling events? If they do not, what do the spectra reveal about the dynamics recorded in the accelerometer data?

REFERENCES

Abdel-Ghaffar, A. M. and R. H. Scanlan. Ambient Vibration Studies of Golden Gate Bridge: I. Suspended Structure. *Journal of Engineering Mechanics, ASCE*, 11:463 (1985).

Bingham, C., Godfrey, M. D., and J. W. Tukey. Modern Techniques of Power Spectrum Estimation. *IEEE Transactions on Audio and Electroacoustics*, AU15:56 (1967).

Bloomfield, P. *Fourier Analysis of Time Series: An Introduction*, John Wiley & Sons, New York (1976).

Brigham, E. O. *The Fast Fourier Transform and Its Applications*, Prentice Hall, Englewood Cliffs, NJ (1988).

Brody, H. The Sweet Spot of a Baseball Bat. *American Journal of Physics*, 54:640 (1986).

Cooley, J. W. The Re-discovery of the Fast Fourier Transform Algorithm. *Mikrochimica Acta (Wien)*, III:33 (1987).

Cooley, J. W. and J. W. Tukey. An Algorithm for the Machine Calculation of Complex Fourier Series. *Mathematics of Computation*, 19:297 (1965).

Cross, R. The Sweet Spot of a Baseball Bat. *American Journal of Physics*, 66:772 (1998).

Dodge, F. T. Engineering Study of Flexible Baffles for Slosh Suppression. NASA Contractor Report, CR-1880, September 1971 (1971).

Knott, C. G. On Lunar Periodicities in Earthquake Frequency. *Proceedings of the Royal Society of London*, 60:457 (1897).

Lamb, H. *Hydrodynamics*, 6th edition, Dover Publications, New York (1945).

Magnuson, C. F. and L. T. Wilson. Shock and Vibration Environments for Large Shipping Containers on Rail Cars and Trucks. Sandia Laboratories, SAND76-0427 (1977).

Miles, T. J., Lucas, M., Halliwell, N. A., and S. J. Rothberg. Torsional and Bending Vibration Measurement on Rotors Using Laser Technology, *Journal of Sound and Vibration*, 226:441 (1999).

Rockmore, D. N. The FFT—An Algorithm the Whole Family Can Use. Department of Mathematics and Computer Science, Dartmouth College (1999).

Schuster, A. On Lunar and Solar Periodicities of Earthquakes. *Proceedings of the Royal Society of London*, 61:455 (1897).

Shust, W. C. Shock and Vibration in Rail and other Transport Modes. PPT presentation, Objective Engineers, Inc. (2007).

Slusher, G. R. Reciprocating Engine and Exhaust Vibration and Temperature Levels in General Aviation Aircraft. *FAA Report NA-68-27* (1968).

Tukey, J. W. An Introduction to the Calculations of Numerical Spectrum Analysis. In: *Spectral Analysis of Time Series* (B. Harris, editor), John Wiley & Sons, New York (1967).

Warpinski, N. R., Du, J., and U. Zimmer. Measurements of Hydraulic-Fracture-Induced Seismicity in Gas Shales. SPE 15197, SPE Hydraulic Fracturing Technology Conference (2012).

11

AN INTRODUCTION TO THE CALCULUS OF VARIATIONS AND THE FINITE-ELEMENT METHOD

SOME PRELIMINARIES

You may recall that we previously made use of some concepts from the *calculus of variations (COV)* in our discussion of the *variational iteration method (VIM)*. We will now expand on that very brief treatment to illustrate some other applications of COV. We will mainly follow the conventions employed by Spiegel (1971) and Weinstock (1974). It is important to note that our approach will focus on *conservative* fields for which the force, F, is related to a potential energy function, ϕ, such that $F = -\nabla \phi$.

Consider a function, $y = f(x)$, in the x-y plane that joins the two points, (x_1, y_1) and (x_2, y_2), together. We recall from elementary calculus that the length of the pathway connecting these two points is

$$L = \int_{x_1}^{x_2} \sqrt{1 + \left(\frac{dy}{dx}\right)^2}\, dx. \qquad (11.1)$$

To reveal the nature of the problem we wish to contemplate more clearly, let us select two specific points:

$$(x_1, y_1) = (1, 4) \quad \text{and} \quad (x_2, y_2) = (4, 9).$$

Now suppose the function, $f(x)$, connecting these two points is a straight line:

$$y = \frac{5}{3}x + \frac{7}{3}. \qquad (11.2)$$

We want to evaluate L from eq. (11.1), recognizing, of course, that the value produced by eq. (11.2) will be the minimum possible length. But we also want to have a couple of comparisons available, so in addition, we will let $y(x)$ be given by the two equations:

$$y = \frac{1}{3}x^2 + \frac{11}{3} \quad \text{and} \quad y = \frac{7}{x} - \frac{797}{240}x + \frac{77}{240}x^3. \qquad (11.3a,b)$$

Now we will compute the arc length from point 1 to point 2 using eq. (11.1) for all three cases; the results are 5.83, 5.91, and 8.98, respectively. The straight line provided the minimum length as we expected. The second-degree equation exhibits a little curvature, so its value for L is a bit longer, and the third-degree equation—which is a good deal more complicated—provides a significantly longer L as we would expect. Imagine now a situation in which we wanted to identify the minimum value of the integral (eq. 11.1) but had no idea of the form of the function, $y(x)$. This certainly sounds like a much more difficult task and we have an additional complication: In technical problems, our objective will almost certainly be far more complicated than the mere distance between two points in the x-y plane as given by eq. (11.1).

As we suggested previously, our real interest is the more general case where we need to identify some function, $y(x)$, where $y_1 = y(x_1)$ and $y_2 = y(x_2)$, but with the stipulation that the integral

Applied Mathematics for Science and Engineering, First Edition. Larry A. Glasgow.
© 2014 John Wiley & Sons, Inc. Published 2014 by John Wiley & Sons, Inc.

$$I = \int_{x_1}^{x_2} F(x, y, y')dx \qquad (11.4)$$

be a minimum or a maximum. The function $y(x)$ that meets this requirement is called an *extremal*, and the integral (eq. 11.4) is referred to as a *functional* with one independent variable, x. We will proceed in the following way: We form a family of comparison functions using a single parameter, ε:

$$Y(x) = y(x) + \varepsilon\phi(x). \qquad (11.5)$$

We require that $\phi(x)$ be a differentiable function that disappears at the endpoints x_1 and x_2 so that $\phi(x_1) = 0$ and $\phi(x_2) = 0$. Now we rewrite eq. (11.4) as

$$I(\varepsilon) = \int_{x_1}^{x_2} F(x, Y, Y')dx \qquad (11.6)$$

and note that by $Y'(x)$, we mean $Y'(x) = y'(x) + \varepsilon\phi'(x)$. Of course, if $\varepsilon = 0$ everywhere, then eq. (11.6) is exactly the same as eq. (11.4). The essential point here is that no matter what form $\phi(x)$ has, the extremizing function, $y(x)$, will be a member of the comparison function family. Let us illustrate this scenario with a graph (Figure 11.1).

If we differentiate eq. (11.6) with respect to ε and set $\varepsilon = 0$, then $dI/d\varepsilon = 0$. Thus,

$$I'(0) = \int_{x_1}^{x_2} \left(\frac{\partial F}{\partial y}\phi + \frac{\partial F}{\partial y'}\phi'\right)dx. \qquad (11.7)$$

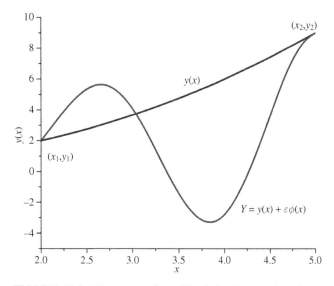

FIGURE 11.1. Illustration of the "family" of comparison functions. If $\varepsilon = 0$, then we get the behavior shown by $y(x)$.

By integrating by parts and noting that the result must be valid for all ϕ, Weinstock (1974) shows that the function we are seeking must satisfy the *Euler–Lagrange differential equation*:

$$\frac{d}{dx}\left(\frac{\partial F}{\partial y'}\right) - \frac{\partial F}{\partial y} = 0, \qquad (11.8)$$

which was derived by Euler in the eighteenth century. We can perform the indicated differentiation to see the equation in its entirety:

$$\frac{d^2 y}{dx^2}\frac{\partial^2 F}{\partial y'^2} + \frac{dy}{dx}\frac{\partial^2 F}{\partial y'\partial y} + \frac{\partial^2 F}{\partial y'\partial x} - \frac{\partial F}{\partial y} = 0.$$

The Euler–Lagrange equation is a necessary—but not sufficient—condition for $y(x)$ to be an extremal; at this stage, we do not know if we have found a minimum, a maximum, or a stationary point (at a stationary point, the derivatives of the function vanish). For a simple example of a stationary point, consider the behavior of $f(x) = x^3$ with $f'(x) = 3x^2$; clearly, at $x = 0, f' = 0$, but this point does not correspond to an extremum. The differential equation (eq. 11.8) may be written in an equivalent form that is often useful:

$$\frac{d}{dx}\left[F - y'\frac{\partial F}{\partial y'}\right] = \frac{\partial F}{\partial x}. \qquad (11.9)$$

Should x *not appear* in F explicitly, then

$$F - y'\frac{\partial F}{\partial y'} = C. \qquad (11.10)$$

Equation (11.10) is known as the *Beltrami identity* (named for Eugenio Beltrami, an Italian mathematician of the nineteenth century). And should y *not appear* explicitly, then $\partial F/\partial y = 0$, and eq. (11.7) is simply $(d/dx)(\partial F/\partial y') = 0$, such that

$$\frac{\partial F}{\partial y'} = C. \qquad (11.11)$$

NOTATION FOR THE CALCULUS OF VARIATIONS

Let us return to eq. (11.4) to retrieve the function $F(x, y, y')$, but now we take the independent variable, x, to be constant. We define the *variation of F* as

$$\Delta F = F(x, y + \varepsilon\phi, y' + \varepsilon\phi') - F(x, y, y'). \quad (11.12)$$

We can employ a truncated Taylor series expansion for $F(x, y + \varepsilon\phi, y' + \varepsilon\phi')$:

$$\cong F(x, y, y') + \frac{\partial F}{\partial y}\varepsilon\phi + \frac{\partial F}{\partial y'}\varepsilon\phi' + \cdots. \quad (11.13)$$

This means that the variation of F is

$$\Delta F \cong \frac{\partial F}{\partial y}\varepsilon\phi + \frac{\partial F}{\partial y'}\varepsilon\phi' + \text{neglected terms.} \quad (11.14)$$

In the literature of the COV, this variation of F is written as

$$\delta F = \frac{\partial F}{\partial y}\varepsilon\phi + \frac{\partial F}{\partial y'}\varepsilon\phi'. \quad (11.15)$$

The reason this is important is because a necessary condition for the *integral of the function*, $F(x, y, y')$, to be an extremum is

$$\delta \int_{x_1}^{x_2} F(x, y, y')dx = 0; \quad (11.16)$$

that is, the first variation of the integral must vanish. Though conceptually attractive, we bear in mind an observation made by Courant and Hilbert (1989): "in the calculus of variations the existence of an extremum for a particular problem cannot be taken for granted. A special existence proof is needed for the solution of each problem or class of problems." Courant and Hilbert provide examples of COV problems that, while appropriately developed, *do not* have solutions.

BRACHISTOCHRONE PROBLEM

The utility of the COV is often demonstrated with the "brachistochrone" problem, which was posed by Johann Bernoulli as a challenge to European mathematicians in 1696. This example serves another purpose as well since it conveys a clearer picture of Sir Isaac Newton's formidable powers; it is extremely difficult three centuries later to grasp how far Newton had surpassed his contemporaries in mathematics and physics. It was reported by Newton's biographers (much of what we know about Newton has come down to us from John Conduitt and William Stukeley) that Newton received the brachistochrone problem in the evening after returning from work at the Royal Mint (Newton served as Master of the Mint for 10 years); Newton developed concepts for the COV *and* solved the brachistichrone problem, completing his work at 4 a.m. the next morning. He insisted that the solution be published anonymously (and it was, by the Royal Society in January 1697), but it has been reported that when Bernoulli saw it, he remarked that "we recognize the lion by his claw."

The physical picture for the brachistochrone problem is as follows: We consider two points in a vertical plane connected by a ribbon or wire that can assume any shape desired. An object of mass, m, slides down the wire under the influence of gravity. The process is frictionless so that at any point in time, the sum of potential and kinetic energies is constant; there are *no dissipative forces in play*, only gravity. The question to be addressed is what shape should the wire have such that the sliding mass reaches the bottom position in the least possible time?

Let the initial point be the origin and the lower point be P corresponding to position (x_2, y_2). Since there are no dissipative forces in this process, we note by energy balance, $PE_0 + KE_0 = PE_2 + KE_2$. We assume that the mass was initially at rest, so the kinetic energy term on the left-hand side is zero:

$$mgy_0 = mg(y_0 - y) + \frac{1}{2}m\left(\frac{ds}{dt}\right)^2, \quad (11.17)$$

and therefore, $ds/dt = \sqrt{2gy}$. Of course, s is the length of the arc connecting the initial and final points. The total time required for the mass to slide from the origin to its final position can be obtained from the quotient of the path length and the object's velocity (keep in mind that we are starting at the origin, so $x = 0$):

$$T_{\text{req}} = \int_{x=0}^{x=x_2} \frac{ds}{\sqrt{2gy}}. \quad (11.18)$$

We already know that the length of an arc is given by eq. (11.1); therefore,

$$T_{\text{req}} = \frac{1}{\sqrt{2g}} \int_0^{x_2} \frac{\sqrt{1 + y'^2}}{\sqrt{y}} dx. \quad (11.19)$$

The objective is to find the function $y(x)$ that minimizes this integral, recognizing that this is exactly the type of COV problem we described previously. One could propose different functions and calculate the total time for each case (as we did in the example at the beginning of this chapter), but we would not know whether the $y(x)$ so identified was really the extremal we are seeking. By comparison with the form for eq. (11.4), we see

$$F = \frac{\sqrt{1 + y'^2}}{\sqrt{y}},$$

and thus F *does not* explicitly depend on x. Consequently, we can make use of eq. (11.10):

$$\frac{\sqrt{1 + y'^2}}{\sqrt{y}} - y'\left[\frac{y'}{\sqrt{1 + y'^2}\sqrt{y}}\right] = C, \quad (11.20)$$

and accordingly,

$$y' = \frac{dy}{dx} = \sqrt{\frac{1/C - y}{y}}. \tag{11.21}$$

We set $A = 1/C$ and change the y variable: $y = A\sin^2\theta$. The reader may wish to show that

$$x = A\left[\theta - \tfrac{1}{2}\sin 2\theta\right]. \tag{11.22}$$

The constant of integration that should appear here is equal to zero since the curve must go through the origin (where $x = y = 0$). The expressions for x and y give us the parametric equations for a cycloid; for example, when $\theta = \pi/6$, $x = 0.0906A$ and $y = 0.25A$, and when $\theta = \pi/4$, $x = 0.2854A$ and $y = 0.5A$. The value for A is selected to ensure that the curve passes through the desired point, (x_2, y_2).

OTHER EXAMPLES

Minimum Surface Area

Again we assume that we have some curve in the x-y plane that connects the two pairs of points, (x_1, y_1) and (x_2, y_2). If we rotate this curve about the x-axis, then a surface is generated that has some particular surface area, as well as some enclosed volume. We know from elementary calculus that this generated surface has an area given by

$$A = \int_{x_1}^{x_2} 2\pi y(x)\sqrt{1 + \left(\frac{dy}{dx}\right)^2}\, dx. \tag{11.23}$$

Therefore, if a straight line extended from the origin to the point $(4, 5)$, then $y(x) = (5/4)x$, and if we rotate this line around the x-axis we find

$$A = 2\pi\left(\frac{5}{4}\right)\sqrt{\frac{41}{16}}\left.\frac{x^2}{2}\right|_0^4 = 100.58.$$

Of course, we could confirm this number by use of the mensuration formula for the lateral surface area of a cone. But our goal in this section goes beyond a determination of surface area; we want to determine the *minimum surface area* formed by the revolution of $y(x)$ about the x-axis. Therefore, we need to identify the *minimum* value of the integral,

$$I = 2\pi \int_{x_1}^{x_2} y\sqrt{1 + y'^2}\, dx. \tag{11.24}$$

Observe once again that the integrand does *not* depend explicitly on x; thus, we can turn immediately to eq. (11.9), $F - y'(\partial F/\partial y') = C$:

$$y\sqrt{1 + y'^2} - \frac{yy'^2}{\sqrt{1 + y'^2}} = C. \tag{11.25}$$

Now we isolate dy/dx to show

$$\int \frac{dy}{\sqrt{\dfrac{y^2}{C^2} - 1}} = x + B. \tag{11.26}$$

We substitute $z = y/C$ to find

$$C\int \frac{dz}{\sqrt{z^2 - 1}} = x + B,$$

and we notice that the denominator of the integral is a quadratic irrationality that would normally call for trigonometric substitution. However, in this case, we might recognize the appropriate antiderivative since the integral is one that appears in every elementary calculus book, and therefore,

$$\cosh^{-1}(z) = \frac{x}{C} + B_1, \quad \text{resulting in } y = C\cosh\left[\frac{x}{C} + B_1\right]. \tag{11.27}$$

The intriguing feature of this *catenary* is that, if we force the curve through the initial point, (x_1, y_1), we *might not be able* to pass through the endpoint, (x_2, y_2); that is, the endpoint might lie outside the envelope of this catenary family. Weinstock discusses this particular case which reveals a limitation of the theory we have employed earlier. Though it is beyond the scope of our discussion, the actual minimum surface area for this situation is given by the *Goldschmidt discontinuous solution*—the catenary family will provide relative or absolute minima for surface area *only if* the endpoint is sited appropriately (in general, the endpoint must be above or to the left of the catenary envelope).

Before we leave this discussion of the catenary, we should make note of the fact that the COV can be used to show that an inverted catenary (the catenary arch) has the lowest possible internal stress. Of course, catenary arches have been used in buildings for centuries and a modern example is the Sheffield Winter Garden in South Yorkshire. The very well-known Gateway Arch in St. Louis, Missouri, is an example of a flattened catenary.

Systems of Particles

We are concerned here with a collection of n-particles in a conservative system for which the potential energy function,

V, can be used to obtain the three force components; for the nth particle (located at positions x, y, and z), we would have

$$F_x^n = -\frac{\partial V}{\partial x_n}, \quad F_y^n = -\frac{\partial V}{\partial y_n}, \quad \text{and} \quad F_z^n = -\frac{\partial V}{\partial z_n}. \quad (11.28)$$

The total kinetic energy, T, for this system can be obtained from

$$T = \frac{1}{2}\sum_{i=1}^{n} m_i \left[\left(\frac{dx}{dt}\right)^2 + \left(\frac{dy}{dt}\right)^2 + \left(\frac{dz}{dt}\right)^2 \right]. \quad (11.29)$$

In classical mechanics, we would describe the behavior of a particle with Newton's law, $F = ma = m(d^2s/dt^2)$. We imagine that at two different times, say, t_1 and t_2, we have two different configurations—the particles have moved around subject to $F = ma$. The *principle of least action** states that the integral of the Lagrangian function over this time interval must have a minimum value. Since the Lagrangian function for this system is merely the difference between kinetic and potential energies, we require that

$$I = \int_{t_1}^{t_2} (T - V)dt = \int_{t_1}^{t_2} L\,dt \quad (11.30)$$

be an extremum (minimum). The problem thus posed is one of determining the paths traveled by the particles that make this integral a minimum. We let the particle velocities be represented by $\dot{q}_1, \dot{q}_2, \ldots \dot{q}_n$ and the coordinate locations be given by $q_1, q_2, \ldots q_n$. Naturally, the kinetic energy of the system is a function of the velocities, the positions, and time, while the potential energy is a function of the coordinates and time. The requirement that the integral (eq. 11.30) be a minimum is often referred to as *Hamilton's principle*, which is applicable to the motion of any *conservative* system. This principle leads to Lagrange's general equations of motion, which we write as

$$\frac{d}{dt}\frac{\partial T}{\partial \dot{q}_i} - \frac{\partial}{\partial q_i}(T - V) = 0 \quad \text{or} \quad \frac{\partial L}{\partial q_i} - \frac{d}{dt}\left(\frac{\partial L}{\partial \dot{q}_i}\right) = 0$$
$$(\text{for } i = 1, 2, \ldots, n), \quad (11.31)$$

where the q_is are generalized coordinates. These equations (eq. 11.31) constitute a set of simultaneous second-order

*The philosophical underpinning of the principle of least action was given by Euler: "As the construction of the universe is the most perfect possible, being the handiwork of an all-wise Maker, nothing can be met with in the world in which some maximal or minimal property is not displayed. There is, consequently, no doubt but that all of the effects of the world can be derived by the method of maxima and minima from their final causes as well as from their efficient ones."

differential equations whose solutions reveal the generalized coordinates as functions of time. The solution, of course, requires that the initial values for both q and $\dot{q}$ be known.

Vibrating String

In the previous section, we considered a situation in which points of mass (particles) were distributed in space. However, Hamilton's principle can also be applied to cases in which mass is continuously distributed. To illustrate, let us consider a string anchored at positions $x = 0$ and $x = L$. The string is flexible, and if it is physically perturbed, it can oscillate normal to its equilibrium (or undisturbed) axis. We will let the string's displacement be ϕ and its mass per unit length be m. The total kinetic energy of this string of length L is then

$$T = \frac{1}{2}\int_0^L m\left(\frac{\partial \phi}{\partial t}\right)^2 dx. \quad (11.32)$$

The string is perfectly elastic, so any disturbance will result in an increase in length. We will assume that the potential energy of the string is directly related to its elongation, and therefore by eq. (11.1), this change in length is given by

$$\int_0^L \sqrt{1 + \left(\frac{\partial \phi}{\partial x}\right)^2}\, dx - L. \quad (11.33)$$

If the increase in string length caused by the displacement is sufficiently small, then

$$\sqrt{1 + \left(\frac{\partial \phi}{\partial x}\right)^2} \approx 1 + \tfrac{1}{2}\left(\frac{\partial \phi}{\partial x}\right)^2$$

such that

$$V \cong \frac{k}{2}\int_0^L \left(\frac{\partial \phi}{\partial x}\right)^2 dx,$$

and therefore,

$$T - V = \frac{1}{2}\int_0^L \left[m\left(\frac{\partial \phi}{\partial t}\right)^2 - k\left(\frac{\partial \phi}{\partial x}\right)^2 \right] dx. \quad (11.34)$$

By the principle of least action, the first variation of I should be zero:

$$I(\phi) = \frac{1}{2}\int_{t_1}^{t_2} \int_0^L \left[m\left(\frac{\partial \phi}{\partial t}\right)^2 - k\left(\frac{\partial \phi}{\partial x}\right)^2 \right] dx\,dt. \quad (11.35)$$

Note that the functional depends *only* on the two derivatives (with respect to t and x); it does not involve x or t or even ϕ explicitly. The Euler–Lagrange equation for this case (we will use subscripts to indicate derivatives with respect to x and t) is

$$\frac{\partial F}{\partial \phi} - \frac{\partial}{\partial x}\left(\frac{\partial F}{\partial \phi_x}\right) - \frac{\partial}{\partial t}\left(\frac{\partial F}{\partial \phi_t}\right), \qquad (11.36)$$

which means that the function ϕ, which makes the first variation of I disappear, must be governed by

$$\frac{\partial^2 \phi}{\partial t^2} = c^2 \frac{\partial^2 \phi}{\partial x^2}, \qquad (11.37)$$

where the characteristic velocity, c, is obtained from the quotient of k/m. Equation (11.37) tells us that the string's displacement must follow the one-dimensional wave equation, exactly as we expected.

Laplace's Equation

We have seen previously that problems concerned with conservative fields (no dissipative effects) can often be described with Laplace's equation, for example, in two dimensions:

$$\nabla^2 \phi = 0 \quad \text{or} \quad \frac{\partial^2 \phi}{\partial x^2} + \frac{\partial^2 \phi}{\partial y^2} = 0. \qquad (11.38)$$

We may also remember that eq. (11.38) has a broad range of applications in electrostatics, heat transfer, gravitation, and hydrodynamics. For cases in which the field variable is specified around the boundary, we have what is known as a Dirichlet problem. Such problems can be expressed alternatively in variational form:

$$I(\phi) = \iint F(x, y, \phi, \phi_x, \phi_y)dxdy, \qquad (11.39)$$

and for the two-dimensional Laplace equation, the variational principle is known:

$$I(\phi) = \iint \frac{1}{2}\left[\left(\frac{\partial \phi}{\partial x}\right)^2 + \left(\frac{\partial \phi}{\partial y}\right)^2\right]dxdy. \qquad (11.40)$$

Our goal in this case is to identify the potential function that causes the first variation of I to disappear. The solution procedure is analogous to the Ritz method, which is discussed in the next section. The main problem with this approach is that the required workload is significant; thus, we would normally use this technique for the solution of the Laplace equation only when a closed-form solution cannot be found, but we absolutely must obtain an analytic approximation.

Boundary-Value Problems

In a typical variational problem, we try to identify a function, $y(x)$, that leads to a minimum value of the integral just as we saw at the beginning of this chapter:

$$I(y) = \int_{x_1}^{x_2} F(x, y, y')dx. \qquad (11.4)$$

A *direct approach* to this problem involves the construction of a sequence of functions, $y_1, y_2, y_3, \ldots, y_n$ such that as $n \to \infty$, we obtain an extremum for I. The set of functions that achieves this objective is called a *minimal* (or *minimizing*) *sequence*. There is a powerful technique that can be used for this purpose and we will look at a detailed example later. However, as Smith (1953) indicates, we need to keep two very important points in mind: First, the function sequence must be selected with care to obtain a suitably rapid rate of convergence, and second, the minimal sequence—even as n becomes very large—may *not necessarily* converge to the actual solution of the variational problem.

The COV can be used to solved certain boundary-value problems through the use of a technique we discussed previously in this course, the *Rayleigh–Ritz* (or simply Ritz) *method*. We can best illustrate this with a simple example for which the analytic solution is easily determined. Suppose we have a differential equation,

$$\frac{d^2 \phi}{dx^2} = -f(x) = a + bx, \qquad (11.41)$$

with $f(x = 0) = 0$ and $f(x = 1) = 7$; and further, we presume a form for the solution, $\phi = 11x - 17x^2 + 13x^3$. This means that $a + bx = -34 + 78x$. The behavior of the function $\phi(x)$ is illustrated in Figure 11.2.

In the application of the Ritz method to this boundary-value problem, we set out to find a function that minimizes the integral:

$$I(\phi) = \int_0^1 \left[\frac{1}{2}\left(\frac{d\phi}{dx}\right)^2 - f(x)\phi(x)\right]dx. \qquad (11.42)$$

We proceed with the understanding that our objective is to identify an analytic approximation for $\phi(x)$ as quickly and easily as possible. We let the trial function be

$$\phi \cong x\left[C_1 + C_2(1-x) + C_3(1-x)^2 + C_4(1-x)^3 + \cdots\right]. \qquad (11.43)$$

The boundary conditions are automatically satisfied if we take $C_1 = 7$. The trial function is substituted into eq. (11.42) and the integration is carried out to obtain I, which depends

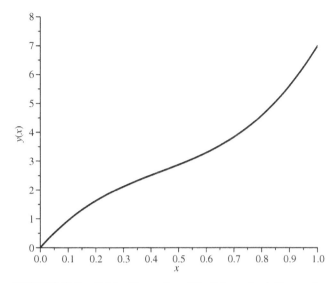

FIGURE 11.2. $\phi(x)$ for illustration of the Ritz method. The reader should integrate eq. (11.41) and verify the solution.

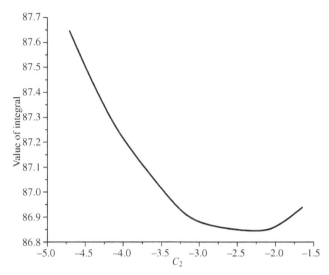

FIGURE 11.3. Calculated values for the integral (eq. 11.42); the minimum occurs when C_2 is approximately equal to -2.25.

on the undetermined constants, C_2 through C_n. Values are obtained for these Cs by differentiating I and setting the derivatives equal to zero:

$$\frac{\partial I}{\partial C_2} = 0, \quad \frac{\partial I}{\partial C_3} = 0, \quad \ldots, \quad \frac{\partial I}{\partial C_n} = 0. \quad (11.44)$$

Naturally, the more terms that are retained in the trial function, the better the result is likely to be. This example provides a terrific opportunity for the reader to explore the Ritz method without expending too much time or effort, as we shall now see. We will begin by truncating our approximation (we know the result will not meet critical inspection, but it will be simple enough to be executed rapidly): $\phi \approx x[7 + C_2(1 - x)]$. Of course, this quadratic form cannot duplicate the point of inflection we see in Figure 11.2, but it will reveal the essence of the technique. Now we compute the value of the integral (eq. 11.42) using plausible values for the parameter, C_2. The results are illustrated in Figure 11.3.

Using $\phi = x[7 - 2.25(1 - x)]$, we obtain

x	0.0	0.2	0.4	0.6	0.8	1.0
$\phi(x)$	0	1.04	2.26	3.66	5.24	7

We can compare the results in this abbreviated table with the behavior of the function illustrated in Figure 11.2. For examples, at $x = 0.4$, the correct value is about 2.5, while at $x = 0.8$, we should have gotten about 4.5 or 4.6. The reader is urged to verify this result and then to add the C_3 term and repeat! How much better is the approximation when the polynomial is third degree?

This direct approach to the solution of variational problems can be applied to more difficult cases if we use a little ingenuity, and we will illustrate this using an example developed by Ritz and discussed in detail by Smith (1953). Suppose we are interested in a problem of the type

$$I = \int_{x_1}^{x_2} \left(p_1 y'^2 + 2 p_2 y y' + p_3 y^2 \right) dx. \quad (11.45)$$

Our objective, of course, is to find $y(x)$ that makes $I(y)$ an extremum (a minimum); we stipulate that p_1, p_2, and p_3 be either constants or functions of x (but they *do not* depend on y). We also require that y' be finite and continuous for the interval $x_1 \le x \le x_2$, and we are given that $y(x_1) = \beta_1$ and $y(x_2) = \beta_2$. It is quite useful in a case like this to employ a transformation that simplifies the boundary conditions. Consider the function

$$y(x) = \frac{x - x_2}{x_1 - x_2} \beta_1 + \frac{x - x_1}{x_2 - x_1} \beta_2 + \phi(x) \quad (11.46)$$

and observe that $\phi(x)$ must disappear (be equal to zero) at both endpoints. This is extremely convenient and the required behavior suggests a function type that is suitable for the minimizing sequence; for example, let the approximations for $\phi(x)$ have the form

$$\phi_1 = C_1 \sin \frac{\pi(x - x_1)}{x_2 - x_1}. \quad (11.47)$$

Therefore,

$$\phi_n = C_1 \sin \frac{\pi(x - x_1)}{x_2 - x_1} + C_2 \sin \frac{2\pi(x - x_1)}{x_2 - x_1} \\ + \cdots + C_n \sin \frac{n\pi(x - x_1)}{x_2 - x_1}. \tag{11.48}$$

Of course, the final step is to determine the constants C_1, C_2, and so on, that make $I(y)$ a minimum. The precise forms for p_1, p_2, and p_3 would have to be known before this process could be carried out.

There is another direct method for determining the minimizing sequence that is described in chapter IV of Courant and Hilbert (1989). It involves discretizing the integral as given by eq. (11.4) using finite-difference approximations and dividing the interval (x_1, x_2) into m pieces such that

$$\int_{x_1}^{x_2} F(x, y, y')dx \approx \sum_{i=0}^{m} F\left(x_i, y_i, \frac{y_{i+1} - y_i}{\Delta x}\right)\Delta x.$$

This is now simply an ordinary minimum problem handled in piecewise fashion. Should higher derivatives appear in the integral, they would be replaced by forward difference approximations.

A CONTEMPORARY COV ANALYSIS OF AN OLD STRUCTURAL PROBLEM

Flexing of a Rod of Small Cross Section

We will begin this part of our discussion by considering a thin rod with a length that extends from $x = 0$ to $x = L$; this rod will be subject to flexing (transverse vibrations). It is easy enough to visualize this situation—imagine a strand of uncooked spaghetti, for example, that is oriented vertically between your fingertip and a tabletop. By pressing down on the end of the pasta, you can observe the lateral deformation and, ultimately, failure as the load is increased. It is this transverse flexing that we wish to consider. We will let $u(x, t)$ represent transverse displacement such that the transverse velocity is $\partial u/\partial t$, and we will take the mass of the rod per unit length to be m. Therefore, the total kinetic energy of the transverse motions can be written as

$$T = \frac{1}{2}m\int_{0}^{L}\left(\frac{\partial u}{\partial t}\right)^2 dx. \tag{11.49}$$

The total potential energy due to strain is determined from the second derivative (with respect to x) of the displacement:

$$V = \frac{1}{2}EJ\int_{0}^{L}\left(\frac{\partial^2 u}{\partial x^2}\right)dx, \tag{11.50}$$

where J is the moment of inertia. Thus, by Hamilton's principle, we have

$$I = \frac{1}{2}\int_{t_1}^{t_2}\int_{0}^{L}\left[m\left(\frac{\partial u}{\partial t}\right)^2 - EJ\left(\frac{\partial^2 u}{\partial x^2}\right)^2\right]dxdt. \tag{11.51}$$

The task confronting us is to choose the function $u(x, t)$ that yields the extremum of I and at the same time satisfies the conditions imposed at the ends of the rod. For a structural member (like our strand of spaghetti), the ends can be *free*, *hinged*, or *clamped*. By free, we mean that neither u nor the slope, $(\partial u/\partial x)$, is constrained to a particular value at $x = 0$ (or $x = L$). For the hinged condition, u is fixed, but the slope is arbitrary (this is the likely situation when we load the pasta strand by pressing down), and for the clamped condition, both the displacement, u, and the slope have particular values. The integral is extremized by creating a set of comparison functions such as $U = u(x, t) + \varepsilon\phi(x, t)$ and it is necessary that the function $\phi(x, t)$ be zero at both times, t_1 and t_2 (you should recognize that this is exactly the process that is illustrated in Figure 11.1). If the rod is hinged at the ends, it is necessary that $\phi(x = 0, x = L)$ be zero, but the slope, $\partial\phi/\partial x$, is arbitrary. Our interest is the case where $\varepsilon = 0$ and, thus, $I'(0) = 0$. For the sake of compactness, we will let the integrand in eq. (11.51) be represented by F (which includes the prefactor of ½), and therefore,

$$I'(0) = \int_{t_1}^{t_2}\int_{0}^{L}\left(\frac{\partial F}{\partial \dot{u}}\dot{\phi} + \frac{\partial F}{\partial u_{xx}}\phi_{xx}\right)dxdt = 0. \tag{11.52}$$

Weinstock (1974) shows that, through repeated integration by parts, one can obtain the differential equation that governs the displacement of the rod (you may recall the analogous development of the Euler–Lagrange differential equation at the beginning of this chapter):

$$m\frac{\partial^2 u}{\partial t^2} + EJ\frac{\partial^4 u}{\partial x^4} = 0 \quad \text{for} \quad 0 \leq x \leq L. \tag{11.53}$$

This is the equation of motion for transverse motions of the thin rod. If the ends of the rod are hinged, then it is necessary that $u = 0$ and $\partial^2 u/\partial x^2 = 0$. We can use the product method to seek a solution for eq. (11.53) as we will now demonstrate by proposing that $u = f(x)g(t)$; this results in two ordinary differential equations:

$$g'' = -\lambda^2 g \quad \text{and} \quad f'''' - \frac{\lambda^2}{\beta}f = 0, \text{ where } \beta = \frac{EJ}{m}. \tag{11.54}$$

The solution for the first of this pair is just: $g = A\sin\lambda t + B\cos\lambda t$, and we should recognize that what we have here is an eigenvalue-eigenfunction problem. Furthermore, to obtain this result, we have assumed that there can be no negative values for λ. To solve this problem, we would need to find the eigenfunctions (the fs) for the sequence of identified λs that satisfy the necessary conditions at the ends of the rod. We will leave this problem for the end of the chapter as a student exercise. What we want to do now, though, is to transition from the transverse displacement of a loaded strand of spaghetti to a much more fundamental question. If we need to use a column as a building support, what kind of column will give us the maximum strength while minimizing the amount of material required? In other words, is there an optimal column *shape*?

The Optimal Column Shape

Let us consider the design of a supporting column subjected to a vertical (axial) load; such columns were a staple of classical architecture (think of the Parthenon in Athens). If such a column is subjected to an extreme compressive load, it will fail or *buckle*. Structural engineers use Euler's formula to estimate the critical buckling load,

$$F_c = \frac{EJ\pi^2}{L^2}, \tag{11.55}$$

where E is Young's modulus of elasticity, J is the moment of inertia of the cross section, and L is the column length. Typically, the modulus of elasticity for carbon steel is about 30×10^6 psi; for concrete, it is about 3×10^6 psi; and for Teflon™, about 75,000 psi. The second moment of an area A (moment of inertia) about the x-axis is $J = \int_A y^2 dA$. The load that a column can carry also depends on how it is secured at the ends (at $x = 0$ and $x = L$); as we noted earlier, columns are commonly *free*, *hinged*, or *clamped* at the ends. These boundary conditions play a critical role in determining the form of the first buckling mode. Letting y represent the transverse displacement, the simple linear model for the buckling of a long, straight column is

$$\frac{d^2y}{dx^2} + \frac{F}{EJ}y = 0, \tag{11.56}$$

and the solution for eq. (11.56) is just $y = A\sin\sqrt{\beta}x + B\cos\sqrt{\beta}x$, where $\beta = F/EJ$. If $y = 0$ at both ends, then $B = 0$ and $\sqrt{\beta} = n\pi/L$; that is, the shape of the buckling mode is revealed and the eigenvalues yield $F = EJ(n^2\pi^2/L^2)$. Such an analysis works well enough for *slender* columns (for steel columns, *slender* means $L/R \geq 140$), but it fails to address the more complete design question, namely: *If the total volume of material and the column length are fixed, what shape or profile can support the greatest load?* In other

words, we want to evaluate columns where the cross-sectional area, A, can vary with length but with the requirement that $V = \int_0^L A(x)dx$. We should think of this enterprise as finding the $A(x)$ that meets this volume requirement and *maximizes* the smallest eigenvalue of F.

In the eighteenth century, Euler worked on the problem of finding the form of $y(x)$ that would lead to minimum stored energy. With the assumption that E and J were constant, Euler was able to minimize

$$\int_0^L \frac{EJ|y''|^2}{\left(1+|y'|^2\right)^{9/4}}\,dx - \lambda\int_0^L \left(1+|y'|^2\right)^{1/2}\,dx. \tag{11.57}$$

The second part of this expression is the length constraint with the Lagrange multiplier, λ. Cox (1992) notes that Euler was also able to determine the critical load for certain non-uniform columns for which $J(x) = (a + bx/L)^q$. The exponent, $q = 2$, corresponds to cylindrical columns for which A varies linearly over the length, L (a tapered cylinder).

It turns out that the question of the strongest column shape (for fixed volume and length) has been revisited repeatedly over the 200 years that have passed since Lagrange "proved" that the answer was a right-circular cylinder. Many have suspected that Lagrange's cylinder was incorrect; Cox makes the very practical observation that an optimal column shape should have large A, where bending might be expected, and reduced A, where little bending would occur. Keller (1960) revisited the problem and found that if both ends of the column were hinged, the optimal shape is actually a "stunted" cycloid (a bit like a sausage that is overly plump in the middle). Because the column is hinged at the ends, A at both $x = 0$ and $x = L$ is reduced; that is, the column does not have to resist bending at the ends. This profile is illustrated in Figure 11.4 and Cox notes that this shape is stronger than a cylinder of the same length and volume by more than 30%. But what about a column that is clamped at both ends? Tadjbakhsh and Keller (1962) examined that case and found a solution for which $A(x) \to 0$ at two interior locations, $x = \frac{1}{4}$ and $x = \frac{3}{4}$. Of course, this means that $y''(x)$ has two singularities and these two points can be thought

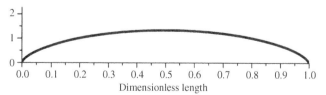

FIGURE 11.4. Optimal column shape if both ends are hinged (profile shown in horizontal orientation) as determined by Keller (1960).

of as interior hinges. For a time, it was generally accepted that the strongest clamped-clamped column would have two interior locations where the cross-sectional area would vanish; however, Olhoff and Rasmussen (1977) discovered that the Tadjbakhsh–Keller solution was incorrect. In the earlier work $y'(x)$ was assumed to be continuous (it does not need to be). Moreover, when Olhoff and Rasmussen used the $A(x)$ profile from the 1962 paper in the conventional eigenvalue formulation, they discovered that the critical compressive load was actually *less* than that for a uniform column.

Olhoff and Rasmussen (1977) developed the appropriate equations, recognizing that a bimodal formulation was necessary, and solved the problem numerically. They point out that the equation set provides a coupled, nonlinear, integro-differential eigenvalue problem. Their results showed that a column with clamped end conditions would have two interior regions where the cross section would diminish significantly but $A(x)$ *would not approach zero*; these contractions occur at fractional length positions of ¼ and ¾ (an excellent illustration is provided by Cox, 1992 as his figure 5). Olhoff and Rasmussen found that a column thus designed is almost 33% stronger than a uniform (cylindrical) column of the same height and volume. Cox and Overton (1992) used nonsmooth analysis to *prove* that A need not be continuous in the sup-norm topology but that it would have a lower (and nonzero) bound. Their results, which are also summarized by Cox, confirmed the numerical calculations of Olhoff and Rasmussen: The strongest clamped-clamped columns have *two* contractions where A is reduced significantly relative to a uniform cylindrical column.

SYSTEMS WITH SURFACE TENSION

Surface tension, σ, is a property of interfaces; it acts as a negative pressure in such a way as to minimize interfacial area. Work must be performed against surface tension to expand the interface and if the area is increased by an amount, dA, then the required work is $dW = \sigma dA$. Values for σ are usually reported as force per unit length, typically dyne per centimeter, and for the air–water interface, $\sigma = 72$ dyne/cm. As we noted earlier, surface tension creates a pressure difference across a curved interface, and we can think of this in the following way: The tendency for a bubble to contract, that is, to decrease its surface area, must be countered at equilibrium by an increase in pressure on the concave (interior) side. For a spherical bubble, the pressure difference is given by $\Delta p = 2\sigma/r$ such that for a 0.5-mm diameter bubble in water, $\Delta p = (2)(72)/(0.025) = 5760$ dyne/cm^2 (576 Pa), or 5.76×10^{-3} bar. Since the action of surface tension is to minimize surface area, it is clear that the situation is very similar to that posed by eq. (11.23) and eq. (11.24). Let us illustrate this point.

Suppose we take two circular loops of wire, separated by a distance, $2h$, but fixed so that the two circles are parallel. We let the circles be located at $x = \pm h$ with $y^2 + z^2 = R^2$. Now, we dip this wire framework into a basin of soapy water and then remove it. What shape will the resulting film have? The surface we are describing is a surface of revolution created by rotating a curve, $f(x)$, about the x-axis such that $f(-h) = R$ and $f(+h) = R$. For this case,

$$I = 2\pi \int_{-h}^{+h} f\sqrt{1 + f'^2}\, dx, \qquad (11.58)$$

and our task is to minimize I. This is equivalent to solving the Beltrami identity:

$$\frac{f'^2 f}{\sqrt{1 + f'^2}} - f\sqrt{1 + f'^2} = C, \qquad (11.59)$$

or consolidating,

$$\frac{-f}{\sqrt{1 + f'^2}} = C.$$

We can do a little rearranging and then integrate, yielding

$$\frac{1}{C}x + a_1 = \cosh^{-1}\left|\frac{f}{\alpha}\right|,$$

such that

$$f = C \cosh\left(\frac{x}{C} + a_1\right). \qquad (11.60)$$

You will see immediately that this is the catenary discussed previously (eq. 11.27). The validity of this result is demonstrated beautifully in the little book of interfacial experiments by Boys (1959; see figure 27, p. 55, of the Dover reprint of the "new and enlarged" edition of 1911). Boys used two large glass rings to draw out the soap film, and he notes that "the film is so far curved as to have a most elegant waist." He also points out that the surface thus formed is a *catenoid*—just as we demonstrated earlier.

THE CONNECTION BETWEEN COV AND THE FINITE-ELEMENT METHOD (FEM)

The finite-element method (FEM) as it is used in engineering applications had its origin in the analysis of structures. In particular, FEM developed in response to structural problems that involved a continuum (as opposed to those, e.g., that concerned a truss with a finite number of beams and connecting points). Hrenikoff (1941) proposed discretizing the continuum by dividing it into a finite number of elements. Initially, many applications of FEM were focused on structures, but now the technique is employed by virtually all branches of science working with continuum mechanics. The COMSOL™ website is an ideal place for the reader to get a sense of the breadth of these uses.

Previously in this chapter, we used the Ritz method to find an approximate solution for a boundary-value problem and you will recall that the trial function was valid over the *entire* domain (from $x = 0$ to $x = 1$). This is a distinguishing characteristic with regard to the FEM where the trial functions (or interpolating functions) are valid only in a piecewise manner. Our intent with FEM is to break the domain into a collection of pieces and then use the trial function(s) to represent the solution over each piece separately. So an important difference between the Ritz method and FEM is now clear—in FEM, the trial function does *not* have to satisfy the boundary conditions.

Suppose, for example, that we have a continuous function in the x-y plane: $\psi = f(x, y)$. Our goal is to represent this function in an approximate fashion over the entire domain but to do so with a collection of small pieces. We will use triangular elements and, for the moment, assume that the field variable is known at the three vertices, which we will label 1, 2, and 3. The simplest feasible representation (simplest in the sense that it allows for the linear variation of ψ with changes in the two independent variables) for ψ will be taken as

$$\psi \cong C_1 + C_2 x + C_3 y. \tag{11.61}$$

You can see immediately that the problem confronting us is one of choosing the "best" values for C_1 through C_3 for our small triangular piece. We will place the vertices at (1, 1), (2, 1), and (2, 2), and we observe that the area of this triangle is ½. We will also arbitrarily take $\psi = x^2 y^2$ such the node values for the dependent variable will be 1, 4, and 16, respectively. Since we know that value of the field variable at each of the vertices (or nodes), we write

$$\psi_1 = 1 = C_1 + C_2(1) + C_3(1), \tag{11.62}$$

$$\psi_2 = 4 = C_1 + C_2(2) + C_3(1), \tag{11.63}$$

and

$$\psi_3 = 16 = C_1 + C_2(2) + C_3(2). \tag{11.64}$$

We use these three equations, solving them simultaneously, to determine the unknown parameters. The result is $C_1 = -14$, $C_2 = 3$, and $C_3 = 12$; these choices provide the correct values at the vertices (or nodes) as we required. But what about the value of ψ at other points within the triangu-

lar element? Our approximation is $\psi \cong -14 + 3x + 12y$, so obtaining a comparison should be easy:

x	y	$\psi_{\text{estimated}}$	ψ_{actual}
1.5	1.0	2.5	2.25
1.5	1.25	5.5	3.5156
1.5	1.5	8.5	5.0625
1.75	1.0	3.25	3.0625
1.75	1.25	6.25	4.7852
1.75	1.5	9.25	6.8906
2.0	1.5	10.0	9.0
2.0	1.75	13.0	12.25

Although the values for ψ are correct at the vertices, some of the estimated values in the interior of the triangular element are in error by as much as 60% (the average error for the points tested in the previous table is 27.6%). This performance is not acceptable, and we have two options: We could increase the degree of the trial function polynomial (e.g., we could make it quadratic with respect to x and y) or we could refine (subdivide) the triangular element. In the case of the latter and continuing with this example, we simply draw a line from (1, 2) to (1.5, 1.5), producing *two* triangles of equal size. Naturally, we have doubled our workload as we must now determine values for two sets of parameters, but this is of little consequence for an automated calculation. If necessary, we can continue subdividing the element until the linear trial (or interpolation) function gives satisfactory agreement over each element. When we are solving elliptic partial differential equations (PDEs), it is usually obvious where the steepest gradients will be located, so the analyst often has a very good idea where the element size might need to be reduced.

We determined the three needed parameters (the C_ns) from eq. (11.62), eq. (11.63), and eq. (11.64) by Gaussian elimination, but equivalently,

$$C_1 = \frac{\psi_1(x_2 y_3 - y_2 x_3) + \psi_2(x_3 y_1 - x_1 y_3) + \psi_3(x_1 y_2 - x_2 y_1)}{2A},$$
$$\tag{11.65}$$

$$C_2 = \frac{\psi_1(y_2 - y_3) + \psi_2(y_3 - y_1) + \psi_3(y_1 - y_2)}{2A}, \tag{11.66}$$

and

$$C_3 = \frac{\psi_1(x_3 - x_2) + \psi_2(x_1 - x_3) + \psi_3(x_2 - x_1)}{2A}. \tag{11.67}$$

The subscripts 1, 2, and 3 refer to the three vertices and we are proceeding from the (1, 1) vertex in the counterclockwise sense; therefore, vertex 2 is located at (2, 1) and vertex 3 is at point (2, 2). A is the area of the triangular element,

$(1/2)bh = (1/2)(1)(1) = 1/2$, so the denominator in eq. (11.65), eq. (11.66), and eq. (11.67) is 1. We will introduce eq. (11.65), eq. (11.66), and eq. (11.67) into eq. (11.61) and write the result in general form:

$$\psi(x, y) = \frac{a_i + b_i x + c_i y}{2A}\psi_i + \frac{a_j + b_j x + c_j y}{2A}\psi_j \qquad (11.68)$$
$$+ \frac{a_k + b_k x + c_k y}{2A}\psi_k,$$

where

$$a_i = x_j y_k - x_k y_j, \quad b_i = y_j - y_k, \quad \text{and} \quad c_i = x_k - x_j. \qquad (11.69)$$

Following Huebner (1975), we define the linear *shape functions* as

$$N_n = \frac{a_n + b_n x + c_n y}{2A}. \qquad (11.70)$$

Therefore, for *each* triangular element, we have three vertex values and three shape functions:

$$\psi = \begin{matrix} \psi_i \\ \psi_j \\ \psi_k \end{matrix} \quad \text{and} \quad N = [N_i, N_j, N_k], \qquad (11.71)$$

and thus, $\psi(x, y) = N_i \psi_i + N_j \psi_j + N_k \psi_k$. Our approximation for the variation of the field variable over the complete domain is then

$$\psi(x, y) = \sum_{s=1}^{M} [N^{(s)}]\{\psi^{(s)}\}, \qquad (11.72)$$

where M is the total number of triangular elements. The remaining task is to identify the nodal (vertex) values of the dependent (or field) variable, ψ, so that the functional $I(\psi)$ is an extremum, where

$$I(\psi) = \sum_{s=1}^{M} I^{(e)}\psi^{(e)}. \qquad (11.73)$$

The first variation of the functional, I, must be zero, and therefore for each element, e, we have

$$\frac{\partial I^{(e)}}{\partial \psi_j} = 0, \qquad (11.74)$$

and the j index refers to the nodes (or vertices) of each element. Equation (11.74) provides a set of simultaneous equations that, when solved, yield the problem's solution in terms of the nodal values of ψ.

Now we will explore the use of this method for an elementary problem in steady-state heat conduction in two dimensions, where the governing equation is

$$\frac{\partial^2 T}{\partial x^2} + \frac{\partial^2 T}{\partial y^2} = 0. \qquad (11.75)$$

We will use a single triangular element so that the process is completely transparent with a linear (with respect to x and y) trial function. Let the vertices of the triangle be located at (x, y) positions, $(0, 0)$, $(2, 1)$, and $(1, 2)$. We let the top correspond to vertex 1 where the temperature is $100°$, so $T(x = 1, y = 2) = 100$. Both bottom vertices, and the bottom edge, are maintained at $50°$. Now we proceed in clockwise fashion from vertex 1, writing the three equations:

$$a_1 + b_1(1) + c_1(2) = 1 \quad a_1 + b_1(2) = 0 \quad a_1 = 0. \quad (11.76)$$

For vertex 2,

$$a_2 + b_2(2) + 0 = 1 \quad a_2 = 0 \quad b_2 + 2c_2 = 0. \quad (11.77)$$

And for vertex 3,

$$a_3 = 1 \quad a_3 + b_3 + c_3(2) = 0 \quad a_3 + b_3(2) = 0. \quad (11.78)$$

In each case, you will note that the equation for the "home" vertex is set equal to one, and it is zero at the other two vertices. Thus, for the first set of equations where $(x = 1, y = 2)$, we find $c_1 = 1/2$; for the second set, $(x = 2, y = 0)$, we find $b_2 = 1/2$ and $c_2 = -(1/4)$. The reader should verify that for vertex 3, $a_3 = 1$, $b_3 = -(1/2)$, and $c_3 = -(1/4)$. We can now write out our approximation:

$$T \approx 100\left(\frac{1}{2}y\right) + 50\left(\frac{1}{2}x - \frac{1}{4}y\right) + 50\left(1 - \frac{1}{2}x - \frac{1}{4}y\right). \quad (11.79)$$

Looking at a few selected points, we obtain $T(x = 1, y = 1) = 75°$, $T(x = 1/2, y = 1/2) = 62.5°$, $T(x = 2/3, y = 4/3) = 83.33°$, $T(x = 5/3, y = 2/3) = 66.67°$, and $T(0 \le x \le 2, y = 0) = 50°$. Should we wish to improve the quality of our estimated solution, we could place an inverted triangle inside the original domain with vertices at $(1, 0)$, $(1.5, 1)$, and $(0.5, 1)$; the result is four equal-sized triangles inside the original figure. If we subsequently divided each of those four triangles into four equal pieces using the same procedure, we would end up with 16 elements—all identical triangles. By refining the mesh, we can improve the quality of the solution at the cost of additional simultaneous equations (to be solved).

There are *many* examples in the literature in which the FEM is applied to elliptic PDEs; de Vries and Norrie (1971) is especially useful as they illustrate how the variational problem is posed for the Laplace equation (with

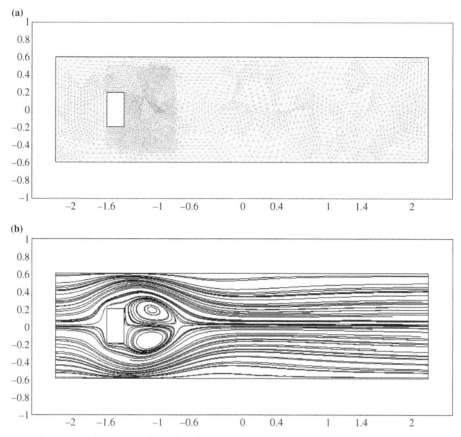

FIGURE 11.5. Flow around a rectangular obstruction at low Reynolds number: (a) discretization (triangular elements) and (b) streamlines. The inflow boundary condition (left edge) was set to constant velocity and the horizontal boundaries had the slip/symmetry condition applied. Notice how the element size was reduced near the obstruction. These results were obtained with the finite-element method using COMSOL.

emphasis on potential flow) for different types of boundary conditions. There are also many examples of simple FEM codes for readers wishing to start from the ground up; examples are included in Huebner (1975) and Reddy and Gartling (1994). For complex problems requiring FEM—particularly if the analyst has time constraints—commercial codes are probably worth the cost of acquisition. One such program that is very well known is COMSOL, which was originally marketed as FEMLAB™. This software was designed to find solutions for "multiphysics" problems such as buoyancy-driven flows arising in free convection. The user interface allows one to rapidly explore the effects of changing the boundary conditions, refining the mesh, and so on. Such programs can be used to solve quite difficult problems with complicated boundaries; an example (viscous flow around a rectangular obstruction) computed with COMSOL is shown in Figure 11.5a,b. Please keep in mind that one should fully test some benchmark cases related to any problem of interest before relying on a solution produced by such software.

CONCLUSION

The COV is extremely useful for problems in which an *extremal* is being sought and the *functional* is simple enough to be dealt with analytically. Unfortunately, in practice, this often turns out to be quite restrictive because, for a problem of interest, (1) the variational principle may not be known, (2) the existence of an extremum may not be guaranteed, (3) solution for a properly posed problem may not even be possible, and (4) the problem may be highly nonlinear such that identifying the minimizing sequence requires solution of large sets of nonlinear algebraic equations. To illustrate these points, consider the integral

$$I = \int_{-1}^{+1} x^4 \left(\frac{dy}{dx}\right)^2 dx. \qquad (11.80)$$

Now suppose we wanted to identify a continuous function, $y(x)$, with a piecewise continuous derivative that would minimize eq. (11.80); we require that $y(-1) = -1$ and that $y(+1) = +1$. Courant and Hilbert (1989) indicate that, although the integral can be made small, no appropriate function $y(x)$ can be identified that would cause the variation of eq. (11.80) to vanish.

Nevertheless, COV is an important tool of historical significance that can be quite valuable in the right context; as we have seen, for example, COV ideas are used routinely in the application of the FEM to the solution of PDEs. And FEM has emerged as an invaluable asset for the analysis of problems in acoustics, electromagnetics, fluid dynamics, geophysics, heat transfer, mass transfer, optics, quantum mechanics, structural mechanics, wave propagation, and so on. The interested reader can get a sense of the breadth of possibilities by reviewing the proceedings from any of the recent COMSOL conferences.

PROBLEMS

11.1. In "Variational Methods for the Solution of Problems of Equilibrium and Vibration," Courant (1943) points out that the equivalence between boundary-value problems involving PDEs and COV has been studied for a very long time. He notes that both Lord Rayleigh and Walther Ritz suggested the variational approach to such problems might be reduced to a much simpler extremum problem in which the number of undetermined parameters would be manageably finite. In part II of his paper, Courant describes how one goes about solving a variational problem numerically by constructing a minimizing sequence (this is what is often referred to as the Rayleigh–Ritz method, as we described previously). He points out that the convergence of this process can be improved by adding a "sensitizing" functional to I. Read Courant's paper and determine how the addition of the "sensitizing" term will impact the nature of the simultaneous equations that must be solved. Courant suggests that an important objection to the Rayleigh–Ritz method is the difficulty the analyst has in assessing the accuracy of the approximation. Explore this topic and determine if substantive progress has been made in the years that have passed since Courant's original presentation in Washington in the spring of 1941.

11.2. Examine the eigenvalue-eigenfunction problem described in our discussion in "Flexing of a Rod of Small Cross Section." What determines the values assumed by λ? What boundary conditions would be applied for the solution of the fourth-order ordinary differential equation, $f'''' - (\lambda^2/\beta)f = 0$? Solve this fourth-order equation numerically using the first two appropriate values for λ.

11.3. The variational principle for the Laplace equation was given earlier in this chapter as eq. (11.40). Suppose we wanted to explore the finite-element solution of a two-dimensional, steady conduction problem in an *anisotropic* medium that includes a thermal energy source term; that is,

$$\frac{\partial}{\partial x}\left(k_x \frac{\partial T}{\partial x}\right) + \frac{\partial}{\partial y}\left(k_y \frac{\partial T}{\partial y}\right) + S = 0,$$

where S might be a function of position, $S = f(x, y)$. We approximate the dependence of T on position with

$$T(x, y) \approx \sum_{j=1}^{n} T_j^e \phi_j^e(x, y),$$

where the T_j^es are the node values of T and the ϕ_j^es are the approximating functions for each finite element. For a weighted-residual approach to this problem, we could write

$$0 = \iint W\left[\frac{\partial}{\partial x}\left(k_x \frac{\partial T}{\partial x}\right) + \frac{\partial}{\partial y}\left(k_y \frac{\partial T}{\partial y}\right) + f(x, y)\right]dxdy.$$

The Ts that appear here are the *approximations*, of course. Our intent is to use this weighted-residual form to generate a set of simultaneous algebraic equations that, when solved, will yield the approximate nodal values for T. You will notice, however, that we have said nothing about the boundary conditions that must accompany the original PDE. Investigate how *both* Dirichlet and Robin's-type boundary conditions are implemented in this weighted-residual treatment of this conduction problem. You will find chapter 2 in Reddy and Gartling (1994) to be helpful in this exercise.

11.4. Courant and Hilbert (1989) describe other direct means by which a minimizing sequence might be determined. Suppose we have

$$I = \int_0^b \int_0^a \left[\left(\frac{\partial \phi}{\partial x}\right)^2 + \left(\frac{\partial \phi}{\partial y}\right)^2\right]dxdy,$$

which is to be minimized subject to the following condition:

$$1 = \int_0^b \int_0^a \phi^2 dxdy.$$

The allowable comparison functions must disappear on the rectangular boundary, of course. Suppose we represent ϕ using a Fourier series:

$$\phi \cong \sum_{m,n=1}^{\infty} B_{mn} \sin \frac{m\pi x}{a} \sin \frac{n\pi y}{b}.$$

The task now is to identify the coefficients, B_{mn}s, that produce the minimum of I. Show that

$$I = \frac{ab\pi^2}{4} \sum_{m,n=1}^{\infty} B_{mn}^2 \left(\frac{m^2}{a^2} + \frac{n^2}{b^2} \right),$$

then use the condition that $1 = \int_0^b \int_0^a \phi^2 dx dy$ to show that the only nonzero B_{mn} is B_{11}, which is equal to $2/\sqrt{ab}$. Finally, show that the minimum value of the integral is

$$\pi^2 \left(\frac{1}{a^2} + \frac{1}{b^2} \right).$$

11.5. In the previous exercise, we explored a direct solution technique for a problem that included an auxiliary condition. We will now examine a very specific case with an appended condition. Suppose we need an extremum for the integral

$$I = \int_{x_1}^{x_2} F(x, y, y') dx$$

but subject to the condition that

$$\int_{x_1}^{x_2} G(x, y, y') dx = C.$$

The variational problem described previously is equivalent to finding the extremum for

$$I^* = \int_{x_1}^{x_2} \left[F(x, y, y') + \lambda G(x, y, y') \right] dx,$$

where λ must be determined from the auxiliary condition. Now let us focus on the x-y plane, where we seek the *minimum* length curve between points (x_1, y_1) and (x_2, y_2) that yields a *specific area* under the curve represented by A. We already know that the length of the curve was given at the beginning of this chapter by eq. (11.1): $L = \int_{x_1}^{x_2} \sqrt{1 + y'^2} dx$. The area under the curve (that we seek) is, of course, $\int_{x_1}^{x_2} y dx = A$. Suppose that the specific points of interest to us are $(0, 0)$ and $(5, 3)$; if we connect these points with a straight line, we form a triangle with an area of $15/2$. But for our problem, we will require that $A = 15$, and this makes it apparent that the curve we seek must be convex up. It is also clear from the preceding description that we need to identify the extremum for

$$I^* = \int_0^5 \left[\sqrt{1 + y'^2} + \lambda y \right] dx.$$

Smith (1953) shows that, to make the first variation of I^* disappear, it is necessary that

$$\frac{d}{dx} \left[\frac{y'}{\sqrt{1 + y'^2}} \right] = \lambda.$$

Determine the exact nature of the curve connecting $(0, 0)$ with $(5, 3)$ that yields an integration an area of 15. What is the numerical value of λ?

11.6. *Isoperimetric* problems appear frequently in the literature of COV. The origin is found in the tale of *Dido* (who may or may not have been an actual historical figure). By legend, Dido fled to North Africa to escape her brother, Pygmalion. On arrival, she asked the Berber *Iarbas* for a spot of land for her party. They agreed that she could claim whatever land might be encircled by a single oxhide, which she cut into very thin strips as to make a very long cord, hence the phrases "cut a dido" or "cutting didoes." In this manner, she was able to claim a small hill that, according to legend, became Carthage. For the Dido problem, we wish to maximize the area given by $I = \int_a^b y dx$, while the total length, $L = \int_a^b \sqrt{1 + y'^2} dx$, must equal some particular value. Of course, the function $y(x)$ must produce a closed path that contains the desired area. Let $H = y + \lambda\sqrt{1 + y'^2}$ and use the Euler equation, $(\partial H/\partial y) - (d/dx)(\partial H/\partial y') = 0$ to find the solution for this problem (we should get the equation for a circle, of course).

REFERENCES

Boys, C. V. *Soap Bubbles: Their Colours and the Forces Which Mold Them*, Dover Publications, New York (1959).

Courant, R. Variational Methods for the Solution of Problems of Equilibrium and Vibration. *Bulletin of the American Mathematical Society*, 49:2165 (1943).

Courant, R. and D. Hilbert. *Methods of Mathematical Physics*, Vol. 1, John Wiley & Sons, New York (1989).

Cox, S. J. The Shape of the Ideal Column. *The Mathematical Intelligencer*, 14:16 (1992).

Cox, S. J. and M. L. Overton. The Optimum Design of Columns against Buckling. *SIAM Journal on Mathematical Analysis*, 23:287 (1992).

de Vries, G. and D. H. Norrie. The Application of the Finite Element Technique to Potential Flow Problems. *Journal of Applied Mechanics, Transactions of the ASME*, 38:798 (1971).

Hrenikoff, A. Solution of Problems in Elasticity by the Framework Method. *Journal of Applied Mechanics*, 8:169 (1941).

Huebner, K. H. *The Finite Element Method for Engineers*, John Wiley & Sons, New York (1975).

Keller, J. B. The Shape of the Strongest Column. *Archive for Rational Mechanics and Analysis*, 5:275 (1960).

Olhoff, N. and S. H. Rasmussen. On Single and Bimodal Optimum Buckling Loads of Clamped Columns. *International Journal of Solids and Structures*, 13:605 (1977).

Reddy, J. N. and D. K. Gartling. *The Finite Element Method in Heat Transfer and Fluid Dynamics*, CRC Press, Boca Raton, FL (1994).

Smith, L. P. *Mathematical Methods for Scientists and Engineers*, Dover Publications, New York (1953).

Spiegel, M. R. *Advanced Mathematics for Engineers and Scientists*, McGraw-Hill Book Company, New York (1971).

Tadjbakhsh, I. and J. B. Keller. Strongest Columns and Isoperimetric Inequalities for Eigenvalues. *Journal of Applied Mechanics*, 29:159 (1962).

Weinstock, R. *Calculus of Variations with Applications to Physics and Engineering*, Dover Publications, New York (1974).

INDEX

Applied Mathematics for Science and Engineering, First Edition. Larry A. Glasgow.
© 2014 John Wiley & Sons, Inc. Published 2014 by John Wiley & Sons, Inc.